D1243429

TWENTY YEARS
OF COLLOID AND
SURFACE CHEMISTRY

The Kendall Award Addresses

TWENTY YEARS OF COLLOID AND SURFACE CHEMISTRY

The Kendall Award Addresses

EDITED BY

KAROL J. MYSELS
Gulf General Atomic Co.

CARLOS M. SAMOUR
The Kendall Co.

JOHN H. HOLLISTER
The Kendall Co.

1973
AMERICAN CHEMICAL SOCIETY
WASHINGTON, D.C.

QD
549
M92

NMU LIBRARY

Copyright © 1973
American Chemical Society
All Rights Reserved

Library of Congress Catalog Card 73-77341
ISBN 8412-0167-6
PRINTED IN THE UNITED STATES OF AMERICA

Preface

During the twenty years spanned by these Kendall award papers we see the last vestiges of art leave the study of Colloid and Surface chemistry, the technological maturity is seen; mathematics, physics and the biological sciences become home ground for the Colloid and Surface chemists and their territories now extend to the whole world of science.

The Kendall Company has been greatly honored to participate in these awards and its scientists rejoice that this book will help to propogate and preserve the wisdom of the recipients. They have enlarged man's understanding of his universe. For scientists there is no higher communication or, in the giving and receiving, no more important or humbling gift.

Special thanks are due to a long time friend of the Kendall Company, Karol J. Mysels, who in 1951 suggested the Kendall awards and recently through his work on this book has helped to give them tangible form.

September 15th, 1972

Boston, Massachusetts

W. M. Bright
President and Chief Executive Officer
The Kendall Company

Introduction

The Kendall Award was first bestowed in 1953. Its official name has changed slightly over the years from 'Kendall Company Award in Colloid Chemistry' to 'Kendall Company Award in Colloid or Surface Chemistry', and finally to 'The American Chemical Society Award in Colloid or Surface Chemistry sponsored by The Kendall Company'; but it has been the only national award in this field. As part of the bestowal ceremonies, all recipients have delivered addresses. These Award Addresses mirror the personalities and the work of those selected annually from among residents of the USA and Canada, by an anonymous and changing jury, for "outstanding contributions" to the field with the proviso that "special consideration will be given to the independence of thought and the originality shown".

It may therefore be of interest on the twentieth anniversary of the Award to present the collection of these original addresses along with some Addenda by their authors to update them when feasible.

Several of these addresses were based on subsequently published papers but others were prepared only for the occasion or even delivered from notes or presented extemporaneously. This created some problems in assembling the present volume. Fortunately, we were able to obtain several manuscripts of the Award presentations or a publication which was closely related to the actual addresses, one of them along with an introduction and conclusion written now but "in the spirit of the award time". Due to these manuscripts and to the Addenda, much of this book is original, unpublished material.

In most cases, especially recently, the awardees have used the award address as an occasion to review their best work and place it in perspective. Others presented a specific new contribution which excited them at the time. A few have preferred to make it a personal rather than scientific occasion and offered an insight into what made them 'tick' as scientists.

Some of the flavor of the original occasion is of course lost in print. Movies and color slides could not be included in this book. Nevertheless the variety of personalities, areas of interest, approaches and points of view, as well as the gradual evolution of the whole field are manifest from the printed page.

The variety of topics covered and techniques discussed is enormous. Even when nominally the subject is the same, such as the contact angle, the earlier approach of Bartell has little in common with that of Zisman and, still later, Zettlemoyer touches upon it again from a different point of view. The solid surface has also been treated by Emmett, Brunauer, Halsey and Zettlemoyer but with emphasis on various aspects of its interaction with gases: its function in catalysis, the importance of proper area determinations, the mutual interaction of adsorbed molecules, and the characteristics of hydrophobic behavior. The liquid-gas surfaces are the subjects of Mysels and of Hansen, but one is concerned with thin films and the other with ripples on bulk liquid. Aqueous dispersions are studied by Matijevic and serve him as a tool to solve problems of inorganic chemistry. The first monodisperse aerosols were discussed by La Mer, and Kerker reported recently on advances made by light scattering and other modern techniques in their study. The theory of radiation scattering in general, and especially of X-rays, was the subject of Debye. Williams discussed the role of the ultra-centrifuge and, later, Vinograd its improvements and uses in studying DNA. Muscular contraction was the concern of Hill and biological problems also were very much on Scatchard's mind. Holmes ranges, in his review, from Vitamin A to the contraction of the silica gels which others use as model solid surfaces. Ferry and Mason both discuss rheology, one from the microscopic and the other the macroscopic point of view.

Thus there are some interweaving threads in these Award Addresses, but the diversity may appear overwhelming. Yet there seem to be two questions to which all the authors are addressing themselves, though each in his own way and to differing degrees and with different emphasis. These are: How can our basic knowledge of nature be used to better understand this world of neglected dimensions of surface layers and colloidal dispersions? And, perhaps more important, how can the study of this special world enlarge our basic understanding of nature? This is the effort that the Kendall Award has tried to recognize and encourage.

Warner Eustis and Horace Secrist, Directors of Research, both now retired, of The Kendall Company saw the importance of providing this recognition and encouragement and made the Award possible over these years.

The Editors

Table of Contents

The Kendall Award—A Background Account

Karol J. Mysels

That an award in colloid and surface chemistry would find its place in the array of prizes administered by the American Chemical Society was inevitable. That this prize happened to be established quite early and that it is the Kendall Award is due to special circumstances of which it may be of interest to tell the story although, by necessity, it will be an incomplete and a very personal one indeed.

In 1938 I came to the USA and began my graduate work at Harvard with Arthur B. Lamb, whose main research concern was the physical chemistry of Werner complexes and who had been past president of the American Chemical Society and editor since 1917 of its Journal. Like most US chemists of his generation, Lamb had studied in Germany and there roomed in the same house with James W. McBain, a Canadian who established his reputation in colloid chemistry at Bristol and then joined Stanford University in 1927.

Soon after school started, the traditional departmental "smoker" was held and I found that I could attend without being a smoker myself. There I remember having a lengthy discussion about the deteriorating international situation with Professor Kistiakovsky and another new graduate student named Willard M. Bright, who was also starting his research with Lamb. As I got to know Bright better, I learned that he had studied at Toledo University on a football scholarship until his fourth year, when the chemistry department finally decided that their prize student should not be spending most of his energies in team workouts and gave him a teaching fellowship. He was also very different from most other graduate chemistry students in that while the rest of us aimed somewhere in the direction of Nobel Prizes and pure scientific research, his ambition was to do industrial research and to direct it. In the meantime he was earning his way through school and thought that this was particularly easy at Harvard because of the limited competition for part-time jobs.

My thesis work on cobalt complexes took me into the problem of surface conductivity through which I became acquainted with McBain's work in this area. After taking a job with the Shell Development Company in San Francisco I discovered that McBain was not in Stamford, Connecticut, as I had thought, but at Stanford University some thirty miles away. Hence, after getting my thesis and my naturalization out of the way, I went there to ask him if I could work in his laboratory on weekends to try to prove out some ideas on surface conductivity which were quite contrary to his publications. He agreed readily and I worked with him for about a year that way, although it turned out soon that my original idea was not right. After a brief enlistment in the Army, I then worked full time with him on the aluminum soaps which at that time formed Napalm and I learned to appreciate his remarkable scientific insight and leadership qualities.

Through the following years I kept in touch with Bright, who in the meantime joined The Kendall Company and, after investigating the theory of adhesion, developed the first polyethylene adhesive tape and saw it through successful commercialization under the Polyken name. The company then decided to open the Theodore Clark laboratory devoted to long-range research and named him its first director. I also kept in touch with Lamb and often had lunch with him during ACS national meetings. At one of these, about 1949, he mentioned that he had been thinking of getting some public recognition for McBain's many contributions to science but found that none of the existing awards fitted his achievements. I later checked and found that this was indeed the case and it became clear that a prize recognizing contributions to colloid chemistry would fill a real lacuna. After some further discussions with several friends, I brought the matter up at the business meeting of the Colloid Division during the Fall 1950 meeting in Chicago. The response was rather lukewarm as I remember, especially when the financial problems were raised, but a committee, composed of Desirée Le Beau, a former division chairman, Herbert Davis, a councillor of the division, and me as chairman, was appointed to examine the question more thoroughly. A request for suggestions was included in the divisional newsletter, but none came in. The committee, however, corresponded during the year to discuss the desirability of the award, and examined a number of possibilities ranging from a scroll to an ACS award and including a specially dedicated issue of the Journal of Physical Chemistry. Editor Lind poured some cold water on this one! We finally reached the conclusion that the award was definitely desirable and should be substantial.

At the next business meeting of the Division, in the Fall of 1951, I presented our report pointing out that there were general big prizes in chemistry ranging from the Nobel Prize to the Gibbs Medal, and that colloid chemists occasionally did receive them. There were also more specialized prizes but only few colloid chemists, such as those below 36 years of age or women, were at all eligible for them. Yet the advantages of such a specialized prize are large: To the recipient it can be a well-deserved reward for his efforts and an encouragement to continue; to the outsider it points out the best work in the field and attracts attention to it more effectively than this work could by itself. It thus brings the whole field to the attention of non-specialists and does it by its most deserving aspects, and in so doing improves its eligibility for other awards. In addition, such an award may be an encouragement for young men to enter a field in which recognition can be earned. Its bestowal could be an occasion for those interested in the field to gather together, and for the Division it could be an important yearly activity and accomplishment. It seemed, however, that any worthwhile award would be beyond the means of the Division itself or those of the Colloid Symposium so that outside support should be sought to establish a significant one.

After some discussion, the Division went on record unanimously in favor of such an award and a committee, under L. H. Reyerson's chairmanship, was appointed to explore ways of securing financial support. Soon after the meeting I spoke to Will Bright and told him about this need. As I remember he was quite noncommittal about the whole idea, and it was a delightful surprise when a few months later came a copy (reproduced here) of his letter of January 10, 1952, to Emery, the secretary of the ACS, offering to establish the Kendall Award.

The formalities of activating the award took over a year so that the first one was presented in 1954. Unfortunately, McBain passed away early in 1953 soon after returning from India, where he had gone at Nehru's request to organize the first National Chemical Laboratory in Poona. The Division of Colloid Chemistry held a symposium in his memory at the time when the first Kendall awardee was announced.

One should mention perhaps some of the others who are not included in this volume despite their outstanding contributions to the field. Harkins had already passed away in 1951. Langmuir died in the Summer of 1957, but it should be noted that the Kendall award did not originally include surface chemistry explicitly, the field in which he made his Nobel Prize-winning contributions. Finally, the man who perhaps did most to make colloid chemistry a science by his explanation of Brownian motion, diffusion, light scattering, and intrinsic viscosity—Einstein, was probably never nominated, perhaps just because he seemed so much greater than the field. I know that it never occurred to me that he would be eligible until after I first met him a few months before his death in 1955. I can still only guess whether he would have been interested.

In the years that elapsed since, many changes have taken place. Will Bright left The Kendall Company in 1952 and, after a distinguished career with Lever Brothers, R. J. Reynolds and Warner-Lambert, returned to it as President and Chief Executive Officer two years ago. Clearly, his action on the award was not his only wise decision. The Division recognized his contribution by a special certificate in 1972 at the divisional luncheon during the Spring Meeting of the American Chemical Society in Boston.

The Kendall Award and the Symposium organized annually in honor of the recipient have become focal points of the activities of the Division of Colloid and Surface Chemistry. The several sessions of the Symposium are the highlight of the Spring Meeting of the Division and bring together some of the best work in the field, both in this country and abroad. It is the place to hear about important current work and to meet both the recognized authorities and the rising stars of colloid and surface chemistry. The culmination of the Symposium is of course the award address, usually an hour long, which gives the recipient a chance to put his best foot forward. The resulting footprints form this volume. Drs. Hollister and Samour, heads of two of the laboratories of The Kendall Company, who have made personally most of the presentations of the awards over the years, hope, together with me, that these convey some of the spirit of the research recognized by the Kendall Awards.

Karl — I think this will interest you. Bill

January 10, 1952

Mr. Alden H. Emery
Executive Secretary
1155 Sixteenth Street, N.W.
Washington 6, D. C.

Dear Mr. Emery:

The Kendall Company wishes to initiate and sponsor for at least five years a research award in the field of colloid chemistry. The purpose, nature, and rules of eligibility for the award are as follows:

The Kendall Company Award in Colloid Chemistry

Purpose - To recognize and encourage outstanding contribution to the science of colloid chemistry in the United States and Canada.

Nature - The award consists of $1,000 and a certificate. An allowance of not more than $150 is provided for traveling expenses to the meeting at which the award will be presented.

Establishment and Support - The award was established in 1952 and has been supported since that date by The Kendall Company

Rules of Eligibility - A nominee must be a resident of the United States or Canada and must have made an outstanding contribution to the science of colloid chemistry. Special consideration will be given to the independence of thought and the originality shown.

It is the Company's desire that the American Chemical Society accept responsibility for the administration of the award. It is our understanding that a committee of experts chosen by the Society in the field of colloid science will select the recipient, that the award will be presented at a national meeting of the

2.

American Chemical Society, and that the award recipient will be expected to present an address in person on his work in colloid chemistry before the colloid division of the Society at a session of the division during the meeting at which the award is presented.

The Company hopes that the first presentation of this award can be made in 1953.

We realize that any remaining details must be quickly settled. We believe that this letter covers the facts pertinent to the award, and we look forward to an early receipt of your instructions regarding the next steps necessary to carry forward the arrangements.

Yours very truly,

THE KENDALL COMPANY

Willard M. Bright,
Director, Theodore Clark Laboratory

WMB/ces

cc:
Mr. R. M. Warren
Prof. L. H. Reyerson

Kendall Award Rules—1972

■ ACS Award in Colloid or Surface Chemistry sponsored by The Kendall Company

Purpose. To recognize and encourage outstanding scientific contributions to colloid or surface chemistry in the United States and Canada.

Nature. The award consists of $2,000 and a certificate. An allowance of not more than $350 is provided for traveling expenses to the meeting at which the award will be presented.

Establishment and Support. The award was established in 1952 by The Kendall Company.

Rules of Eligibility. A nominee must be a resident of the United States or Canada and must have made outstanding scientific contributions to colloid or surface chemistry. Special consideration will be given to the independence of thought and the originality shown.

1954

HARRY N. HOLMES

OBERLIN COLLEGE
OBERLIN, OHIO

. . . an authority on colloid chemistry.

The following Kendall Award Address of the late Prof. Holmes is reproduced by permission from
The Journal of Chemical Education, **31**, 600 (1954)

THE GROWTH OF COLLOID CHEMISTRY IN THE UNITED STATES[1]

HARRY N. HOLMES
Oberlin College, Oberlin, Ohio

It was heart-warming to be selected as the first recipient of the Kendall Company Award in Colloid Chemistry, for I labored long—and early—in the vineyard. I feel certain that this honor comes to me, in part, as a recognition of my pioneering efforts in this field.

As long ago as 1910 I gave some laboratory instruction in the preparation of colloids (the first in this country) at Earlham College, and in 1913, bursting with enthusiasm but very inadequate knowledge, I published a sixteen-page review of the general subject in *School Science and Mathematics*.

To the best of my knowledge the second institution to give considerable attention to colloid instruction was Massachusetts Institute of Technology early in 1916.

My first colloid research paper, "Electrostenolysis," published in 1914, got in by the back door under cover of Freundlich's insistence that "Capillarchemie" was a better name than "Kolloidchemie." The theme of this modest paper was that certain metals and metallic oxides could be deposited by a direct current of electricity in very fine capillaries placed between the electrodes.

While teaching at Earlham I listened to a lecture on colloid chemistry by Wolfgang Ostwald, then visiting America, and had stimulating talks with Martin H. Fischer in Cincinnati. These two able chemists strengthened me in my determination to get in on the ground floor of a promising subject.

Prior to 1915 very few papers on colloid chemistry were published in any single year in the journals of this country, although there was much activity in Germany. Here, W. D. Bancroft and Martin H. Fischer were the most active in such research before 1915. As early as 1902 W. R. Whitney was interested, briefly and quietly, in colloidal gold, while in 1903 Cushman published on rock dust and Ashley in 1909 wrote on soils.

Up to 1920 the other authors were Harkins, Langmuir, Gortner, Mathews, Loeb, Alexander, Spence, Brigge, Newman, Bogue, Nagel, Holmes, and a few more. These were joined in the period 1920–30 by a distinguished group including Sheppard, Weiser, Bingham, Taylor, Bartell, McBain, Patrick, Miller, Bradfield, Hauser, Wilson, and Thomas. The increase in papers published since 1930 has been phenomenal.

In 1923 J. H. Mathews and I planned the first Colloid Symposium and, with the aid of an enthusiastic committee, made it a reality, at the University of Wisconsin. The success of this meeting (and its monograph) inspired other groups in various fields to follow suit.

NATIONAL RESEARCH COMMITTEE CAMPAIGN

In 1919 the National Research Committee appointed me chairman of its Committee on Colloid Chemistry, a position that I held for six years. At the request of this committee, in 1921 I wrote the first laboratory manual of colloid chemistry used in this country (three editions). Later I wrote a small text, but there were other, better ones available.

Later I contributed chapters to "Medical Physics," "Collier's Encyclopedia," and the comprehensive books by Alexander and by Bogue.

The Committee wished to spread the gospel in every way possible, so some of us were urged to lecture on our subject whenever invited. I did more than my share, for I addressed a large number of sections of the A. C. S. and spoke at nearly all our larger universities as well as at eight universities in Europe, all the time learning more than I taught.

My own research dealt with dialysis, gels, silicic acid gels, silica gels, catalysts, adsorption, emulsions, and some odds and ends. Finally, in a very logical way, I moved over into the vitamins and biochemistry in general.

DIALYSIS

Soon after moving to Oberlin in 1914 I published a paper on "Washing precipitates for peptization" and a note on "An inexpensive dialyzer." Dialysis seems to have held my fancy, for I came back to it from time to time. For my laboratory manual I developed the idea of "hot dialysis"—as faster than the usual method. With A. L. Elder in 1931 a paper was published on "A new cell for electrodialysis." This device speeded up removal of ionic impurities. Again for the manual, I introduced the use of artificial sausage skins in dialysis.

Of more interest to biologists was my paper, "A practical model of the animal cell membrane," 1939. In this work a membrane (quickly and cheaply prepared) permitted passage of either water-soluble or oil-soluble material, quite similar to the cell walls of animal bodies.

GELS

Gels (or jellies) occupied my attention for several years. Three papers, 1916, 1918, and 1919, with Rindfuss, Arnold, and Fall, dealt with the peptization, dialysis, and gel formation of ferric arsenate and ferric

[1] Presented as the Kendall Award address at the 125th Meeting of the American Chemical Society, Kansas City, Mo., March, 1954.

phosphate. The ferric arsenate jelly was beautiful and tempting but obviously had no future as a table jelly. "Jellies by slow neutralization," with Paul H. Fall, 1919, developed the theme that when quick mixing of two solutions yields only a gelatinous precipitate, slow mixing by diffusion through a membrane may yield an excellent jelly.

In the first "Colloid Symposium Monograph," 1923, I published a somewhat general paper, "Gel formation," in which I offered the "brush-heap" theory of gel structure. This resulted from the observation that many beautiful gels were composed of minute crystals with adsorbed and entangled liquid. Also in this paper was the observation that addition of various dehydrating agents to pectin sols caused formation of good pectin jellies.

"The influence of a second liquid upon the formation of soap gels," with R. N. Maxson, 1928, brought out the fact that although anhydrous alkali soaps in general are dispersed with difficulty in hot solvents such as turpentine and benzene, they readily form sols when a relatively small amount of water, oleic acid, or similar polar solvent is present. There are narrow limits to the percentage of polar additive. On cooling, gels (not of long life) are formed, showing increased solvent holding power. "Silicic acid gels," 1918, was the first of a long series of related papers dealing with hydrated silica, dry silica, adsorption, and catalysis. Contrary to previous belief, it was shown that at high concentrations of the acid mixed with sodium silicate gels do form. The dehydrating influence of nonionized molecules of the acid is the determining factor in time of set.

"The formation of crystals in gels," 1917, used the theory that the influence of gels on crystal formation is largely that of regulated diffusion of reacting substances. Although silicic acid was generally used in the experiments, fine powders of sulfur, etc., also served, thus explaining many geologic phenomena. Basic silicic acid gels permitted formation of crystals not possible in the usual acid gels. Gold salts in an acid gel (made with dilute sulfuric) were reduced by inward diffusion of oxalic acid to yield gleaming crystals of gold.

In the next paper, "Rhythmic banding," 1918, controlled conditions permitted formation of beautifully colored bands of colloidal gold, or sharp bands of copper chromate, or of basic mercuric chloride, etc. A new theory of these "Liesegang rings" was advanced.

"The vibration and syneresis of silicic acid gels," 1919, with Kaufman and Nicholas, described the special preparation of silicic acid gels which, half-filling a test tube, vibrated musically on sharp tapping of the tube. They vibrate as rigid solids, rigid because of adhesion to the glass and a great tendency to contract. In vaselined test tubes such gels contracted visibly and pulled away from the walls. A fascinating stunt was tapping a pint milk bottle less than half-filled with the gel. The resulting sound was reminiscent of cathedral bells.

SILICA GELS

The general interest in Patrick's "silica gel" led us at Oberlin into certain experiments that resulted in a paper, "A new type of silica gel," 1925, with J. A. Anderson. We added a solution of ferric chloride to a solution of sodium silicate and obtained as a gelatinous precipitate a molecular mixture of hydrated ferric oxide and hydrated silica. After drying it to a rigid structure we dissolved out the ferric oxide with dilute hydrochloric acid. After washing and drying, a very porous, chalk-white gel of pure silica (slightly hydrated) resulted. In the next paper, "Increasing the internal volume of silica gels by moist heat treatment," 1926, with Sullivan and Metcalf, we checked the initial drying at 60 per cent water content, allowed the partly dried gels to sweat or synerize for a week or two in closed vessels, and then boiled them for a few hours in dilute acid before washing and final drying. Structure was set to a noncollapsing state so that capillaries resulted both from later loss of water and from removal of ferric oxide (as ferric chloride).

From a stream of air saturated with benzene vapor at 30°C. the Patrick type of silica gel adsorbed 32 per cent of its own weight, while our best heat-treated gel adsorbed 126 per cent. However, each has its advantages and disadvantages.

CATALYSTS

It was obvious that such a porous solid could serve as effective support for active catalysts, platinum, for example. A paper, "Platinized silica gels as catalysts for the oxidation of sulfur dioxide," 1929, with Ramsay and Elder, resulted. A high degree of efficiency was obtained.

About this time Jaeger announced a promoted vanadium catalyst rather crudely mixed with a silicious support. We sought to improve this product by mixing solutions of potassium metavanadate and ferric chloride to obtain a precipitate which we peptized, or colloidally dispersed, with sufficient excess of ferric chloride. This colloidal solution was then added to sodium silicate solution to obtain a precipitate composed of catalyst, promotor, and silica support—all molecularly mixed.

After proper drying and testing at 500°C. with a gas flow of 8 per cent sulfur dioxide, an efficiency of 98 per cent was observed. Details are given in the paper, "Vanadium compounds as catalysts for the oxidation of sulfur dioxide," 1930, with A. L. Elder.

Since the usual precipitation of catalysts throughout porous solids may be far from uniform it seemed advisable to try to secure uniformity by first soaking the solid in a solution of the salt desired, salts of platinum, iron, etc., drying, admitting a suitable water-soluble gas (NH_3, H_2S, etc.), and finally immersing the porous solid containing the dry salt and adsorbed gas in water. Delayed reaction then resulted. The idea of delayed reaction was extended so as to reduce platinum salts by carefully moistening the dried gel containing sodium chloroplatinate with a formalin solution. Fortunately, reduction in this instance is slow at room temperatures. Upon raising the temperature to 100°C. reduction is completed. ("Uniform distribution of catalysts in porous solids," 1928, with R. C. Williams, was the resulting paper.)

My final excursion into the field of catalysis followed isolation of vitamin A in the Oberlin laboratories. Other workers had observed orange and red color bands while filtering solutions of fish-liver oils through columns of fine white powders. They could not prove their suspicions that these bands were due to changes in the vitamin A known to be present. As soon as Corbet and I obtained the vitamin without confusing impurities—a great advantage for us—we tried it in seven different solvents on seven different adsorbents. Catalytic formation of orange-red bands was observed in several instances. Evidently care is called for in such chromatographic treatment of sensitive organic compounds. Details are given in "Catalytic effects of porous powders on pure vitamin A," 1939, with Ruth Corbet.

ADSORPTION

Bancroft once told me, "Adsorption is the heart of the subject." With porous gels at hand, it was natural for me to delve in this field. I questioned Freundlich's extension of Traube's rule to nonaqueous solution, that adsorption of organic substances from aqueous solution increases strongly and regularly as we ascend the homologous series.

Since Freundlich's studies dealt with carbon, a nonpolar solid, and water, a polar liquid, it occurred to me that the order of adsorption in a homologous series of fatty acids must be reversed when the acids are adsorbed on silica, a polar solid, from toluene, a nonpolar liquid. This surmise proved to be correct, as shown in a paper, "The reversal of Traube's rule of adsorption," 1928, with J. B. McKelvey.

"The removal of sulfur from petroleum distillates," 1932, with Elder and Beeman, was a report on experiments with silica gels and with our own type of silica gels impregnated with copper sulfide. With careful preparation, adsorption was high. In the above paper an improved form of the sulfur lamp for the oil industry was described.

Thiophene-free benzene has some importance in the manufacture of certain dyes, so it seemed worth while to inquire why treatment of ordinary benzene with anhydrous aluminum chloride had never proved successful. Reaction alone was not enough, for the colored product was somewhat soluble and at high temperatures removal by adsorption was poor. We secured complete removal of thiophene by successive treatments at about 35°C., a temperature high enough for reaction yet sufficiently low for effective adsorption. The details are given in the paper "Removal of thiophene from benzene," 1934, with N. Beeman.

My determination to attempt the isolation of fat-soluble vitamin A seemed to make necessary a preliminary research, "The adsorption of fats from volatile solvents," 1930, with J. B. Thor. Polarity influence was evident.

The next effort, "Comparative studies on the adsorption behaviors of crude vitamin A, carotene and cholesterol," 1933, with Lava, Delfs, and Cassidy, emphasized the importance of polarity in solvents and in the adsorbents used, carbon, our silica gel, and eight types of alumina. A by-product of this research was "The isolation of carotene," 1932, with H. M. Leicester.

Near, but not complete, success in isolation of vitamin A from the nonsaponifiable fraction of fish-liver oils was reported in the paper, "Preparation of a potent vitamin A concentrate," 1935, with Cassidy, Manly, and Hartzler. In Europe, Karrer had been reviving use of adsorption columns, as described by Tswett in 1906, so I adopted and varied this valuable technique for separation of impurities from vitamin A. It is probable that I was the first to introduce chromatography into this country. Karrer extruded the wet columns and cut them into sections, while I preferred to wash the adsorbed bands through, collecting the filtrate in fractions. Carbon had not been reported for Tswett columns, partly because it obscured the colors of bands and, in vitamin A research, had caused serious oxidation loss. The latter objection was removed in the Oberlin laboratory by thorough removal of air in carbon activation and replacement with nitrogen.

Filtering a solution of a concentrate, first through carbon and then through magnesium oxide, we obtained finally a much more potent concentrate of vitamin A than any previously reported from Europe. However, since we realized that probable catalytic effects of the porous powders would prevent isolation of a pure product, the plan of attack was greatly changed. Crystalline vitamin A containing one molecule of methyl alcohol of crystallization was obtained, as reported in "The isolation of crystalline vitamin A," 1937, with Ruth Corbet. Later, other workers recrystallized this product from methyl formate without any solvent of crystallization.

EMULSIONS

My interest in emulsions began rather early and was heightened by Bancroft's adsorption-film theory. A series of papers resulted: "Gelatin as an emulsifying agent," 1920, with W. C. Child; "Cellulose nitrate as an emulsifying agent," 1922, with D. H. Cameron; "Polar emulsifying agents," 1925, with H. A. Williams; "Chromatic emulsions," 1922, with D. H. Cameron; "Emulsion films," 1925; "Iodine as an emulsifying agent," 1925; and an extended review of my work in a French journal made up the list.

The beautiful and spectacular chromatic emulsions with all the colors of the rainbow excited most interest. In this research it was shown that transparent emulsions result when the refractive index of the two liquid phases is the same while the optical dispersive power is also the same. However, when the optical dispersive power of one phase is much greater than that of the other, beautiful color effects are obtained.

1955

JOHN W. WILLIAMS

UNIVERSITY OF WISCONSIN
MADISON, WISC.

. . . for fundamental research achievements in the science of colloid chemistry, particularly for improving the theory and practice of the methods of sedimentation analysis and for applying them in the solution of problems in biology and medicine.

The following Kendall Award Address is reproduced by permission from *The Journal of Chemical Education,* **32,** 579 (1955). The addendum has been written by Prof. Williams for this volume.

THE USES OF THE ULTRACENTRIFUGE[1]

J. W. WILLIAMS
University of Wisconsin, Madison, Wisconsin

In MAKING an appearance of this kind there are two general thoughts that I might develop. I could attempt to treat the subject "Thirty years of colloid chemistry" or I could seek to trace my own experiences with the title "Thirty years in colloid chemistry." This is the period of time which has elapsed since I received my final university degree. Actually, in the time available I shall be able to do neither, but I do want to make a short and simple analysis of what appear to me to be the trends in colloid chemistry during this era before speaking in more detail about a subject within the general discipline with which I have become associated, namely, that of sedimentation analysis and the uses of the ultracentrifuge.

The theoretical and experimental methods for the study of colloids and macromolecules have been vastly improved and developed since the days of Graham and more recently of Einstein, and Zsigmondy and Perrin. Thus, the electron microscope has replaced the ultramicroscope as the tool for the study of particle size. With it not only colloidal particles but also the larger macromolecules are now made visible, and one can determine by direct observation the size and shape of individual virus, hemocyanin, and certain polysaccharide macromolecules. Again, size distribution curves for the classical inorganic colloids are made readily available from measurements made on the electron microphotographs.

For the organic macromolecules of the protein chemist and the high polymer chemist, the ultracentrifuge has become an important instrument for the absolute determination of molecular weight and weight distribution. Along with its development have come greatly improved procedures for the observation and description of linear and rotary diffusion phenomena.

With the passage of these 30 years the emphasis in colloid chemistry has shifted from the inorganic to the organic systems. With the new tools it has become possible to contribute to the understanding of the functions of these organocolloids in life processes and in technology. It is largely through them that our subject has become closely allied to biochemistry and medicine on the one hand, and high polymer chemistry on the other.

As we come to the consideration of my own participation perhaps I may be permitted to remark that two courses which I took at the University of Wisconsin made a deep and lasting impression upon me. One of them, given by Professor T. Svedberg in 1923, described the subject "colloid chemistry" and the other, presented by Professor P. Debye in 1927, treated the topic "polar molecules." I am sure you are familiar with the two well known monographs which form the printed record of these lectures.[2,3]

Superficially the two subjects might have seemed to be far apart; actually, it has turned out that they have many points of close relation. The title of one of my earlier articles[4] "Dipole theory and the size of molecules," is suggestive of one important point of connection, especially when I remark that were I now to assign the title I would use the word "macromolecules" instead of "molecules." At this time and with the use of the dipole theory, remarkably close estimates were made of the molecular weights of such protein molecules as ovalbumin, serum albumin, hemoglobin, and zein, the object having been stated to be "the hope the results obtained in this way may supplement those made available by using the more common sedimentation, diffusion, and osmotic pressure methods."

One interesting result of this particular endeavor was the fact that the zein, which we now know must have formed a highly polydisperse system, gave an average molecular weight of 22,000 from its time of relaxation, while the minimum molecular weight found in textbooks of biochemistry was 100,000. It was a direct result of my desire to prove what I considered to be the essential correctness of my own result that I became interested in sedimentation analysis. It did not matter that we discovered later that the higher (so-called minimum) value was based solely upon an analytical amount of an amino acid in the protein which was eventually shown to be completely absent; we were already embarked on a program of research built around the ultracentrifuge.

The subject, "The uses of the ultracentrifuge," is a broad one. In ultracentrifugal analysis there are two distinct approaches. The machine itself must be one in which convection currents due to temperature inequalities along the column and vibrations due to rotor unbalance are eliminated, so that proper mathematical analyses can be made either of the movements (sedimentation velocity) or of the redistributions (sedimentation equilibrium) of the components of a solution as the rotor revolves at a constant angular velocity. Relatively low speeds of a few thousand revolutions per

[1] Presented as the Kendall Award address at the 127th Meeting of the American Chemical Society, Cincinnati, March, 1955.

[2] SVEDBERG, "Colloid Chemistry," Chemical Catalog Co., Inc., New York, 1924.
[3] DEBYE, "Polar Molecules," Chemical Catalog Co., Inc., New York, 1929.
[4] WILLIAMS, Trans. Faraday Soc., 30, 723 (1934).

minute are sufficient to give sedimentation rates of the same order as the diffusion rates of proteins and organic high polymers, and so to produce a measurable redistribution of concentration at sedimentation equilibrium. High speeds of some 50,000 to 75,000 revolutions per minute are required to overcome diffusion and produce a moving boundary, the rate of movement of which can be said to be observed in sedimentation velocity experiments.

For this reason an ultracentrifuge laboratory should possess equipment to study solutions and suspensions at centrifugal forces from a few times gravity up to the highest attainable forces. There are several types of successful ultracentrifuge on the market, so they need not be now described. The machine itself must be provided with a suitable optical system for the process under observation. The particular choice seems now to be shifting from a system based upon ray optics (Lamm, Philpot, Svensson) to one which makes use of wave optics (Rayleigh, Gouy, Jamin), because of the greater inherent accuracy of the latter.

THEORY

The concentration distribution of small, monodisperse particles settling through a viscous medium in various force fields has been the subject of a number of theoretical investigations.[5] The attempt, limited in scope, will be made here to develop what we shall call an "equation of the ultracentrifuge," and to show that from it there may be obtained suitable equations for use with the actual data of sedimentation equilbrium and sedimentation velocity experiments.

Before going ahead we wish to observe that there are two general approaches in an effort of this kind—the kinetic and the thermodynamic analyses of the problem. The former utilizes a microscopic picture of the colloidal system and provides the simplest derivations of equations to describe the observed phenomena. The equations derived by using the kinetic analysis approach validity only very near to infinite dilution. On the other hand, thermodynamics considers the over-all energy of a mass of material rather than the molecular motions in it, and can be applied rigorously to any system at (or nearly at) equilibrium. Very important, it provides a means for extending the laws of infinite dilution to high concentrations.

It is our present opinion that not enough attention has been paid to the fact that the equations as ordinarily printed and used are limiting laws. They have usually come from a kinetic analysis, so they do not make allowances for the effects of solute-solvent and solute-solute interactions. Some of them have led to a vast literature devoted to determinations of the shape of macromolecules and what is called their extents of solvation. This is a subject upon which we shall later comment.

[5] MASON AND WEAVER, *Phys. Rev.*, **23**, 412 (1924); WEAVER, *ibid.*, **27**, 499 (1926); FAXEN, *Arkiv. Mat., Astron., Fysik*, **21B**, No. 2 (1929); LAMM, *ibid.*, **21B**, No. 2 (1929); ARCHIBALD, *Phys. Rev.*, **54**, 371 (1938); SVEDBERG AND PEDERSEN, "The Ultracentrifuge," Clarendon Press, Oxford, **1940**.

In the thermodynamic analyses there are a few basic notions which we shall use. The motion of an uncharged macromolecule in an ultracentrifuge is defined by the sum of the centrifugal potential, $\dfrac{-M\omega^2 x^2}{2}$, and the chemical potential, μ_2, i. e., the total potential, $\bar{\mu}_2$, and the frictional resistance with which the large molecule opposes the motion. If the macromolecule is charged, the potential of the electrical field must be included with the others to give the total potential. Since our equations *apply only to a two-component system* we shall use the subscripts 1 and 2, the former for the solvent and the latter for the solute. We need not have been so restricted in the treatment of the sedimentation equilibrium problem.

For sedimentation equilibrium the underlying thought is that the total Gibbs potential for the solute of the system is constant throughout the cell, i. e.:

$$\bar{\mu}_2 = \mu_2 - \frac{M\omega^2 x^2}{2} + z_2\phi$$

In this completely general equation M_2 is the molecular weight of the solute, z_2 is its valence, ϕ is the electrical potential, ω is the constant angular velocity of the ultracentrifuge rotor, and x is the distance from the center of rotation.

For sedimentation velocity the negative gradient of total potential measures the driving force on the macromolecule. At constant velocity of fall of the macromolecule this driving force becomes equal to the frictional force, or:

$$f_2\left(\frac{dx}{dt}\right)_2 = -\left(\frac{\delta\bar{\mu}_2}{\delta x}\right)_t$$

where f_2 is a frictional coefficient.

Where a molecule is caused to sediment in a centrifugal field, neither diffusion (with gradient of chemical potential as driving force) nor the frictional resistance of the molecule to motion can be neglected. It is this thought which leads to the development of the equation of the ultracentrifuge. It contains cases of both sedimentation equilibrium and sedimentation velocity. In the latter case we are still restricted to a two-component noncompressible system. We shall consider the behavior of an uncharged macromolecule. Then the total potential of this component is:

$$\bar{\mu}_2 = \mu_2 - M_2\frac{\omega^2 x^2}{2}$$

The symbol μ_2 represents the chemical potential of the constituent at the actual pressure p. At constant temperature and composition the variation of μ_2 with pressure is described by the equation:

$$\mu_2 = (\mu_2)_{p_o} + \int_{p_o}^{p} \frac{\delta\mu_2}{\delta p}\, dp$$

where $\dfrac{\delta\mu_2}{\delta p} = \bar{V}_2 =$ partial molal volume $= M_2\bar{v}_2$, $p_o =$ pressure at the meniscus in the cell, and $dp = \omega^2 x \rho\, dx$, where $\rho =$ density of the *solution*. The quantity M_2

is the molecular weight of the solute, and $\bar{v}_2$ is its partial specific volume.

Thus:

$$\bar{\mu}_2 = (\mu_2)_{po} - \frac{M_2\omega^2}{2}\left(x^2 - 2\int_0^x \bar{v}_2\rho x\, dx\right)$$

On the molar basis, the driving force on the solute molecule is:

$$F_2 = -\frac{\delta\bar{\mu}_2}{\delta x} = M_2\left(\frac{d^2x}{dt^2}\right)_2 + f_2\left(\frac{dx}{dt}\right)_2$$

This is the equation of motion.

Since the acceleration $(d^2x/dt^2)_2$ is negligible, and since $(\mu_2)_{po} = (\mu^o{}_2)_{po} + RT\ln y_2 c_2$, we write for the equation of the ultracentrifuge:

$$\left(\frac{dx}{dt}\right)_2 = -\frac{RT}{c_2 f_2}\left(1 + c_2\frac{d\ln y_2}{dc_2}\right)\frac{\delta c_2}{\delta x} + \frac{M_2\omega^2 x}{f_2}(1 - \bar{v}_2\rho)$$

In this equation c_2 is the concentration of solute (volume basis) and y_2 is its activity coefficient.

Sedimentation Equilibrium. At equilibrium $\left(\dfrac{dx}{dt}\right)_2 = 0$. For the ideal case, where $y_2 = 1$, and with a homogeneous solute,

$$M_2 = \frac{RT}{(1 - \bar{v}_2\rho)\omega^2}\frac{dc_x/dx}{x\cdot c_x} = \frac{dc_x/dx}{c_x\cdot 2Ax}$$

where $\dfrac{1}{2A} = \dfrac{RT}{(1 - \bar{v}_2\rho)\omega^2}$, $\bar{v}_2$ is the partial specific volume of the solute, and c_x is its concentration at a distance x from the center of rotation.

If the system is not ideal, the molecular weight as a function of solute concentration may be expressed by the formula:

$$\frac{1}{M_2} = \left(\frac{1}{M_2}\right)_o + Bc$$

where $(1/M_2)_o$ is the reciprocal of the true solute molecular weight, i. e., at infinite dilution. In an equivalent expression:

$$(M_2)_o = \frac{dc_x/dx}{c_x(2Ax - B(dc/dx))}$$

The constant B is used as a measure of the nonideality of the solute.

In nonideal and heterogeneous systems the expressions become still more complicated. However, the experiment does make available several of the different types of average molecular weight and thus the means to describe the heterogeneity. More complete thermodynamic theory and the application of some molecular weight-distribution functions are found in two recent publications from this laboratory.[6]

Sedimentation Velocity. The quantity $\delta c_2/\delta x = 0$ over much of the cell. The sedimentation coefficient may be a function of concentration but not of pressure.

For the ideal case with the two component system, then:

$$M_2(1 - \bar{v}_2\rho)\omega^2 x = f_2(dx/dt)_2$$

with the sedimentation coefficient being defined as $s_2 = (dx/dt)_2/\omega^2 x$. When the sedimentation rate $(dx/dt)_2$ has been observed, there is still an unknown quantity, the frictional coefficient f_2. It is ordinarily evaluated by an independent observation of the diffusion coefficient, since according to kinetic theory, $f_2 = RT/D_2$, at infinite dilution. The familiar Svedberg equation has been obtained by substitution to give $M_2 = \dfrac{RTs_2}{D_2(1 - \bar{v}_2\rho)}$.

We note now that the density ρ becomes that of the solvent in which the macromolecule is suspended, and $\bar{v}_2$ is the partial specific volume of the solute. The values of s_2, D_2 and $\bar{v}_2$ are extrapolated to infinite dilution of the solute after correction of s and D for the viscosity effects of the solvent. The procedure of extrapolation is an arbitrary but practical one.

The customary derivation of the Svedberg equation is based upon kinetic theory and is unsatisfactory from several points of view. An outline of the thermodynamic argument is here given, again with the caution that it applies strictly only to a two-component system, one in which any volume changes which occur as a result of the transport process are negligible. The frictional force, a frictional coefficient multiplied by a velocity, is set equal to the negative of the gradient of total potential, thus:

$$f_2\left(\frac{dx}{dt}\right)_2 = -\left(\frac{\delta\bar{\mu}_2}{\delta x}\right)_t = -\left[\frac{\delta}{\delta x}\left(\mu_2 - M_2\frac{\omega^2 x^2}{2}\right)\right]_t$$

$$= -\left[\left(\frac{\delta\mu_2}{\delta x}\right)\left(\frac{\delta p}{\delta x}\right) - M_2\omega^2 x\right]$$

$$= -\omega^2 x[M_2\bar{v}_2\rho - M_2]$$

$$s_2 = \left(\frac{dx}{dt}\right)_2\bigg/\omega^2 x = \frac{M_2(1 - \bar{v}_2\rho)}{f_2}$$

The usual sedimentation velocity experiments are performed in three-component systems, but their evaluation and the common computations for the size, shape, and degree of solvation of the dissolved macromolecule are based upon equations such as those given above which apply strictly only to two-component, incompressible systems.

The equations descriptive of the sedimentation velocity behavior in three-component systems are still pretty much a matter of the future, but progress in this direction has begun. Hooyman and associates[7] *claim* to have derived the Svedberg equation in the framework of the thermodynamics of irreversible processes which permits a rigorous definition of all the quantities involved, putting the nonequilibrium case on a basis which meets the same requirements of exactness as the equilibrium equation. Gosting, Baldwin, and Dunlop[8] have initated what seems to be a very significant program in the study of interacting flows in liquid diffusion.

The quantity s is described as a coefficient, not a

[6] GOLDBERG, *J. Phys. Chem.*, 57, 194 (1953); WALES, ADLER, AND VAN HOLDE, *J. Phys. Colloid Chem.*, 55, 145 (1951).

[7] HOOYMAN, HOLTAN, MAZUR, AND DEGROOT, *Physica*, 19, 1095 (1953).

[8] GOSTING, BALDWIN, AND DUNLOP, forthcoming publications.

constant. It is somewhat variable with solute concentration. In addition, it changes with the pH and ionic strength of the solution. It is independent of the field strength, $\omega^2 x$.

In the case of a polydisperse solute one speaks of a distribution of sedimentation coefficients, $g(s)$, which specifies the weight fraction of the sample with sedimentation coefficients:

$$g(s) = \frac{1}{c_o}\frac{dc}{ds}$$

Here c_o is the total concentration and $\frac{dc}{ds}$ is the derivative of the plot of s versus concentration of material having a sedimentation coefficient less than or equal to s. When the effects of diffusion are negligible and the sedimentation coefficients are constant, independent of concentration, $g(s)$ is obtained directly from the boundary gradient curve of $\frac{\delta c}{\delta x}$ versus x, according to the equation:

$$g(s) = \frac{\delta c}{\delta x}\,\omega^2 t x^3 / c_o x^2_o$$

The relationship between x and s is given by the integrated form of the definition of sedimentation coefficient, i. e., $x = x_o e^{s\omega^2 t}$. Thus, x_o is the distance between the meniscus and the center of rotation. In general, it is necessary to make allowance for the effects of diffusion and of variation of sedimentation coefficient with concentration in finding $g(s)$.

USES OF THE ULTRACENTRIFUGE

The use of centrifugal force in many important roles in industry is known to all. As a person representing the University of Wisconsin, it is natural that I should mention the Babcock cream separator as an example. But the instrument here involved might be termed a supercentrifuge as contrasted with the ultracentrifuge of the scientific laboratory, the machine of vibration-free and connection-free sedimentation with which we are now concerned. The quantitative measurement of sedimentation (and diffusion) behavior has provided means for the assignment of molecular characteristic constants to proteins and other polymeric materials.

The existence of well defined macromolecules in solution was proven by physical-chemical studies of protein solutions. The colloidal character of proteins was early recognized, but it was actually a matter of surprise when sedimentation analysis proved that many, if not most of the soluble proteins were built up of well defined molecular units. With the establishment of this fact there followed a rapid development of the tools themselves and of the extensions to a wide variety of the macromolecules of biology and medicine and of technology. More gradually, but none the less surely has come the realization that the sedimentation methods may be also used with success in investigations with smaller molecules. Some suggestions are made in the following paragraphs of their wide variety of applications. Since the sedimentation velocity experiment is the more familiar one it will be given preference in the discussion.

Sedimentation Velocity. The utility of the ultracentrifuge is nowhere better demonstrated than in the determination of the molecular weight for the soluble globular proteins by using the familiar Svedberg equation. That this is a whole subject in itself will be at once evident if reference is made to the Svedberg and Pedersen monograph or to more recent reviews of Lundgren and Ward[9] or of Edsall.[10] It is of course the familiar application.

There are times when one wonders if the application of this basic equation has become too familiar. We have noted that many users of the excellent machines now in use apply this equation in the evaluation of experiments with complicated systems (solvent, buffer and supporting electrolyte, protein). In addition, there has been built up a voluminous literature devoted to the assignment of shapes and extents of hydration to protein molecules. These molecular characteristics come from operations with the frictional force in the velocity problem. Obviously the sedimentation coefficient is a function of both of them, but at present there is no way to separate the effects of each in the modification of the frictional coefficient. The kinds of trouble into which one can get are well stated in the introductory paragraphs of a recent article in which the hydrodynamic properties of proteins are considered.[11] In a few words, the true hydrodynamically effective volume of the protein molecule in solution is inaccessible to exact definition. To this extent, at least, these authors have performed a valuable service even though to test the validity of their alternate approach seems at the moment to be quite out of the question. Sedimentation and diffusion coefficients are not known to within anything like the accuracy which would be required for the equations.

The familiar equations of Herzog, Illig, and Kudar and of Perrin have been put to use to compute the molecular characteristic dimensions of many highly elongated and even flexible macromolecules, but an examination of the assumptions underlying the development of these equations should have caused restraint. The type of difficulty into which one can be led is well illustrated by the literature pertinent to the shape of gelatin molecules in saline solution. It had been assumed, as before in the case of the globular proteins, that one dealt here with molecules whose form could be approximated as ellipsoids of revolution, so dimensions were assigned and much was said and even written about the mechanism by which plasma extender gelatin was lost after infusion into human veins, and how these molecules could be modified chemically to give them longer retention times per given molecular size. Actually, the gelatin molecules in solution take

[9] Lundgren and Ward, in Greenberg, Editor, "Amino-acids and Proteins," Thomas, Springfield, 1951.

[10] Edsall, in Neurath and Bailey, Editors, "The Proteins," Vol. I, Part A. Academic Press, Inc., New York, 1953.

[11] Scheraga and Mandelkern, *J. Am. Chem. Soc.*, **75**, 179 (1953).

the form of randomly-kinked chains;[12] thus the mechanism of elimination from the body is entirely different from that which had been envisioned. In this way there is explained the fact that gelatins, whether or not they have been subjected to chemical treatments, show little if any differences in retention times provided the simple average molecular weights are alike.

Now we are getting away from the purpose of the report. What I am trying to express is my belief that more attention should be devoted to the development and extension of the basic theory of the sedimentation velocity method, to put it to use in the determination of quantities which can be made the subject of exact definition. Molecular weight is one of them. It now can be said that even though the protein cannot be obtained in what the physical chemist would call pure form, the work of Drs. Gosting and Baldwin has made it possible to obtain the molecular weight of a homogeneous protein which is present as a major constituent of the system under observation. The experiments must be performed in dilute solution to eliminate or at least minimize the effects of any flow interaction of diffusion.

In polydisperse systems the distributions of sedimentation coefficient have been or are being evaluated for protein systems such as casein, gelatin, the gamma globulins, and the beta lipoproteins; and for polysaccharide systems such as the dextrans, glycogens, starches, and the pneumococcus polysaccharides. There are difficulties inherent in all of them. For example, in the case of the beta lipoproteins there is superimposed a distribution of sedimentation coefficients which is based upon solute density differences upon that which goes with variations in size and shape. One might proceed by making separations of the lipoproteins into classes of like densities prior to the analytical sedimentation velocity experiments, to give sedimentation coefficient distributions of size and shape in each density class. This appears not to have been done in the experiments of Gofman and associates and one wonders about the interpretation of their experiments.

The usual methods for the determination of sedimentation coefficients are based upon observations of a "moving boundary." It is also possible to obtain such data by using experiments in which accumulations of solute are determined. Thus, in the case of smaller molecules, often of great biological interest, the procedure is to determine the flow of solute through an arbitrary plane in the region of uniform concentration in the ultracentrifuge cell. This approach has been successfully utilized in the assignment of sedimentation coefficients to a number of polysaccharides of low molecular weight, and even to glucose. These are again two-component systems, and an adequate theory is available.

Biochemists are now beginning to use an experiment based upon the idea of accumulation, to provide sedimentation coefficient data for a protein constituent of some characteristic activity which has not been separated from other constituents. The system is now a multicomponent system, so that very carefully controlled experimental conditions are now required to insure a reasonable result.

The sedimentation-velocity method is beginning to be applied in the study of organic high-polymer systems, but many problems of theoretical and practical kinds remain to be worked out before the results can have fundamental and quantitative significance. At the pressures known to exist in the ultracentrifuge cells, the organic solvents become more dense and more viscous but the pressure dependence of these quantities is not accurately described in the literature and proper corrections cannot be made of observed sedimentation rates for reduction to standard conditions.

The low centrifugal force machine has the advantage that it is effective, not only in connection with sedimentation-equilibrium studies of the kind now to be listed, but also for operation as a sedimentation-velocity instrument to describe the particle size distributions in finely divided materials which extend from the finest of inorganic colloids up to particles large enough to measure by means of an ordinary good microscope. As examples, mention may be made of paint pigments, carbon-black suspensions, emulsions of neoprene latex, colloidal gold, etc. The particle size distribution is often an important variable in determining the ultility and merit of the powdered material. For instance, the covering power and tensile strength of a paint depends upon this distribution.

Sedimentation Equilibrium. Substantial progress has been made in the detailed study of the molecular weight, state of aggregation in solution, and molecular-weight distribution of a number of organic, long-chain high polymers, particularly cellulose, cellulose derivatives, rubber, rubber substitutes, polysaccharides, lignins, etc. In some cases there has been established the relationship between viscosity and molecular weight, in others the correlation between sedimentation constant and molecular weight, the purpose being to give valuable information about molecular form in solution. The unique description of the molecular mass heterogeneity has been worked out in principle, but in practice, using the optical systems based upon ray optics, the result is ordinarily subject to some arbitrariness. At the present time the equilibrium experiment is more satisfactory than the velocity one for observations of the heterogeneity in the systems containing flexible, long-chain molecules.

The sedimentation equilibrium method also offers new possibilities for the study of the inorganic electrolytes and polyelectrolytes. In the case of the former the experiment provides data for the determination of the activity coefficients of the solute, since at equilibrium the total potential (the sum of the chemical, electrical, and centrifugal potentials) does not vary with distance in the cell.

For the polyelectrolytes (sometimes called "colloidal electrolytes") new possibilities are provided for the study of micelle formation or solute association in

12 WILLIAMS, SAUNDERS, AND CICIRELLI, *J. Am. Chem. Soc.,* **58,** 774 (1954).

detergent and dye systems. In the case of the detergents an unfavorable density difference between solute and solvent (water) makes the experiments difficult. Certainly studies will eventually be made of the condition of the solute in solutions of the alkali silicates, stannates, tungstates and molybdates as it depends upon solute concentration and the presence of other electrolytes.

Combination of Sedimentation Velocity and Sedimentation Equilibrium. During recent years much has been learned about the means by which corrections for the effects of nonideality of solution can be made in both sedimentation-velocity and sedimentation-equilibrium experiments. The importance of these corrections cannot be overemphasized. With them, the results of sedimentation-equilibrium and sedimentation-velocity experiments can be combined to give two kinds of new information. For polydisperse systems in the equilibrium case, mention has been made of the fact it is practically necessary to assume a general form of the distribution function.[6] This means that any minor details of the distribution will fail to make their appearance. This is not true with the velocity diagrams and their evaluation, because any arbitrary distribution of sedimentation velocities can be handled. In the combination of the two experiments the latter distribution can be converted to a distribution of molecular weights to obviate the difficulty. Furthermore, there can be evaluated the relationship between sedimentation coefficient and molecular weight, $s = kM^\alpha$, at infinite dilution, to provide information about shape of the macromolecules in solution.[13]

[13] WILLIAMS, *J. Polymer Sci.*, 12, 351 (1954); WILLIAMS AND SAUNDERS, *J. Phys. Chem.*, 58, 854 (1954).

By this combination there may be achieved a sedimentation analysis, which was Svedberg's original purpose. It has been a long time in development, the chief reason being that there have been difficult problems in taking apparent values of sedimentation coefficient and molecular weight and converting them to data which are valid at infinite dilution.

ACKNOWLEDGMENT

For advice and help in connection with the development of a research program in sedimentation analysis I am greatly indebted to many teachers, associates, and graduate students. Their number is large, so that mention by name is not practical; nonetheless, the distinction of the 1955 Kendall Company award is conferred upon them as well as upon me.

I do wish to record here the fact that it was Dr. Warren Weaver of the Rockefeller Foundation and President E. B. Fred of the University of Wisconsin who saw to it that an earlier small effort of mine was enriched by the purchase, installation, and maintenance of two ultracentrifuges, nearly 20 years ago.

Over the years the scientific program has received substantial financial aid from the University of Wisconsin, the Wisconsin Alumni Research Foundation, the Rockefeller Foundation, the National Institutes of Health, the Office of Naval Research, and other institutions. I am greatly indebted to them.

To two of my most recent associates, Drs. L. J. Gosting and R. L. Baldwin, I must make an especial acknowledgment. High loyalty and devotion on the one hand, and unusual scientific ability on the other, are combined to an extraordinary degree in them.

ADDENDUM 1972

John W. Williams

University of Wisconsin
Madison, Wisconsin

Beginning in the mid-1920's with the famous researches of Svedberg (with Rinde and with Fåhraeus) the uses and purposes of the ultracentrifuge have remained today substantially unaltered. The real changes beyond the publication date of our Kendall Award Address have come largely in extensions of area coverage, in the enrichment of theory and practice for the operations with the instrument, and in the development of more sophisticated apparatus.

The determination of the detailed double-spiral structure of DNA in 1953 was the achievement which brought on the dramatic development of the new molecular biology, or perhaps more specifically, the molecular genetics of today. So just as an analytical ultracentrifugation early had become one of the most important techniques for the investigation of protein and organic high polymer molecules, it has since assumed a parallel position in nucleic acid physical chemistry. Thus a great new domain has been added to the two areas of application considered in the 1955 survey. (cf. Vinograd, J. and Hearst, J. E., *Fortschr. d. Chem. Org. Naturstoffe,* **XX**, 372 [1962].)

Enrichment of the basic theory for ultracentrifugal analysis to include polydisperse systems both at sedimentation equilibrium and in transport really had begun two years prior to the preparation of our report; but only brief mention was made of it.

Gibbsian thermodynamics provided the means to derive the completely general mathematical statements which are necessary to describe the multicomponent, or polydisperse, nonideal systems at sedimentation equilibrium. Then with the use of certain mathematical approximations, the equations have been put into forms for practical application in the interpretation of the several types of experiment; "low-speed", meniscus depletion, etc. With the use of these extended equations and different methods the ultracentrifuge has had a record of substantial achievement in the treatment of polydisperse, nonideal systems, in protein self-association reactions, in the density gradient sedimentation equilibrium procedures of the nucleic acid chemist, and so on. (cf. Williams, J. W., Van Holde, K. E., Baldwin, R. L. and Fujita, H., *Chem. Revs.,* **58**, 715 [1958]. Fujita, H., "Mathematical Theory of Sedimentation Analysis", Academic Press, Inc., New York [1962].)

In an interesting development it has been shown that the equations descriptive of a ternary system, when formulated in terms of certain properly defined refractive index and density derivatives, reduce in form to that of the simple equation descriptive of the two component system, much simpler in use. When applied to the macromolecular solute, the molecular weight obtained for it will be very close to reality. These analyses are especially advantageous in connection with systems at sedimentation equilibrium. (cf. Casassa, E. F. and Eisenberg, H., *Adv. Protein Chem.,* **19**, 287 [1964].)

Transport in the cell under the influence of the centrifugal field is a typical rate process, one complicated by diffusion. As such, it has now been shown to be possible to describe a component flow in terms of the thermodynamics of irreversible processes, in this way to remove the restriction to the idealized two-component, incompressible system of the classical theory. One now derives a set of generalized flow equations of the Lamm type; it is upon this set that the extended sedimentation velocity equations have been built. It appears that even the simple Svedberg molecular weight equation really derives from thermodynamics. This equation had its source in statements which were not sufficiently precise in their representation of either the volume, $M_2 \bar{v}_2$, or the density, $\dfrac{1}{\bar{v}_2}$, of the molecule in solution. (cf. Hooyman, G. J., *Physica,* **22**, 751, 761 [1956]. Fujita, H., "Mathematical Theory of Sedimentation Analysis", Academic Press, Inc., New York [1962], Lamm, O., *Kgl. Tekn. Högskolans Handl.,* No. 134, 1 [1959]. Williams, J. W., "Ultracentrifugation of Macromolecules", Academic Press, Inc., New York [1972].)

The shift in the refractive index optical methods for the ultracentrifuge from systems based on ray optics to those which utilize wave optics has continued. The Rayleigh system has become quite popular; the Gouy method, while extremely useful in diffusion, is not adaptable to the ultracentrifuge. (cf. Richards, E. G. and Schachman, H. K., *J. Phys. Chem.,* **63**, 1578 [1959].)

Vastly improved accessories have been provided for the ultracentrifuge. Temperature controls, speed regulation devices, and automatic scanners are particularly deserving of mention. New instruments have been designed and assembled for the measurement of partial specific volume. And for the present regular user of the ultracentrifuge, the computer has become an indispensible instrument.

1956

VICTOR K. LA MER

COLUMBIA UNIVERSITY
NEW YORK, N. Y.

. . . for his many contributions to the science of colloid chemistry, especially for the formulation of the theory of the formation of monodispersed hydrosols and for improvements in the theory of the production and use of aerosols for military and other purposes.

The following article selected with the advice of Prof. Milton Kerker, a student of the late Prof. La Mer, as closest to his Kendall Award Address is from *Air Pollution* by Louis C. McCabe. Copyright 1952 by McGraw-Hill, Inc. Used with permission of McGraw-Hill Book Company.

THE PREPARATION, COLLECTION, AND MEASUREMENT OF AEROSOLS

Victor K. La Mer

Department of Chemistry, Columbia University, New York

Most of the important properties of aerosols, such as optical obscuration, inertial deposition, evaporation or condensation, thermal and electrical precipitation, filtration, insecticidal toxicity, etc., are strongly dependent upon the particle size of the disperse phase; hence, the measurement of size is of paramount importance in characterizing an aerosol. To describe size accurately with a single parameter, it is necessary that the particles be either spheres or cubes, a requirement which is not always met by solid aerosols. Unless all the particles are of the same size, it also becomes necessary to know the form of the distribution function and the parameters specifying it. The scanty evidence available today indicates that most aerosols, like the larger dust suspensions, conform better to the Hatch-Choate logarithmic distribution function than to the symmetrical Gaussian distribution[54, 55].

All of these difficulties can be avoided by working with monodisperse liquid aerosols, where a single parameter, the radius of the droplet, suffices for a description of their properties. By the term monodisperse we shall imply preparations whose size distribution does not deviate by more than 10 percent from the mean value.

PREPARATION OF MONODISPERSE AEROSOLS

In 1941, Dr. David Sinclair and the writer[56, 57] succeeded in preparing for the first time such monodisperse aerosols from a variety of substances, e.g., oleic and stearic acids, lubricating oils, menthol, etc. It now appears that almost any substance, liquid or solid, which can be vaporized without appreciable decomposition and which has a narrow range of boiling point, can be made into a monodisperse aerosol, by the carefully regulated condensation of the vapor upon suitable nuclei to form supercooled liquid droplets. When the stable phase at room temperature is a solid, the droplets of liquid are metastable, but nevertheless they usually persist in that state for considerable periods of time (e.g., sulphur aerosols[58], and supercooled natural water fogs) unless artificially seeded or cooled below $-38°$ C.

Figure 40 shows a diagrammatic sketch of a working model of a monodisperse aerosol generator made of pyrex glass. The essential parts consist of a boiler, d, in which the substance, c, is vaporized and mixed with a stream of well-filtered air through a, containing a regulated number of condensation nuclei produced in b. The gaseous mixture containing superheated vapor and nuclei goes into a reheater, e, where any residual spray is converted into molecular vapor. It then passes up through an air jacketed cooling column, f, several feet in height. It is essential that the apparatus be enclosed in a well-insulated box, since small fluctuations in the existing temperature gradients destroy the monodisperse character of the product.

The source of nuclei in b may be gaseous ions obtained from a spark discharge,[59, 60] sodium chloride vapor[61] produced by heating a plug of the solid just below red heat on a nichrome wire, or a very fine dust[62] of sodium chloride produced by atomizing a dilute aqueous solution in a nebulizer and evaporating to dryness by passage over sulphuric acid.

When no nuclei are present, condensation proceeds exclusively on the walls of the condensing column; when nuclei are present, the vapor condenses to a fog which, when illuminated from various directions, exhibits beautiful colors in passing up the column.

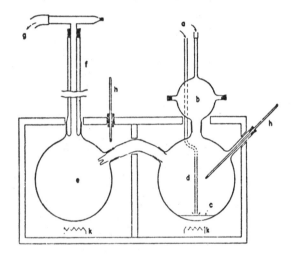

Fig. 40. All-glass monodisperse aerosol generator; a, inlets for diluting air (dust free); b, chamber for production of nuclei; c, liquid or solid to be vaporized; d, boiler; e, reheater; f, double-wall glass chimney; g, outlet; h, thermometers.

Brilliant colors appear only when the preparation is quite monodisperse; less monodisperse preparations exhibit pastel shades and polydisperse preparations are white or opalescent. This observation has furnished the key to improving the monodisperse character of a preparation.

The control of particle size is achieved in part by regulating the number of condensation nuclei produced but more particularly by regulating the temperature of the boiler. This temperature, along with the extent of dilution with air, determines the degree of supersaturation of the vapor when cooled to a given temperature in the column. Monodisperse growth is achieved by slow uniform cooling leading to uniform condensation and growth upon the nuclei. Our laboratory apparatus produces aerosols at the rate of about four liters per minute with radii within a limit of ± 10 percent. Higher rates of production yield less monodisperse preparations. Particle radii as small as 0.005μ and as large as 20μ have been made, but the customary useful range is from 0.02 to 2.0μ. Since the rate of coagulation is bimolecular and decreases as the square of the dilution, the aerosol should be diluted promptly with filtered dry air to preserve its monodisperse character.

OPTICAL PROPERTIES OF MONODISPERSE AEROSOLS

Light of a given wave length, incident upon an aerosol droplet, is scattered in a manner which can be predicted quantitatively knowing the angle of observation, θ, to incident light, the index of refraction of the substance, and the radius of the droplet. Conversely, knowing the radiation pattern one can compute the radius of the droplet.

Rayleigh (1871) showed that when the radius of the sphere was less than one-tenth of the wave length of incident natural light, the polar scattering diagram was symmetrical and of the angular form $(1 + \cos^2 \theta)$; that blue light was scattered more intensely than red; and that at $\theta = 90°$ to the incident beam, the scattered light was completely polarized since it consisted entirely of the component J_1, corresponding to light polarized in the vertical direction, as epitomized in the equation (see reference 56 for definition of symbols):

$$\frac{J_1}{J_2} = \frac{9\pi^2}{2R^2} \left| \frac{m^2-1}{m^2+2} \right|^2 \frac{V^2}{\lambda^4} \frac{1}{\cos^2 \theta} \quad (1)$$

As the particle radius increases, more and more light is scattered in the forward direction, the angle of maximum polarization is displaced from 90°, and the degree of polarization at 90° is decreased. Calibration of the polarization has furnished an excellent method of measuring particle size in the range 0.1 to 0.2 micron (see Fig. 41).

When the particle radius increases over 0.2 micron, Sinclair and the writer discovered (1941) the beautiful new optical effect referred to above, and now known[63] as Higher Order Tyndall Scattering Spectra (H.O.T.S), which is characterized by the appearance of alternate red and green bands in the scattered light when observed over the angle, θ, equal to 180° to the direction of the incident light.

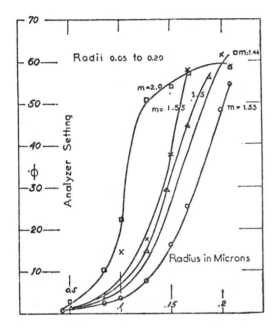

Fig. 41. Calibration diagram using the Cornu analyzer for the determination of the state of polarization ($\tan^2\phi = J_1/J_2$) for 90° scattering in the "Owl" as a function of the radius; m is relative index of refraction.

The particle radius in tenths of a micron is roughly proportional to the number of red bands observed in the angle 0 to 180° to that of the incident light. This ready method can be made more quantitative by determining to ± 1° the angle of maximum red intensity for each band (Fig. 42) from observations on aerosols whose size is known from measurement of the rate of sedimentation using the Stokes-Cunningham law for the calculation of the radius. In the case of monodisperse aerosols the upper boundary of the sedimenting cloud is sharp if convection currents are eliminated by careful temperature control.

In our initial researches[59] the various optical properties were calibrated solely by empirical methods, such as rate of sedimentation. Later they were checked with the theoretical predictions of the Mie theory of light-scattering, by using the complicated functions of that theory calculated (1941-1945) for us by Dr. Arnold Lowan and his associates of the Applied Mathematics Panel of N.D.R.C.

Mie Theory. In 1908, Gustav Mie gave a general analytical solution for the scattering of light by an optically isotropic sphere of any size in terms of the index of refraction and the angle of observation. The result is expressed in a complicated series of terms involving Bessel, Hankel, and Legendre functions[56]. Mie's theory has received quite in-

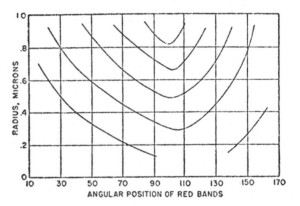

Fig. 42. Angular position θ of the maximum intensity of red scattered light as function of radius of droplet. Valid for water and glycerine — water aerosols.

adequate attention previously because the extensive tables of the functions, necessary for numerical checks of the theory, have become available only since 1941.

It suffices to state that Mie's equation reduces to a single term identical with the Rayleigh result (Equation 1) when $r < \lambda/10$. When $r >> \lambda$, the Mie equation approaches asymptotically the results of classical geometric optics. It therefore constitutes a complete solution for the problem of the scattering of light by spheres of any size by furnishing the functions describing the behavior in the complicated intermediate region in which the radius is comparable with the wave length of the light.

NMU LIBRARY

The writer and co-workers[64, 65] have subjected Mie's theory to exhaustive tests using monodisperse sulphur hydrosols as well as aerosols and have found the theory to be accurate in all details. It may be used with confidence for the prediction of all optical effects arising from scattering by spheres.

A portable apparatus[60] for measuring the angles of the H.O.T.S. and the degree of polarization at a selected angle bears the code name "Owl." A sample of aerosol is pumped into the circular observation chamber and the light scattered from a parallel beam is observed as a function of angle in the telescope. The semicircular chamber is fitted with a bipartite strip of polaroid film, one-half of which passes light perpendicularly polarized and the other half parallely polarized in respect to the plane of observation. Since the red and green bands appear at different angles for the two polarized components, the optical effect is intensified by contrast, when they appear in juxtaposition. The angles of maximum red intensity of one of the components are then plotted on a calibration diagram like Fig. 42. If the preparation is not strictly monodisperse, some of the inner orders may be blurred or obliterated leading to an incorrect result, when calculated from the total number of observable spectra. Recourse is then taken to measuring accurately the angles of the observable outer spectra. The results for the various spectra should plot horizontally on this diagram, otherwise one of the orders has been incorrectly identified. This method yields satisfactory results between 0.2 to 1.2 microns.

For larger sizes the classical corona method[66] is applicable or, better still, one may collect the aerosol on a glass slide and observe it under a light microscope. When no scattering orders are observed ($r < 0.2$ micron), the telescope is set at exactly 90°, and an analyzing polaroid inserted in the eyepiece. The degree of polarization is then measured by finding the angular setting, ϕ, of the analyzing polaroid which produces an equal illumination in the bipartite field (see Fig. 42, $i_1 / i_2 = \tan^2 \phi$).

Although the "Owl" works quite satisfactorily in the laboratory where the preparations deviate only slightly from strict monodispersity, it requires an experienced observer if estimates of the average particle size of the more polydisperse preparations encountered in the field are attempted. The transmission method (see below) tolerates somewhat wider deviations from strict monodispersity and is consequently a more practical method in such cases.

Recently, we have extended the methods based upon the Higher Order Tyndall Spectra–which may be described as based upon the dispersion of wave length as a function of angle–to a study of the maxima and minima in polarization and of the phase angle of elliptically polarized light as a function of angle of observation[65]. The characteristic patterns of polarization maxima and phase angle are similar to those for the H.O.T.S. They furnish alternate and equally reliable methods of size determination.

Transmission Methods. For transparent nonabsorbing materials, S, the scattering cross section of the particles is related to the transmission, T, by the equation:

$$T = I/I_0 = e^{-Snl} = e^{-Jcl}$$

Here I_0 is the incident, and I the emergent intensity of light of wave length, λ, n is the number of particles per cm.[3], c the mass concentration, J the scattering cross section per gram, and l the length of sample traversed by the light. It is convenient to define a scattering area coefficient $K = S / \pi r^2$ equal to the ratio of the optical, S, to the geometric scattering areas. K, shown in Fig. 43, has been computed from the Mie theory in terms of the dimensionless variable $a = 2\pi r / \lambda$. The exact shape of the K curve depends upon the relative refractive index of the aerosol substance as illustrated in Fig. 1 of reference 56. K is related to the optical density as $\log T = \log I_0 / I = K\pi r^2 nl$. When the transmissions are computed to a unit path length of $l = 1$ cm., only two variables, r and n, remain. They can be evaluated by making two measurements of T at two wave lengths, λ_1

and λ_2, and solving the equations simultaneously for r and n.

Rather than relying upon two measurements which may fall upon one or more of the "wiggles" in the K curve, it is safer to determine the entire transmittance curve, by making a large number of measurements over as wide a range of different wave lengths as is practical and then compare the observed curve

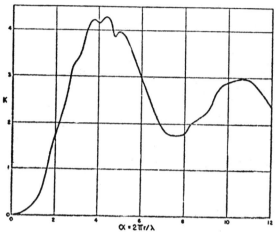

Fig. 43. Optical scattering coefficient $K = \dfrac{S}{\pi v^2}$.

S is the total scattering expressed as an optical obscuring area per gram of material and having the geometric cross-sectional area π_{r2}. The abscissa alpha is the dimensionless variable $2_\pi r/\lambda$.

with the theoretical K curve. This is accomplished best by plotting $\log K$ as ordinate against $\log \alpha = \log 2\pi r /\lambda$ as a standard curve. The experimental values of $\log \log I_0 / I$ are then plotted against $\log \lambda$ on a separate paper, and the curves superimposed. The transposition above the abscissa which is necessary to superimpose a characteristic part of the curve such as the first minimum or first maximum, permits the computation of $\log r$. Similarly, the transposition in the ordinates necessary to achieve superposition of the ordinates yields the value of n.

If the preparation is strictly monodisperse, the characteristic form of the K curve is reproduced even to the subsidiary wiggles, as has been shown by extensive studies on the sulphur hydrosols.[67,68,69] If the preparation is not strictly monodisperse, these "wiggles"

are smoothed out but the characteristic maximum and minimum remain. Figure 8, reference 62, shows some examples. In polydisperse preparations, only very shallow maxima and minima are perceptible, and it becomes difficult to determine a value of r. This is to be expected since the radius, r, has lost its precise meaning in such cases (see Fig. 6, reference 68).

With a Beckmann spectrophotometer, it is possible to determine the transmission curve in a quarter of an hour, and in a few more minutes the concentration and average radius of the particles can be calculated by the graphical procedure just outlined.

In the case of small particles, the characteristic maxima and minima occasionally do not fall within a convenient range of wave lengths. The following modification is helpful. Instead of plotting $\log K$ as ordinate, one plots $\log K/\alpha^z$, which is easily calculable since K is tabulated as a function of α; z may be 1, 2, or 3. Instead of $\log \log I_0 /I$, one then plots $\log \lambda^z \log I_0 /I$. The effect of dividing K by α^z is to shift the position of the maxima and minima into more easily accessible experimental regions of wave length. Although this procedure makes the maximum and minimum less pronounced, and thus renders the determination r less precise, this defect is offset by the easier determination of the transposition of the ordinate with the flatter curve. Consequently, there is an increase in the precision of determining n, the number of particles per unit of volume.

Optical Obscuration by Monodisperse Fogs. Several interesting and important conclusions may be drawn from an inspection of the K curve.

(*a*) The Rayleigh region where the intensity of scattering is proportional to r^6 /λ^4 is restricted to the beginning of this curve; namely, $\alpha < 0.03$, or $r < \lambda /20$.

(*b*) The intensity of the scattering (or decrease in transmission) increases rapidly with increasing values of r until a sharp maximum value is reached. Since the transmission is proportional to $r^2 K$, the maximum value (minimum transmission) occurs at

$r \cong 3/5\lambda$, the exact value being dependent upon m, the relative index of refraction of disperse phase to medium. See Fig. 1, reference 56, for values of water, oils, and sulphur. The optimum particle size for the obscuration of green light ($\lambda = 0.524$) is about 0.33 micron radius for $m = 1.5$.

(c) In the ascending branch of the curve, blue light is scattered better than red, and the transmitted light will be reddish, as is customarily observed at sunset. In the descending branch of the curve the Rayleigh equation loses completely any remaining qualitative significance for the scattering now actually decreases with increasing radius! This is shown most clearly in Fig. 5 of reference 56. This portion of the curve corresponds to a better transmission of blue than red light. At the maximum of the K curve, the transmitted light is a magenta color (minus green filter) and at the minimum, the transmitted light is green, followed by an ascending branch where red transmission again prevails. By observing the residual color of the sun's disk when almost obscured by a screening smoke, a rapid and sufficiently accurate estimate of particle size could be made by a lay observer for regulating the screening smoke generator.

(d) The absolute value of the scattering at the maximum is over four times greater than that computed for the geometrical obscuring area of the particles, namely, πr^2. The assumption frequently made in many textbooks that the obscuration of medium sized particles is equal to the geometrical obscuring area is without foundation. Axford, Sawyer, and Sudgen[70] have investigated the influence of the inhomogeneity existing in some practical smokes upon the transmission characteristics predicted by Mie's theory.

Filtration of Aerosols. The primary motivation for these studies was to determine the filtration efficiency of aerosols as a function of particle size and nature of filter. In our first experiments (1941), the efficiency was measured by comparing the optical density before and after passage of a monodisperse smoke through the filter. Monodisperse aerosols yielded a penetration curve with a distinct and reproducible maximum penetration value at a radius equal to 0.17 micron, whereas the previous measurements with polydisperse aerosols had yielded irregular and nonreproducible results in respect to radius of maximum filter penetration.

Guided by this finding, W. H. Rodebush and Irving Langmuir made significant suggestions to the Chemical Warfare Service for the improvement of the gas mask filter. So little smoke of the most penetrating size passed through the several layers of these new filter papers in the improved C.W.S. gas mask that a much more sensitive method of detecting aerosol was required than could be obtained by measuring either the light transmitted or scattered at 90°. David Sinclair and Seymore Hochberg in the writer's laboratory accordingly developed in 1941 the very sensitive optical detecting system shown schematically in Fig. 44.

The Forward-angle Tyndallometer. The apparatus operates upon the principle that, according to the Mie theory, the intensity scattered at angles $\theta = 5$ to $30°$ may be as much as one hundred to one thousand times that scattered at 90°, when the particle radius is larger than that demanded for the validity of Rayleigh's law. In this new form of Tyndallometer, all of the light scattered at these small angles from the forward direction is collected and brought to a focus for observation. A small volume of the aerosol drawn into the center of the chamber is illuminated by forming at this point the image of a 16 c.p. automobile headlight bulb by means of two aspheric condenser lenses. The central portion of the wide angle cone of rays is blocked out by a disk one-fourth inch in radius, and the intensity of the scattered light in the remainder of the cone is observed at the other end of the chamber along the axis of the cone. The diameter of the observation window of the sensitive photometer is less than that of the blocked-out region and hence no direct rays fall upon it.

When no particles are present, the observer sees only the black background of the disk.

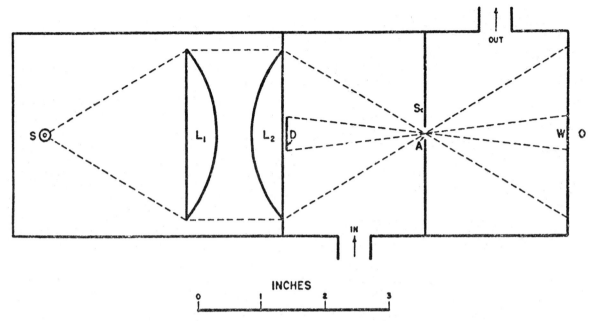

INCHES

Fig. 44. Forward-angle Tyndallometer. S, source of light — automobile headlight bulb. L_1 and L_2, aspheric condenser lenses about 2¼ inch focal length and 2½ inch diameter, furnishing a high-intensity image of the source at the image point A. At this point, the conical beam passes through a 3/16-inch diameter aperture in the screen Sc. The aerosol enters through a glass tube just below A, traverses the conical beam on both sides of the screen, and leaves the chamber just above A. D is an opaque disk of about ¼ inch radius. O is either a visual photometer (Luckiesh brightness meter) or a sensitive photoelement. W is a window of diameter smaller than D.

When an aerosol is present, the background becomes illuminated by the light scattered from the direction of the incident cone of rays. A visual photometer[*] is focused on the image of the filament formed in the smoke. The brightness of the image could .be measured satisfactorily except when the particle size is large and the concentration so low that flickering occurs due to individual particles. The chamber must be airtight and the zero reading made with well-filtered air, as single dust particles will brilliantly illuminate the background and interfere with the measurement.

It is believed that this type of Tyndallometer is the most sensitive ever reported. It was used successfully by Sinclair (1941) for measurements of mass concentration of stearic acid and other oil smokes from 200 down to 10^{-3} microgram (10^{-9} gram) per liter, (10^{-12} gram per cm.^{3}) depending upon particle size. It was instrumental in improving and

[*] A portable Luckiesh brightness meter is satisfactory.

testing the canister in use in 1941. Although we have found visual methods in the hands of a trained investigator to be the most flexible and suitable for exploratory research, the continuous readings demanded in control work fatigue the eye and photoelectric equipment is indicated. Hochberg (1942) accordingly developed a simple balanced photoelectric photometer employing this principle. Gucker and his associates (O'Konski and Pickard), then at Northwestern University, were assigned the problem of perfecting the photoelectric methods. By using improved electrical circuits, as described elsewhere (Gucker, "Determination of Concentration and Size of Particulate Matter by Light Scattering and Sonic Techniques"), they have succeeded in making the apparatus count individual aerosol particles.

Rates of Condensation and Evaporation. When an aqueous salt solution is nebulized and evaporated to dryness, the resulting salt nuclei are far from monodisperse, yet the aerosol resulting from growth upon such nuclei is quite monodisperse. The explanation for this surprising result is contained in the diffusional theory of growth.[62]

The rate evaporation of small spheres has shown by Morse and by Langmuir to follow the law

$$\frac{-dm}{dt} = kr$$

which may be transformed into

$$\frac{-dr}{dt} = \frac{k'}{r}$$

and integrates to $r^2 = r_0^2 + bt$.

Here m is mass; r, radius; t, time; and k, k', and b are constants. We have found that the rate of growth both of aerosols[60] and of sulphur hydrosols[72] follow this law; namely, that the surface area ($4\pi r^2$) increases (or decreases in the case of evaporation) linearly with time; r_0 is the radius of the initial nucleus.

Since the term bt is much larger than r_0^2 and is practically the same for all particles, the final radius after perceptible growth will be almost the same regardless of the initial r_0. Thus if bt has the value $0.16\mu^2$, a nucleus of $r_0 = 0.01\mu$ will yield a particle with a radius $= 0.40\mu$, while a nucleus ten times as large, $r_0 = 0.10\mu$, will produce a particle with $r = 0.41\mu$, or a difference of only 2.5 percent as compared to an initial 1,000 percent difference in the nuclei.

Thermal Repulsion. An aerosol particle in a thermal gradient[73] is subjected to a radiometer force which drives the particle from the hotter to the colder region. It is this force which produces the dust deposits on cold walls near radiators. H. H. Watson[74] has developed an ingenious sampling device based upon this principle. A thin wire is stretched between two glass microscope cover slips separated from each other by a distance of a few millimeters, and the wire is heated electrically to about 100^0 C. Aerosol is then passed slowly between the cover slips. The difference in temperature between the hot wire and cool cover slip drives all the particles to the cover slip where they are deposited in a convenient form for microscopic observation. The deposition is quantitative and the apparatus is generally recognized as an ideal one for preparing samples in the laboratory for all types of microscopic examination and counting. The thermal precipitation suffers from the disadvantage that the distance between wire and plates must be small and the rate of flow past the wire slow. This has restricted its use in field sampling where enormous quantities of air have to be drawn through in order to find a few particles.

Rosenblatt[73] found that the driving force is always proportional to the thermal gradient dT/dx, but that it varied with the pressure, P, the mean free path of the gas molecules, λ, and the radius of the particle, a, in a complicated manner. When L/a is small, the force is inversely proportional to P; when L/a is large, the force varies in direct proportion to P. When L and a are approximately the same magnitude, the force rises to a maximum. It is possible that some improvement in the design of thermal precipitators to make them suitable for higher rates of flow needed in airplane sampling might be accomplished by a further consideration of these principles.

Inertial Deposition. When an aerosol particle flows past a body, the particles tend to follow the streamlines of flow. However, the greater inertia of aerosol particles as compared to the gas molecules causes deviation from the streamline flow with the result that some of the particles come within van der Waals's range of attraction of the body and are deposited. This fundamental behavior is responsible for the removal of aerosol particles when they are passed through the mat of fine fibers constituting a filter; the deposition of an aerosol upon foliage is a manifestation of the same phenomena on a gross scale of observation. The deposition of an insecticidal aerosol upon the bodies of insects from a moving cloud is a further example corresponding to an intermediate scale of body sizes.

Sell[75] investigated the deposition of a cloud of dust, moving in a wind tunnel upon flat plates, cylinders, and spheres. He found that the deposition could be expressed as a function of the product, D^2V, which forms the significant variable in a dimensionless group. Here D is the diameter of the particle, and V, its velocity in the moving cloud. Sell's theory was confirmed by Winsche[76,77] on the penetration of aerosols through foliage, and also by Latta, La Mer, Hochberg, et al.,[61,78] who determined the toxicity of DDT oil aerosols to caged mosquitoes in a wind tunnel as a function of the wind velocity and the particle size of the monodispersed insecticidal aerosol. These experimental data are in excellent agreement with Sell's theory, and can be expressed by the equation, $\log M = K \log (D^2V) +$ constant, where M is the mass deposited, and K is a constant equal to —1 when D^2V is varied between 2 to 200 microns[2] miles per hour, with M ranging between 300 mg. and 1 mg. per square foot of wind tunnel cross section. Between values of D^2V equal to 200 to 800, the value of K decreases to zero. Between values of $D^2V = 800$ to 7,000, K is zero, the value of M then being equal to the constant, 0.8 mg. DDT per square foot. The values of the constants given are valid, of course, only for small bodies of the order of 0.03 cm.[2] but the equations should hold with appropriate constants for bodies of any size.

Cascade Impactor. May[79] describes a metal instrument in which aerosol at the rate of 17.5 l/min. is forced through a series of four jets which are set normal to four glass slides. The cross-sectional area of the jets is made progressively smaller; hence, the linear velocity of the gas increases at each successive jet, and particles of progressively smaller size are impacted out on successive slides.

The device should afford a means for rapid measurement of particle-size distribution upon the successive slides either by weight of the mass deposited, chemical analysis, or by microscopic count. The apparatus requires calibration to determine size distribution on each slide for each given type of aerosol.

Sonkin[80] has described an improved glass apparatus which employs a jet of air approaching sonic velocity and hence represents the maximum velocity that can be imparted. The table gives Sonkin's summary of the two instruments using wet particulates.

Jet No.	Jet speed, mph (May apparatus)	Range of droplet size, microns
1	5	200–10
2	30	20–3
3	50	7–1
4	80	3–0.7
	(Sonkin improved instrument)	
1	60 (approx.)	0–0.7
2	90	1.5–0.25
3	180	1.1–0.15
4	700	0.7–0.1

Opinions are divided on the efficiency and reliability of the cascade impactor. This laboratory tested the May type of instrument early in the war period but abandoned it in favor of other methods. R. M. Ferry, who has had much experience in the counting of bacteria and spores by impingement, states that neither the May nor the Sonkin instrument was satisfactory for his work, and developed a special impinger of his own design[81]. He finds less effective impingement of solids than of liquid droplets. There is a reduction in collection efficiency when jet velocities exceed optimal values, all of which can be explained qualitatively on the assumption that the particles are elastic and bounce. Bacteria may shatter into microscopically invisible fragments.

S. Laskin[82] reported a long series of experiments with the British impactor (May) using uranium dioxide and other uranium dusts in the range of 1 to 17 microns. He finds that with certain modifications in design, it can be used routinely for the size-mass distribution in the range indicated. His modified impactor is made of solid brass, by precision machining, rather than by working sheet metal. Resin-coated slides are necessary to prevent shattering and bouncing. He coats the deposited particles with selenium (index of refraction 3.8) by shadow-casting methods to improve the optical resolution. From his figure 10.9, it is quite evident that the usual methods of size-counting, employing visual microscopy, have failed to count the smaller particles. Thus, at 0.3 micron diameter, the uncoated procedure indicated a frequency of 3 percent, whereas selenium coating makes so many small particles visible that a frequency of 20 percent was registered! Presumably the electron microscope would have resolved still smaller particles which have escaped attention.

The cascade impactor merits reinvestigation with monodisperse aerosols and precise optical methods for the detection of particles which may fail to impinge, or are disrupted on impact.

References

[54]Hatch, T., and Choate, S. P., *J. Franklin Inst.*, 207: 369-387, 1929.

[55]Dalla Valle, J. M., "Micromeritics," Pitman Publishing Corp., New York, 1943.

[56]Sinclair, D., and La Mer, V. K., *Chem. Rev.*, 44: 245, 1949.

[57]*Idem, N.D.R.C. Report No. 57*, July, 1941.

[58]Ford, G., and La Mer, V. K., *J. Am. Chem. Soc.*, in press, February, 1950.

[59]La Mer, V. K., and Sinclair, D., *N.D.R.C. Report No. 57*, July, 1941.

[60]*Ibid., No. 1668*, 1943.

[61]La Mer, V. K., and Hochberg, S., *Chem. Rev.*, 44: 341, 1949.

[62]Wilson, I. B. and La Mer, V. K., *J. Ind. Hyg. Toxicol.*, 30: 265, 1948.

[63]Johnson, and La Mer, V. K., *J. Am. Chem. Soc.*, 69: 1184, 1947.

[64]Kenyon, A. S., and La Mer, V. K., *J. Coll. Sci.*, 4: 163, 1949.

[65]Kerker, M., and La Mer, V. K., *J. Am. Chem. Soc.*, in press.

[66]Humphreys, W. J., "Physics of the Air," p. 547, McGraw-Hill Book Company, Inc., New York, 1940.

[67]La Mer, V. K., and Barnes, *J. Coll. Sci.*, 1: 71, 1946.

[68]Barnes, and La Mer, V. K., *J. Coll. Sci.*, 1: 79, 1946.

[69]Barnes, Kenyon, A. S., Zaiser, and La Mer, V. K., *J. Coll. Sci.*, 2: 349, 1947.

[70]Axford, Sawyer, and Sugden, *Proc. Roy. Soc.*, A195: 13, 1948.

[72]Zaiser, and La Mer, V. K., *J. Coll. Sci.*, 3: 571, 1948.

[73]Rosenblatt, P., and La Mer, V. K., *Phys. Rev.*, 70: 385, 1946.

[74]Watson, H. H., *Trans. Faraday Soc.*, 32: 1037, 1936.

[75]Sell, "*Forschungs arbeiten Verein deutscher Ingenier*," 1: 347 (Berlin: V.D.I., 1931).

[76]Winsche, W. E., *O.S.R.D. Informal Report 10* (March, 1944), pp. 4-55.

[77]Johnstone, H. F., Winsche, W. E., and Smith, L. W., *Chem. Rev.* 44: 353, 1949.

[78]Latta, La Mer, V. K., Hochberg, S., *et al.*, *J. Wash. Acad. Sci.*, 37: 397, 1947.

[79]May, K. R., *J. Sci. Instruments*, 22: 187-195, 1945.

[80]Sonkin, L. S., *J. Ind. Hyg. Toxicol.*, 28: 269, 1946.

[81]Ferry, R. M., Farr, and Hartman, *Chem. Rev.*, 44: 389, 398, 402, 1949.

[82]Laskin, S., "Pharmacology and Toxicology of Uranium Compounds," National Nuclear Energy Series, Div. VI, Vol. I, Book I, 1st ed., 1949.

1957

PETER J. W. DEBYE

CORNELL UNIVERSITY
ITHACA, N. Y.

. . . for his fundamental contributions to the basic theories of colloid chemistry and especially for his theoretical and experimental development of light scattering techniques for the study of solutions of very large molecules.

The following article has been selected with the advice of Prof. J. W. Williams, a student of the late Prof. Debye, as very similar to his Kendall Award Address. It is: P. Debye, *"Scattering of Radiation by Non-crystaline Media"*, from *Non-crystaline Solids*, edited by V. D. Frechette. Copyright © 1960 by John Wiley & Sons, Inc. By permission.

by

P. Debye

Cornell University

Scattering of Radiation by Non-crystalline Media

Formulation of the Interference Principle

One rather powerful, but by no means all-powerful, method for investigating the structure of non-crystalline substances can be based on the observation of the angular distribution of the intensity of scattered radiation. The kind of information about the structure we can obtain in this way, obvious for crystalline media, is not so evident for non-crystalline media. A prime objective of our discussion therefore is a proper definition of such structures. In order to avoid complications unnecessary for this purpose only cases will be considered in which, in the scattering process, the interaction between the radiation and the atoms of the medium is small in magnitude. This certainly covers with a high degree of accuracy the scattering of visible light and of x-rays. Cases of stronger interaction have been considered by van Hove (1954, 1958).

Let us assume a medium with an average dielectric constant ε on top of which local fluctuations $\delta\varepsilon$ of random character are superimposed. Through this medium we send a primary plane light wave which is polarized and in which the electric force has the amplitude E_0. Suppose also that the fluctuations $\delta\varepsilon$ are confined to a small volume which, however, may be large compared to the wavelength λ. By a straightforward application of Maxwell's equations, we then find that this volume is the center of scattered radiations. At a large distance R from this volume (large compared to the dimensions of the volume) and within the medium of dielectric constant ε, the scattered wave has an amplitude and phase (both measured in the usual way by writing for the electric force E of the scattered wave) which lie in the plane defined by the directions of E_0 and R and is perpendicular to R:

$$E = E_0 \frac{e^{-ikR}}{R} \frac{k^2}{4\pi} \sin \vartheta \int \frac{\delta\varepsilon}{\varepsilon} e^{ik(\mathbf{s},\mathbf{r})} \, d\tau. \tag{1}$$

In this relation λ is the wavelength in the medium and $k = 2\pi/\lambda$. The angle between the directions of E_0 and of R is called ϑ. $\mathbf{s}$ is a vector which is the difference between a unit vector $\mathbf{S}$ in the direction of R and another unit vector $\mathbf{S}_0$ in the direction of propagation of the primary wave. If we call the angle between these two vectors θ, the absolute value of the difference vector $\mathbf{s}$ becomes $2 \sin (\theta/2)$. Finally $\mathbf{r}$ is a vector defining the position of the volume element $d\tau$ in the scattering volume with respect to an arbitrary center; from it the large distance R is measured. The integration has to be carried out over the scattering volume; the factor $e^{ik(\mathbf{s},\mathbf{r})}$ measures the interference between the wavelets emitted by the different volume elements.

The Correlation Function

From relation (1) we obtain the scattered intensity by multiplying E and its conjugate complex. We call the primary intensity g_0 and the secondary intensity g, and this process leads immediately to the relation

$$\frac{g}{g_0} = \frac{k^4}{16\pi^2} \frac{\sin^2 \vartheta}{R^2} \iint \left(\frac{\delta\varepsilon}{\varepsilon}\right)\left(\frac{\delta\varepsilon}{\varepsilon}\right)' e^{ik(\mathbf{s},\mathbf{r}-\mathbf{r}')} \, d\tau \, d\tau' \tag{2}$$

This is the instantaneous intensity. Only the average value, which, in the case of fluctuations $\delta\varepsilon$ which vary with time, is a time average. If the fluctuations are frozen in, we can replace the time average by a space average since the position of the sample under investigation obviously has no influence on the observed intensity. In order to be able to perform the integration we have to know the average value of the product

$$(\delta\varepsilon/\varepsilon)_A (\delta\varepsilon/\varepsilon)_B$$

in which the first fluctuation is measured in a point A and the second in another point B. In the case of stationary fluctuations we shall obtain this average by letting the measuring stick AB take up all possible positions and orientations within the scattering medium. We shall consider only the isotropic case; then we can say that the product in question will be a function of the distance AB only, which we shall call r. For $r = 0$ this product takes the value

$$\mathrm{Av}\ \langle (\delta \varepsilon / \varepsilon)^2 \rangle$$

For large values of r the product obviously tends to zero. We now introduce the correlation function $C(r)$ by the definition

$$\mathrm{Av} \left\langle \left(\frac{\delta \varepsilon}{\varepsilon}\right)_A \left(\frac{\delta \varepsilon}{\varepsilon}\right)_B \right\rangle = \mathrm{Av} \left\langle \left(\frac{\delta \varepsilon}{\varepsilon}\right)^2 \right\rangle C(r) \tag{3}$$

which makes $C(r)$ dimensionless and makes it start with 1 for $r = 0$. The same problem has appeared in many different connections, and it leads in all cases to equivalent mathematical formulations. As examples can be mentioned: (a) the theory of turbulence (Taylor, 1920, 1938); (b) the scattering of sound in water (Pekeris, 1947); (c) the scattering of radio waves (Booker and Gordon, 1950). After the introduction of C the integration in (2) can be performed; it leads to the result

$$\frac{g}{g_0} = \frac{k^4}{16\pi^2} \frac{\sin^2 \vartheta}{R^2} V \, \mathrm{Av} \left\langle \left(\frac{\delta \varepsilon}{\varepsilon}\right)^2 \right\rangle \int_0^\infty C(r) \frac{\sin ksr}{ksr} \, d\tau \tag{4}$$

in which V is the volume illuminated by the primary light wave. The right-hand side is dimensionless as it should be, since the integral represents a volume. This volume depends on the angle of observation through the inclusion of s in its definition and measures the interference effect due to the irregularities in the medium. It is obvious that this effect appears as a result of the correlation of the fluctuations in neighboring points. It can easily be seen that, should the primary radiation be unpolarized, it is only necessary to replace $\sin^2 \vartheta$ by $\frac{1}{2}(1 + \cos^2 \theta)$.

The Special Case of X-rays

The scattering of x-rays is due mainly to the electrons dispersed in the medium, and for many practical cases these electrons can be treated as free. Under the influence of a periodic electric field,

$$E e^{i\omega t}$$

the amplitude x of an electron with charge e and mass m will be

$$x = -\frac{eE}{m\omega^2} e^{i\omega t}$$

The polarization P due to n such free electrons per cubic centimeter will be

$$P = -\frac{ne^2}{m\omega^2} E e^{i\omega t}$$

which means that the dielectric constant ε will be

$$\varepsilon = 1 - \frac{4\pi n e^2}{m\omega^2} \tag{5}$$

which is very little different from unity and shows that the phase velocity in the medium is larger than the velocity of light in vacuum. The only thing necessary to describe the x-ray scattering is to replace the dielectric constant in (4) by its value (5). The result is

$$\frac{\mathcal{I}}{\mathcal{I}_0} = \left(\frac{e^2}{mc^2}\right)^2 \frac{\sin^2 \vartheta}{R^2} V \text{ Av } \langle (\delta n)^2 \rangle \int_0^\infty C(r) \frac{\sin ksr}{ksr} d\tau \tag{6}$$

The factor $(e^2/mc^2)^2$ can be called the scattering cross section of the single free electron; it has the value 8.06×10^{-26} cm^2. Owing to the dispersion peculiar for free electrons, the usual characteristic dependence of Rayleigh scattering on the wavelength is wiped out. Equation (6) can be read as saying that the scattering of the $Vn = \mathfrak{N}$ electrons contained in volume V is equivalent to that of

$$\mathfrak{N}^* = V \text{ Av } \langle (\delta n)^2 \rangle \int_0^\infty C(r) \frac{\sin ksr}{ksr} d\tau \tag{6'}$$

independent free electrons. Their number $\mathfrak{N}^*$ varies of course with s, that is, with the angle of observation θ. Since $\delta\varepsilon$ is proportional to δn, the correlation function $C(r)$ can here be defined by the relation

$$\text{Av } \langle \delta n_A \, \delta n_B \rangle = \text{Av } \langle (\delta n)^2 \rangle \, C(r)$$

Calculation of the Correlation Function from the Intensity Distribution

It is clear that for the definition of the structure of a non-crystalline medium the correlation function plays the same role as the lattice structure and the distribution of the atoms within the lattice cell for a crystalline medium. For this reason it is important to see how the correlation function can be derived from observations about the angular distribution of the scattered intensity. It turns out that the solution of this problem is given immediately by the application of a Fourier inversion. We know that if

$$f(x) = 2\int_0^\infty \varphi(u) \sin 2\pi x u \, du$$

the function $\varphi(u)$ can be found by calculating

$$\varphi(u) = 2\int_0^\infty f(x) \sin 2\pi u x \, dx$$

Now we can say, as is seen from the preceding paragraphs, that the experiment provides us with a curve the ordinates of which represent a volume Ω as a function of $s = 2 \sin(\theta/2)$ defined by

$$\Omega(ks) = \int_0^\infty C(r) \frac{\sin ksr}{ksr} \, d\tau \tag{7}$$

Since $k = 2\pi/\lambda$, this relation can also be written in the form

$$\frac{s}{\lambda} \Omega\left(\frac{s}{\lambda}\right) = 2\int_0^\infty rC(r) \sin 2\pi \frac{s}{\lambda} r \, dr \tag{7'}$$

which by application of Fourier's theorem leads immediately to

$$rC(r) = 2\int_0^\infty \frac{s}{\lambda} \Omega\left(\frac{s}{\lambda}\right) \sin 2\pi r \frac{s}{\lambda} d\left(\frac{s}{\lambda}\right) \tag{8}$$

In actual practice it never is possible to match the mathematical elegance of this solution for the simple reason that the range of $s = 2 \sin(\theta/2)$ is limited from 0 to 2 and therefore the curve for $\Omega(s/\lambda)$ is known only for a finite interval instead of the required range from 0 to ∞.

In some important cases it can be shown that the correlation function is a simple exponential,

$$C(r) = e^{-r/a} \tag{9}$$

a relation which defines immediately a correlation length a. If this is so, the angular intensity distribution is represented by a simple formula. The integration by which the volume Ω is defined can be performed readily and leads to

$$\Omega = \frac{8\pi a^3}{(1 + k^2 s^2 a^2)^2} \tag{10}$$

The corresponding intensity distribution can be checked by plotting the reciprocal of the square root of the scattered intensity as a function of s^2. If the correlation function is indeed a simple exponential, this procedure gives a straight line the slope of which is a measure for the correlation length a.

Equation (10) shows, with reference to our general formula for the scattered intensity, that for large enough values of s this intensity decreases in proportion to s^{-4}. This feature is quite general, independent of the special form of the correlation function. Substituting σ for ks, our formula (7) for the definition of the volume Ω, which in essence describes the angular intensity distribution, can be written in the form

$$\Omega = -4\pi \frac{1}{\sigma} \frac{d}{d\sigma} \int_0^\infty C(r) \cos \sigma r \, dr$$

By repeated partial integration the integral can be developed in powers of $1/\sigma$, and the first approximation obtained in this way is

$$\int_0^\infty C(r) \cos \sigma r \, dr = \frac{C'(0)}{\sigma^2} + \cdots$$

which makes

$$\Omega = -8\pi \frac{C'(0)}{\sigma^4} + \cdots = -\frac{8\pi}{k^4} \frac{C'(0)}{s^4} + \cdots \qquad (11)$$

This result is correct provided $C(r)$ vanishes strongly enough for large values of r and, what is more important, provided a finite gradient of $C(r)$ for $r = 0$ exists.

The Analysis of Hole Structures

(See Debye and Bueche, 1949; Debye, Anderson, and Brumberger, 1957; Guinier, Fournet, Walker, and Yudowitch, 1955.) In catalytic processes structures are used which have a high specific surface S/V, which, deviating from the usual custom, we shall define as the surface per unit volume and not per unit weight. Values of, for example, $100 \text{ m}^2/\text{cc}$ are common. Such a specific surface has the dimension of a reciprocal length, and $100 \text{ m}^2/\text{cc}$ corresponds to a length of 100 Å. It is evident that, in case we want to investigate such structures by their x-ray scattering, we shall have to concentrate on small angle scattering if the usual x-rays of wavelengths about 1 Å are used. As soon as we do this, however, we deliberately concentrate on geometrical properties of the hole structure and cannot expect to find any indication of the atomic structure of the solid material, which can appear only at larger angles. For our purpose, then, the structure can be considered as consisting of material parts in which the electron density n is constant and holes in which the electron density is zero. If we take χ to represent the porosity (quotient of volume of holes to total volume) the average electron density will be

$$n_{\text{av}} = n(1 - \chi) + 0\chi = n(1 - \chi)$$

and the local fluctuations Δ_M in material and Δ_H in holes will be

$$\Delta_M = n - n(1 - \chi) = n\chi$$
$$\Delta_H = 0 - n(1 - \chi) = -n(1 - \chi) \qquad (12)$$

which leads to

$$\mathrm{Av}\,\langle\Delta^2\rangle = (1-\chi)\Delta_M{}^2 + \chi\Delta_H{}^2 = n^2\chi(1-\chi) \tag{13}$$

We must now define the correlation function. In the special case of our hole structure the measuring stick AB which we introduced to define the correlation function can appear only in four different positions depending on whether end A or end B is in material or in a hole. In order to calculate the average value of $\Delta_A\,\Delta_B$ we introduce the four probabilities

$$\begin{matrix} p_{00} & p_{01} \\ p_{10} & p_{11} \end{matrix} \tag{14}$$

where p_{01} is the probability that our measuring stick, being with end A in a hole (index 0), will have end B in material (index 1). The other three possibilities are defined in an analogous way. We now can calculate the average value of the product $\Delta_A\,\Delta_B$ for a given length of the measuring stick and obtain

$$\mathrm{Av}\,\langle\Delta_A\,\Delta_B\rangle = \chi p_{00}\,\Delta_H{}^2 + [\chi p_{01} + (1-\chi)p_{10}]\,\Delta_M\,\Delta_H$$
$$+ (1-\chi)p_{11}\,\Delta_M{}^2 \tag{15}$$

The four probabilities we introduced are not independent of each other. Obviously the following three relations must hold.

$$p_{00} + p_{01} = 1 \qquad p_{10} + p_{11} = 1 \qquad \chi p_{01} = (1-\chi)p_{10} \tag{16}$$

and this means that we must have

$$\begin{matrix} p_{00} = 1 - (1-\chi)R & p_{01} = (1-\chi)R \\ p_{10} = \chi R & p_{11} = 1 - \chi R \end{matrix} \tag{17}$$

in which R is an as yet unknown function of the length of the stick. Substituting (17) in (15) gives, by using (12), the result

$$\mathrm{Av}\,\langle\Delta_A\,\Delta_B\rangle = n^2\chi(1-\chi)(1-R)$$

Now according to (13) the factor on the right-hand side represents $\mathrm{Av}\,\langle\Delta^2\rangle$, so we come to the result that our former correlation function $C(r)$ and the new function $R(r)$ are connected by the relation

$$C(r) = 1 - R(r) \tag{18}$$

Since the correlation function $C(r)$ can be derived from the angular intensity distribution, the same is now true for $R(r)$, and we can, according to (17), also evaluate the four probabilities p. It is easier to condense the whole information by asking for the probability p_{diss}, i.e., that the ends of our measuring stick will be dissimilar (hole and material or material and hole). For this probability we find

$$p_{\text{diss}} = \chi p_{01} + (1 - \chi)p_{10} = 2\chi(1 - \chi)R = 2\chi(1 - \chi)[1 - C] \quad (19)$$

The curve for p_{diss} so obtained can be considered, in a qualitative way, as a representation of the distribution of hole sizes. It starts with 0 for $r = 0$ and ends with the ordinate $2\chi(1 - \chi)$ for large values of r.

The Relation between Correlation Function and Specific Surface

(See G. Porod, 1951, 1952; Guinier et al., 1955). If we consider the probability of dissimilar ends p_{diss} for very small lengths of our stick, this stick will have to cut through the surface, and in evaluating the value of the probability we shall have to follow this surface through the whole volume. From this it is evident that a relation should exist which connects p_{diss} with the specific surface. A simple geometrical argument (Debye, Anderson, and Brumberger, 1957) shows that, in the limit for very small lengths r of our stick in a volume V which contains a surface S,

$$p_{\text{diss}} = (S/2V)r \quad (20)$$

On the other hand, relation (19) shows that for small values of r we have

$$p_{\text{diss}} = -2\chi(1 - \chi)C'(0)r \quad (20')$$

Comparison of these two relations then shows that the specific surface S/V can be determined by evaluating the tangent of the correlation curve for $r = 0$ according to the relation

$$S/V = -4\chi(1 - \chi)C'(0) \quad (21)$$

In case the correlation function is a simple exponential function as in (9), this boils down to

$$S/V = 4\chi(1 - \chi)(1/a) \quad (21')$$

In this case the representation of the intensity curve in the manner described on page 5 provides immediately a measure of the specific surface by the slope of the straight line in question.

In an experimental investigation (Debye, Anderson, Brumberger, 1957) Anderson and Brumberger showed that in many practical cases of gel catalysts the straight-line representation of the reciprocal square root of the intensity versus the square of the scattering angle fitted the experimental results. That brought up the question: Why should the special exponential correlation curve appear at all? It could be shown (Debye, Anderson, and Brumberger, 1957) that an absolutely random arrangement of the interfaces between holes and material inside the sample indeed leads to the exponential correlation function. In equation (11) it was shown that for sufficiently large values of s the scattered intensity should decrease proportionately with the reciprocal fourth power of s, and the proportionality was found to be proportional to $C'(0)$. Since in the meantime we have seen that $C'(0)$ is proportional to the specific surface (equation 21), we see that for sufficiently large values of s the observed intensity itself should be a measure of the specific surface. Starting with equation (6) for the equivalent number $\mathfrak{N}^*$ of free electrons, the Av $\langle(\delta n)^2\rangle$ can be expressed by the electron density n of the material and the porosity χ according to equation (13). For the integral Ω (see equation 7) appearing as the second factor which determines $\mathfrak{N}^*$, we can substitute its asymptotic value according to (11). In this way we arrive at

$$\mathfrak{N}^* = -\frac{8\pi}{k^4}Vn^2\chi(1-\chi)\frac{C'(0)}{s^4}$$

Finally $C'(0)$ and the specific surface are connected by relation (21). Thus we have in the range of larger angles

$$\mathfrak{N}^* = \frac{2\pi}{k^4}\frac{n^2S}{s^4} \tag{22}$$

It is interesting to note that in the case of gel catalysts and by a purely experimental approach Van Nordstrand came to the conclusions that (a) the scattered intensity decreased at larger angles in proportion to the reciprocal 4th power of the angle, and (b) in this range the intensity was proportional to the surface as derived from adsorption measurements, as long as samples of the same material were compared (Van Nordstrand and Hach, 1953; Van Nordstrand and Johnson, 1954).

Angular Dissymmetry of Critical Opalescence

Under ordinary circumstances the scattering of liquids is rather small and for visible light does not show any peculiar angular intensity distribution; it is essentially like that of an infinitely small dipole. However, in condensing a simple gas to a liquid and observing it in the vicinity of the critical point or in experimenting with two liquids in the vicinity of their critical mixing temperature, very strong scattering of visible light is observed. At the same time the scattered intensity becomes concentrated more and more in the forward direction (the direction of propagation of the primary beam) the nearer the temperature comes to the critical temperature (Zimm, 1950; Fuerth and Williams, 1954). In the light of our general formula, equation (4), this must mean that under such circumstances an appreciable correlation has to exist between the density or concentration fluctuations or the corresponding fluctuations of the refractive index over distances which are comparable with the wavelength of visible light. This is clear since the main angular dependence of the scattered intensity from the angle of observation is represented by the integral over the correlation function $C(r)$, and this integral will show appreciable variations with s only when $C(r)$ has not yet reached its final value of zero for distances r which make kr of the order of magnitude 1 or r of the order magnitude $\lambda/2\pi$.

From equation (4) only, we cannot yet calculate the intensity of scattering. It will be proportional to

$$\text{Av} \left\langle \left(\frac{\delta \varepsilon}{\varepsilon} \right)^2 \right\rangle \int_0^\infty (Cr)\, d\tau$$

at least under ordinary circumstances, when the correlation distance is small compared with the wavelength of the light, but the value of this factor can be determined only after we have specified the reason for the fluctuations. From now on we shall consider only the case of the condensation of a gas, since this sufficiently illustrates the situation.

Following Einstein (1910), we accept as the reason for the fluctuations the thermal molecular motion. For the actual calculation, however, it will be more appropriate for our purpose to use a line of reasoning followed by Brillouin (1922). He considers the thermal motion in the liquid a superposition of elastic (sonic) waves, the same way as has been found to be practical for the calculation of the specific heat of solids. He starts out by calculating from equation (1) the scattered intensity of a homogeneous liquid through which a plain sonic wave travels which induces small periodic fluctuations η of the dielectric constant ε. Experiments showing the scattering of light by artificial supersonic waves were made by Debye and Sears (1932). Such fluctuations can be represented by

$$\eta = \eta_0 e^{i\omega t} e^{-iK(\mathbf{S},\mathbf{r})}$$

for waves of frequency ω and a direction indicated by the unit vector $\mathbf{S}$. Brillouin then accepts for the number of such waves, dZ, all independent from each other in volume V and in the interval $d\Omega$ (with $d\Omega$ indicating the element of solid angle) the usual relation

$$dZ = \frac{V}{(2\pi)^3} K^2 \, dK \, d\Omega$$

Performing the integration over the whole volume, which is illuminated (taken to be a sphere for the purpose of the actual calculation), he finds:

(a) Of all the thermal waves which are traversing the liquid only one is effective for the scattering. It is a wave with a front such that the scattered light can be considered as primary light reflected at this front.

(b) Between the wavelength λ of the light and the wavelength Λ of the reflecting sonic wave, Bragg's relation must hold. This relation is written

$$s/\lambda = 1/\Lambda \tag{23}$$

If now the amplitude of the dielectric constant fluctuation in this special wave is called η_0, Brillouin's calculations show that the intensity scattered by a volume V is proportional to the square of the amplitude. They lead to the relation

$$\frac{\mathscr{I}}{\mathscr{I}_0} = \frac{k^4}{16\pi^2} \frac{\sin^2 \vartheta}{R^2} V^2 \left(\frac{\eta_0}{\varepsilon}\right)^2 \tag{24}$$

The total energy of such a wave filling volume V is

$$\frac{V}{2\kappa} \left(\frac{\sigma_0}{\rho}\right)^2 \tag{25}$$

if in the usual way κ indicates the compressibility and if the amplitude of the density fluctuations is σ_0 whereas ρ is the average density of the liquid. This must be equal to βT in thermal equilibrium (β is Boltzmann's constant); therefore

$$\left(\frac{\sigma_0}{\rho}\right)^2 = \frac{2\kappa\beta T}{V} \tag{26}$$

Since we can express η_0 in σ_0 by the relation

$$\eta_0 = \sigma_0 \frac{d\varepsilon}{d\rho}$$

we come to the final result

$$\frac{\mathscr{I}}{\mathscr{I}_0} = \frac{k^4}{16\pi^2} \frac{\sin^2 \vartheta}{R^2} V 2\kappa\beta T \left(\frac{\rho}{\varepsilon} \frac{d\varepsilon}{d\rho}\right)^2 \tag{27}$$

This is the result first derived by Einstein (1910). In this formula there is not any indication of an angular dissymmetry. Obviously there is a flaw somewhere in the reasoning, and it comes to the foreground in experiments near the critical point. The next paragraph deals with the necessary correction.

Molecular Energy in an Inhomogeneous Medium

The wavelength of the sound waves on which the scattered light can be considered as reflected changes with the angle θ between primary and secondary ray. According to Bragg's relation it goes from ∞ for $\theta = 0$ to $\lambda/2$ for $\theta = \pi$. In this way a range of frequencies going roughly from 0 to 10^4 megacycles is covered. The observation of the intensity of the scattered light, when θ goes from 0 to π, can be interpreted as a measurement of the square of the amplitude of the thermal supersonic waves over the corresponding range of wavelengths. Einstein's result is equivalent to saying that because of the law of equipartition those amplitude squares are all the same for the sound waves of the whole interval. In scrutinizing the details, it is seen that this statement rests on the assumption that the potential energy involved in the propagation of the sound waves is due to the compressibility. This compressibility gets bigger the nearer we come to the critical point. In this point itself it is infinite, and this has the effect that the energy involved in the compression to a given amplitude approaches zero near the critical point, as shown by equation (25). Under these circumstances it is essential to consider the existence of any other source of energy, even if it is unimportant under ordinary circumstances. The fact that surface tension exists, showing that juxtaposition of two different densities involves an extra energy, indicates that if we want to be exact in calculating the energy in a wave not only the amplitude but also the spatial gradient of the amplitude has to be considered (Roccard, 1933). This remark can be formulated in a quantitative way by the following reasoning.

Suppose that the potential energy between two molecules is solely a function of their mutual distance r. Thinking of universal molecular attraction, we call it $-\varepsilon(r)$. The number of molecules around a central molecule in a shell of radius r and thickness dr is

$$4\pi n r^2 \, dr = n \, d\tau$$

The potential energy of one central molecule due to action of its surroundings therefore is

$$w = -\int n\varepsilon(r) \, d\tau \qquad (28)$$

where in our simplified picture of molecular forces we let the integration

46

go from the distance of contact to infinity. Around the position of the central molecule we can develop the molecular density n in powers of the coordinates around this center. If we break this series off at the second power of these coordinates, the result is

$$w = -I_0 n - \frac{I_2}{6} \Delta n \tag{29}$$

in which Δ stands for the Laplace operator calculated at the position of the central molecule and I_0 and I_2 are defined by the integrals

$$I_0 = \int \mathcal{E}(r)\, d\tau \qquad I_2 = \int r^2 \mathcal{E}(r)\, d\tau \tag{29'}$$

We can define a length l by making

$$l^2 = I_2/I_0 \tag{30}$$

We shall call this length the range of molecular forces. We can now calculate the total potential energy $\mathfrak{u}$ in a volume V containing $\mathfrak{N}$ molecules and find, by integrating over the whole volume,

$$2\mathfrak{u} = -I_0 \int n^2\, d\tau - \frac{I_0}{6} l^2 \int n\, \Delta n\, d\tau \tag{31}$$

We are trying here to discuss a theory in its most simple formulation. For this reason it will be sufficient not to go beyond the classical reasoning of van der Waals. For the potential energy of a liquid he writes

$$\mathfrak{u} = -a/V$$

This shows, according to (31), that the van der Waals constant a can be identified with

$$a = \mathfrak{N}^2 (I_0/2) \tag{32}$$

We can also remark that by partial integration the second integral in (31) can be given the form

$$\int n\, \Delta n\, d\tau = \int n \frac{\partial n}{\partial \nu}\, d\sigma - \int (\mathrm{grad}\ n)^2\, d\tau$$

in which the first integral is to be taken over the boundary surface with the normal ν. Combining all this, we come to the conclusion that in a liquid with densities variable from point to point we not only have per unit volume the van der Waals energy density

$$u_1 = -a/V^2 = -(I_0/2)n^2 \tag{33}$$

47

but also an additional energy density

$$u_2 = (I_0/12)(l \operatorname{grad} n)^2 \tag{33'}$$

Angular Distribution of the Scattered Intensity

If we use the classical van der Waals equation of state for a volume V containing $\mathfrak{N}$ molecules,

$$p = \frac{\mathfrak{N}\beta T}{V - b} - \frac{a}{V^2}$$

the critical parameters are

$$p_c = \frac{1}{27}\frac{a}{b^2} \qquad V_c = 3b \qquad \mathfrak{N}\beta T_c = \frac{8}{27}\frac{a}{b} \tag{34}$$

Using the reduced parameters

$$\varphi = \frac{V}{V_c} \quad \text{and} \quad \tau = \frac{T}{T_c} \tag{35}$$

the reciprocal compressibility calculated from the equation of state comes out to be

$$\frac{1}{\kappa} = 6p_c \left[\frac{4\varphi}{(3\varphi - 1)^2}\,\tau - \frac{1}{\varphi^2} \right] \tag{36}$$

We shall suppose, for the sake of simplicity in the formulas, that the scattering experiment is carried out at constant volume and decreasing temperature and that the density is kept equal to the critical density. In this case $\varphi = 1$ and

$$1/\kappa = 6p_c(\tau - 1) \tag{36'}$$

In equation (25) we calculated the energy, due to the compressibility, contained in a volume V through which a supersonic wave passes. If the value in (36') for the compressibility is substituted, this energy becomes

$$V3p_c(\tau - 1)(\sigma_0/\rho)^2 \tag{37}$$

Half of this is potential energy.

In order to calculate the additional potential energy due to density gradients, we observe that, (a) the relative density fluctuations are the same as the relative fluctuations of the number of molecules per unit volume and (b) the operation gradient performed on a wave of the form

$$A \cos (\omega t - Kx)$$

merely changes this expression into

$$KA \sin (\omega t - Kx)$$

From equation (33′) we then conclude that the additional energy in volume V is

$$\frac{V}{2} \frac{I_0 n^2}{12} K^2 l^2 \left(\frac{\sigma_0}{\rho}\right)^2 \qquad (37')$$

On the other hand, by application of equations (32) and (34) it is found that

$$n^2 I_0 = \frac{2a}{V^2} = \frac{2}{g} \frac{1}{\varphi^2} \frac{a}{b^2} = \frac{6 p_c}{\varphi^2}$$

The ratio of the additional to the compressional potential energy therefore is

$$\frac{1}{6} \frac{K^2 l^2}{\tau - 1} \qquad (38)$$

if again, as before, we assume that the experiment is carried out at $\varphi = 1$. Upon taking the additional potential energy into account one obtains, because of equipartition, the relationship

$$\frac{V}{2\kappa} \left(\frac{\sigma_0}{\rho}\right)^2 \left[1 + \frac{1}{6} \frac{K^2 l^2}{\tau - 1}\right] = \beta T \qquad (39)$$

instead of equation (26), derived previously by Einstein. Remembering that $K = 2\pi/\Lambda$ and that Bragg's relation holds between the wavelength Λ of the supersonic wave and the wavelength λ of the optical wave, we come now, instead of to equation (27), to the final result,

$$\frac{g}{g_0} = \frac{k^4}{16\pi^2} \frac{\sin^2 \vartheta}{R^2} V \left(\frac{\rho}{\varepsilon} \frac{d\varepsilon}{d\rho}\right)^2 \frac{2\kappa\beta T}{1 + \dfrac{2\pi^2}{3} \dfrac{l^2}{\lambda^2} \dfrac{s^2}{\tau - 1}} \qquad (40)$$

Since according to (36′) κ is proportional to $1/(\tau - 1)$, relation (40) represents the familiar fact that, with the approach to the critical point, the intensity increases strongly. More important, because of the denominator, the angular dissymmetry increases at the same time. Assuming $l = 10$ Å and $\lambda = 3000$ Å, a value of $\tau - 1 = 0.73 \times 10^{-4}$ will make the factor of s^2 in the denominator equal to unity, which means a strongly pronounced dissymmetry. On the other hand, for a critical temperature $T_c = 300$ it would only be necessary to approach the critical temperature to a distance of 0.022°C in order to establish that dissymmetry.

It is clear that similar relations will hold for the case of the critical mixing point if the energy relations are discussed, as this is usually done in defining the cohesive energy density. If the foregoing explanation is accepted, this means that the angular dissymmetry in the vicinity of the critical point is a measure for the range of molecular forces. From this point of view measurements of this type become important.

The Correlation Function

We started by showing that in scattering by inhomogeneous media the correlation function is of fundamental importance. At first glance it may seem, from the text of the immediately preceding paragraphs, as if we now had completely lost sight of this point of view. This section is added in order to show what the correlation function is which corresponds to relation (40) and to our physical picture of critical opalescence. Consider two points A and B in the liquid. If a wave passes through the liquid in a direction indicated by the unit vector $\mathbf{S}$, the density fluctuation σ in a point at position $\mathbf{r}$ can be represented by the formula

$$\sigma = \sigma_0 e^{i[\omega t - K(\mathbf{S}, \mathbf{r})]}$$

The average value of the product of the two density fluctuations in points A and B is

$$\text{Av} \langle \sigma_A \sigma_B \rangle = \frac{\sigma_0^2}{2} \cos K(\mathbf{S}, \mathbf{r}_B - \mathbf{r}_A) \tag{41}$$

All the different thermal waves are independent of each other. Remembering that the number of waves in the interval $K^2 \, dK \, d\Omega$ (with $d\Omega$ indicating the element of solid angle) is

$$dZ = \frac{V}{(2\pi)^3} K^2 \, dK \, d\Omega$$

we find immediately

$$\text{Av} \langle \sigma_A \sigma_B \rangle = \frac{V}{(2\pi)^3} \int \frac{\sigma_0^2}{2} \cos K(\mathbf{Sr}) K^2 \, dK \, d\Omega$$

if the vector from A to B now is called $\mathbf{r}$. The integration over the solid angle can be carried out immediately with the result that

$$\text{Av} \langle \sigma_A \sigma_B \rangle = 4\pi \frac{V}{(2\pi)^3} \int_0^{K_m} \frac{\sigma_0^2}{2} \frac{\sin Kr}{Kr} K^2 \, dK \tag{42}$$

The integration starts at $K = 0$, it ends at $K = K_m$, where, as in the case of the specific heats, K_m is determined by the fact that the liquid has a finite number of degrees of freedom. This leads to the relation

$$\frac{V}{(2\pi)^3}\frac{4\pi}{3}K_m{}^3 = 3\mathfrak{N} \tag{43}$$

In our case, where we have taken account of the existence of an energy additional to the compressional energy due to density gradients, we found that equipartition leads to relation (39). Substituting the value for $\sigma_0{}^2$, from this relation gives the result

$$\text{Av}\left\langle \frac{\sigma_A}{\rho}\frac{\sigma_B}{\rho}\right\rangle = \frac{4\pi\kappa\beta T}{(2\pi)^3}\frac{1}{r}\int_0^{K_m}\frac{\sin Kr}{1+K^2/K_0{}^2}K\,dK \tag{44}$$

with

$$K_0{}^2 = 6\frac{\tau-1}{l^2} \tag{44'}$$

For $r = 0$ this leads to

$$\text{Av}\left\langle\left(\frac{\sigma}{\rho}\right)^2\right\rangle = \frac{4\pi\kappa\beta T}{(2\pi)^3}\int_0^{K_m}\frac{K^2\,dK}{1+K^2/K_0{}^2}$$

$$= \frac{4\pi\kappa\beta T}{(2\pi)^3}K_0{}^3\left[\frac{K_m}{K_0}-\arctan\frac{K_m}{K_0}\right] \tag{45}$$

The correlation function is the quotient of (44) and (45). If now we consider first a case in which the experiment is carried out near the critical point, we observe that, since according to (43) and (44')

$$K_m = 2\pi\left(\frac{g}{4\pi}\frac{\mathfrak{N}}{V}\right)^{1/3}\quad\text{and}\quad K_0 = \frac{2\pi}{l}\sqrt{\frac{3}{2\pi^2}(\tau-1)}$$

the quotient K_m/K_0 will be very large. In this case a good approximation will be achieved by substituting ∞ for the upper limit of the integration. Since

$$\int_0^\infty\frac{\sin\alpha x}{1+x^2}dx = \frac{\pi}{2}e^{-\alpha}$$

we find for the correlation function the approximation

$$C(r) = \frac{\pi}{2}\frac{e^{-K_0 r}}{K_m{}^2} \tag{46}$$

An exponential approximation for the correlation function has been mentioned before (Ornstein and Zernike, 1918). In another paper of the same authors (Ornstein and Zernike, 1926), essentially the same formula as our relation (40) for the scattered intensity has been proposed. The reasoning, however, is quite different from that followed here. We are on common ground concerning the importance of correlation, the introduction of which was the outstanding featurre of Zernike's thesis (Zernike, 1916, 1917). It should also be noted that the introduction of the correlation concept provides a quantitative description of what Volmer (1957) calls: "The colloidal nature of liquid mixtures near the critical state."

From the derivation it is clear that the approximation (46) becomes invalid for small values of r. For $r = 0$ we always have $C(r) = 1$.

It is interesting to see what the correlation function expressed by the integral in equation (44) becomes for ordinary cases, for example, in a liquid at a temperature much lower than the critical temperature. In this case it is more appropriate not to try to calculate the compressibility but to use its experimental value. It is easily seen from expression (37′) for the additional energy, in substituting for $n^2 I_0$ its expression in terms of the critical pressure p_c and the reduced volume φ, that $K_0{}^2$ can be defined by the formula

$$K_0{}^2 = \frac{\varphi^2}{\kappa p_c} \frac{1}{l^2} \tag{47}$$

Taking as an example benzene at 20°C, it is found, applying equation (43) for K_m and equation (47) for K_0, that

$$K_m = \frac{2\pi}{5.9} \text{Å}^{-1} \quad \text{and} \quad K_0 = \frac{2\pi}{0.12l} \text{Å}^{-1}$$

K_m and K_0 are now both of the same order of magnitude, although K_m probably is a few times smaller than K_0. Taking, for instance, $l = 10$ Å gives $K_m/K_0 = \frac{1}{5}$. If we introduce now $x = K_m r$ as a new parameter and use in the integral the variable $\xi = K/K_m$ we have according to (44):

$$\text{Av}\left\langle \frac{\sigma_A}{\rho} \frac{\sigma_B}{\rho} \right\rangle = \frac{4\pi\kappa\beta T}{(2\pi)^3} \frac{K_m{}^3}{x^3} \int_0^1 \frac{\sin x\xi}{1 + (K_m{}^2/K_0{}^2)\xi^2} \xi \, d\xi \tag{48}$$

Since $K_m{}^2/K_0{}^2 = 1/25$, it will be a good approximation to neglect the part $(K_m{}^2/K_0{}^2)\xi^2$ in the denominator. So we obtain the approximation

$$\text{Av}\left\langle \frac{\sigma_A}{\rho} \frac{\sigma_B}{\rho} \right\rangle = \frac{4\pi\kappa\beta T}{(2\pi)^3} K_m{}^3 \frac{\sin x - x \cos x}{x^3} \tag{48′}$$

where $x = K_m r$. From this, taking the limiting value for $x = 0$, or from an approximate value calculated from (45) in the limit for $K_m/K_0 = 0$, the correlation function follows immediately as

$$C(r) = 3 \frac{\sin x - x \cos x}{x^3} \tag{49}$$

where $x = K_m r$. This is an interesting theoretical result in view of the well-known experimental results obtained by x-ray scattering for the radial distribution function of the atoms around a single central atom in a liquid, as introduced by Zernike and Prins (1927). Maxima and minima in the scattering diagram of liquids were first observed in 1916 (Debye and Scherrer). They were originally interpreted as the result of interference between the atoms in the molecule. This point of view, however, was soon corrected (Debye, 1925, 1927) and was experimentally proved to be untenable by the observation of a scattering pattern of monatomic liquids (Keesom and de Smedt, 1923).

The effect of approaching the critical temperature can be expressed very simply if, next to a range l of molecular forces, we introduce a range L of correlation by the definition

$$L^2 = \frac{\int C(r) r^2 \, d\tau}{\int C(r) \, d\tau}$$

From equation (4) it is evident that (after corrections for polarization effects) the first variation of the scattered intensity with s will be proportional to s^2 and be represented by a formula of the form

$$\frac{g}{g_0} = 1 - \frac{k^2 s^2}{6} L^2$$

According to equation (40) we also have for small values of s

$$\frac{g}{g_0} = 1 - \frac{2\pi^2}{3} \frac{l^2}{\lambda^2} \frac{s^2}{\tau - 1}$$

Since $k = 2\pi/\lambda$, we see at once that

$$L^2 = l^2/(\tau - 1) \tag{50}$$

We cannot, of course, apply (50) directly to the experiments of Zimm (1950) on the mixture of carbon tetrachloride and perfluoromethylcyclohexane. However, we can determine L by plotting the scattered intensity as a function of s^2 and measuring the tangent of this curve for $s = 0$. From Figure 6 in Zimm's publication it follows, for instance, that for a temperature distance of 0.02°C from the critical temperature we have $L/\lambda = 0.43$, which would make $L = 1300 \, \text{Å}$ for an assumed value of the optical wavelength of $3000 \, \text{Å}$. Moreover, Zimm's curves for different values of $T - T_c$ give values of L which follow relation (50) in their dependence on the temperature distance from the critical temperature, although, strictly speaking, our derivation of that relation does not apply directly to the case of liquid mixtures. We conclude that observations of the angular intensity distribution near the critical temperature can be used to investigate the range of molecular forces since, in order to obtain the correlation range in this vicinity, we have to multiply by such a big factor that the last-named range becomes comparable to the wavelength of visible light.

REFERENCES

Booker, H. G., and Gordon, W. E., *Proc. I.R.E.* **38**, 401 (1950).

Brillouin, L., *Ann. phys.* **17**, 88 (1922).

Debye, P., *J. Math. Phys.* **4**, 133 (1925).

Debye, P., *Physik. Z.* **28**, 135 (1927).

Debye, P., Anderson, H. R., and Brumberger, H., *J. Appl. Phys.* **28**, 679 (1957).

Debye, P., and Bueche, A. M., *J. Appl. Phys.* **20**, 518 (1949).

Debye, P., and Scherrer, P., *Goettinger Nachr.* **16** (1916).

Debye, P., and Sears, F. W., *Proc. Natl. Acad. Sci. U. S.* **18**, 409 (1932).

Einstein, A., *Ann. Physik* **33**, 1275 (1918).

Fuerth, R., and Williams, C. L., *Proc. Roy. Soc. (London)* **A224**, 104 (1954).

Guinier, A., Fournet, G., Walker, G B., and Yudowitch, K. L., *Small Angle Scattering of X-rays*, Wiley, New York, 1955.

Keesom, W. H., and de Smedt, J., *J. phys. radium* [6] **4**, 144 (1923).

Ornstein, L. S., and Zernike, F., *Physik. Z.* **19**, 134 (1918).

Ornstein, L. S., and Zernike, F., *Physik. Z.* **27**, 761 (1926).

Pekeris, C. L., *Phys. Rev.* **71**, 268 (1947).

Porod, G., *Kolloid Z.* **124**, 83 (1951).

Porod, G., *Kolloid Z.* **125**, 51, 109 (1952).

Roccard, Yves, *J. phys. radium* **4**, 165 (1933).

Taylor, G. I., *Proc. London Math. Soc.* **20**, 196 (1920).

Taylor, G. I., *Proc. Roy. Soc. (London)* **164**, 478 (1938).

van Hove, L., *Phys. Rev.* **95**, 249 (1954).

van Hove, L., *Physica* **24**, 404 (1958).

Van Nordstrand, R. A., and Hach, K. M., in a paper presented to the A.C.S. Meeting, Chicago, Sept. 1953.

Van Nordstrand, R. A., and Johnson, M. F. L., in a paper presented to the A.C.S. Meeting, New York, Sept. 1954.

Volmer, M., *Z. physik. Chem.* **207**, 307 (1957).

Zernike, F. (Thesis), *Proc. Acad. Sci. Amsterdam* **17**, 793 (1916).

Zernike, F. (Thesis), *Arch. néerl. sci.* **Ser. 3A, 4**, 74 (1917).

Zernike, F., and Prins, J. A., *Z. Physik* **41**, 184 (1927).

Zimm, B. H., *J. Phys. & Colloid Chem.* **54**, 1306 (1950).

1958

PAUL H. EMMETT

THE JOHNS HOPKINS UNIVERSITY
BALTIMORE, MD.

. . . for devising the first generally applicable method for measuring the surface area of finely-divided solids.

The following Kendall Award Address is reproduced by permission from *The Journal of Physical Chemistry*, **63**, 449 (1959). The Addendum has been written by Prof. Emmett for this volume.

ADSORPTION AND CATALYSIS

By Paul H. Emmett

Chemistry Department, The Johns Hopkins University, Baltimore, Md.

Received October 7, 1958

When Berzelius in 1836 coined the word "catalysis" and used it to describe a number of observations that had already been recorded in the literature, he clearly recognized that the reacting components in catalytic reactions were held to the surface or adsorbed during the period in which they were reacting. In the intervening years since Berzelius gave birth to this new field of science, many studies have been reported under the general title of "Adsorption and Catalysis." This occasion seems to be a suitable one for pointing out the many ways in which adsorption has now become related to catalytic action and to the study of catalysts, and also for calling attention to the general direction along which progress is being made toward establishing a firm scientific explanation for the behavior of catalysts.

The existence of two types of adsorption of gases on solids has been recognized for many years.[1-3] One type, now commonly called physical adsorption, is non-specific and in general occurs between all gases and all solids. It is capable of forming monolayers and even multilayers of adsorbed gas on solids close to the boiling point of a gaseous adsorbate. The other type commonly known as chemical adsorption or activated adsorption is specific in nature and occurs under conditions in which the gaseous adsorbate might be expected to combine chemically with the surface of the solid adsorbent. Both types of adsorption are now considered to be useful and important in the study of catalysts.

Physical Adsorption.—Two distinct uses for physical adsorption in studying catalysts have now been developed. One of them is concerned with the measurement of the surface areas of finely divided catalysts; the other, with the measurement of the pore size of catalytic materials.

The method for measuring surface areas by the physical adsorption of gases is now so well known as to require little comment.[4-6] Briefly, it involves measuring the adsorption isotherms of suitable inert adsorbates at temperatures close to the boiling points of the gases and determining by appropriate plots the point on the isotherms corresponding to the volume of gas required to form a single close-packed layer on the catalyst surface. By assuming a suitable cross-sectional area for the adsorbate molecules, one can then calculate the absolute surface area of the catalyst in square meters per gram. By these techniques, the relative surface areas of a number of catalysts or other solid adsorbents can be determined quite accurately. The absolute value of the surface area is, of course, subject to the accuracy with which one can estimate the area covered by each molecule in a close packed layer of physically adsorbed gas. It is probably safe to assume that the absolute surface area of solids can now be measured within an uncertainty of $\pm 30\%$ by these gas adsorption techniques, but relative surface areas for a series of solids can probably be estimated within a few per cent. The gas adsorption method is now universally used for specifying the surface areas of catalytic materials.

An equally or perhaps more important application of physical adsorption in catalysis has to do with measuring the pore size of the tiny capillaries that are present in most catalysts. Wheeler[7] was the first to point out that by taking into account the formation both of multilayers of adsorbed gas and the capillary condensation of adsorbate in the tiny capillaries one could, in principle, calculate the size distribution of capillaries in solid adsorbents.

(1) I. Langmuir, *J. Am. Chem. Soc.*, **38**, 2221 (1916).

(2) A. F. Benton, *ibid.*, **45**, 887, 900 (1923).

(3) H. S. Taylor, *ibid.*, **53**, 578 (1931).

(4) P. H. Emmett and S. Brunauer, *ibid.*, **59**, 1553 (1937).

(5) S. Brunauer, P. H. Emmett and E. Teller, *ibid.*, **60**, 309 (1938).

(6) P. H. Emmett, "Catalysis," Vol. I, Chap. 2, Reinhold Publ. Corp., New York, N. Y., 1954.

(7) A. Wheeler, Gordon Research Conference, 1945 and 1946.

Barrett, Joyner and Halenda[8] following this suggestion, arrived at a detailed procedure that could be used for making such calculations without making any arbitrary assumptions as to the nature of the size distribution curves for the capillaries in the solid catalysts. Using this procedure, one can now calculate the pore distributions over the range 10 to 2000 Å. diameter with a fair degree of assurance that the distribution so calculated is a good approximation to the true distribution. Confirmation of the pore distribution so calculated has been obtained on a number of solids by use of the mercury porosimeter[9-11] method by which the pressure necessary to force mercury into the various capillaries is used in calculating pore distributions. By a combination of the gas adsorption and mercury porosimeter method, one can obtain a fairly reliable distribution curve for the pore size of a given solid and can obtain an even more reliable value for the average pore size.

Wheeler[12] has also pioneered the development of the theory for interpreting the influence of pore size on many characteristics of the behavior of a catalyst toward various reactants. In particular, he has shown that pore distribution can affect the apparent order of the catalytic reaction, the temperature coefficient, the fraction of the surface participating in and contributing to the catalytic reaction,[13] the specificity of the catalyst, and the behavior of the catalyst after it is exposed to various poisons. In a word, he has shown that for a given catalytic reaction, a particular catalyst can be tailor made by appropriate alteration of the method of preparation to yield a surface area and a pore distribution that will be optimum for a particular gas reaction. This makes it possible to put catalysis on a much more quantitative and scientific basis than has heretofore been possible.

For a good many years, physical adsorption was considered to involve such weak binding forces with the solid adsorbents as not to cause any alteration or change in the adsorbent. After five years ago, Cook, Pack and Oblad[14] pointed out the possibility that even physical adsorption could distort the atomic arrangement in the surface of the catalyst. In 1954, Yates[15] furnished proof of this effect by showing that the adsorption of inert gases such as nitrogen, krypton and argon on a sample of porous glass[16] caused an appreciable expansion in the porous glass. This expansion was detectable even when only a few per cent. of the surface was covered with adsorbed gas. Although these observations are still not completely explained, they leave no doubt that physical adsorption, even at the temperature of liquid nitrogen is capable of altering the surface energy of solid adsorbents and of causing either the expansion or contraction of the solid depending upon the particular gas–solid combination that is being used.

It is generally assumed that chemical and not physical adsorption is a form that is directly involved in catalytic reactions. Even this view may require some modification. For example, there are many catalytic reactions in which only one of two or more reactants apparently has to be chemisorbed as a pre-requisite to the occurrence of a catalytic reaction. It is entirely possible that the other components are merely physically adsorbed on the catalyst at the time of reaction. Furthermore, the very nice work of Becker[17] and Ehrlich[18] on the chemisorption of nitrogen on tungsten seems to indicate that the nitrogen molecules are first physically adsorbed and then are transformed over to a type of chemical adsorption. If this is generally true, then, physical adsorption may be a forerunner to catalytic reactions and to the chemical adsorption that is usually considered to be involved as a basic and essential part of catalytic reactions.

Chemical Adsorption—Chemical adsorption traditionally has found two applications in the study of catalysts. To begin with it has been found very useful in measuring the fraction of a catalyst surface which consists of the catalytically active component compared to the portion which consists of material that either acts as a support or as a promoter to the principal catalyst. For example, it has been found possible to measure the fraction of the surface of an iron synthetic ammonia catalyst that consists of metallic iron,[19] by the study of the low temperature chemisorption of carbon monoxide. Similarly, if these catalysts are promoted with alkaline oxides, one can measure the concentration of such alkaline promoters on the surface of the catalyst by determining the chemisorption of carbon dioxide on the promoter at $-78°$. Properly designed experiments using isotopic forms of the adsorbate can even give information as to the extent to which the surface appears to be homogeneous and the extent to which it appears to be heterogeneous. Such work has been done, for example, using non-radioactive carbon monoxide and radio-active carbon monoxide on iron catalysts.[20,21]

By far the biggest application of chemisorption is related to the part it plays in actual catalytic reactions. Two examples will suffice as illustrations. For many years, the chemisorption of hydrocarbons on standard silica–alumina cracking catalysts could not be detected. Volumetric measurements seemed to indicate that less than 0.1% of the surface of these catalysts was covered by chemisorbed hydrocarbons during cracking reactions.[22] Recently,

(8) E. P. Barrett, L. G. Joyner and P. P. Halenda, *J. Am. Chem. Soc.*, **73**, 373 (1951).

(9) L. G. Joyner, E. P. Barrett and R. E. Skold, *ibid.*, **73**, 3155 (1951).

(10) E. W. Washburn, *Phys. Rev.*, **17**, 273 (1921).

(11) H. L. Ritter and L. C. Drake, *Ind. Eng. Chem., Anal. Ed.*, **17**, 782 (1945).

(12) A. Wheeler, "Catalysis," Vol. II, (P. H. Emmett, ed.), Reinhold Publ. Corp., New York, N. Y., 1955, Chap 2.

(13) E. W. Thiele, *Ind. Eng. Chem.*, **31**, 916 (1939).

(14) M. A. Cook, D. H. Pack and A. G. Oblad, *J. Chem. Phys.*, **19**, 367 (1951).

(15) D. J. C. Yates, *Proc. Roy. Soc. (London)*, **224A**, 526 (1954).

(16) M. E. Nordberg, *J. Am. Ceramic Soc.*, **27**, 299 (1944).

(17) J. A. Becker, "Advances in Catalysis," Vol. VII, Academic Press, New York, N. Y., 1955, p. 135.

(18) G. Ehrlich, *J. Phys. Chem. Solids*, **1**, 3 (1956).

(19) P. H. Emmett and S. Brunauer, *J. Am. Chem. Soc.*, **59**, 310 (1937).

(20) J. T. Kummer and P. H. Emmett, *J. Chem. Phys.*, **19**, 289 (1951).

(21) R. P. Eischens, *J. Am. Chem. Soc.*, **74**, 6167 (1952).

(22) R. C. Zabor and P. H. Emmett, *ibid.*, **73**, 5639 (1951).

(23) D. S. MacIver and P. H. Emmett, THIS JOURNAL, **62**, 935 (1958).

by the use of radioactive techniques, it has been possible to show that chemisorption of such hydrocarbons as isobutane does actually occur on the silica–alumina catalysts but is restricted to a very small fraction (about 0.01%) of the total surface. It now appears therefore that this catalyst as well as all other known catalysts, function through the chemisorption of at least one of the reactants on the catalyst surface.

A second example, the chemisorption of nitrogen on iron is of especial interest because of the enormous amount of work that has been done upon it. The catalytic synthesis of ammonia over iron catalysts became a commercial process forty-five years ago. Since that time a good deal of interest has been shown in attempting to deduce the mechanism of the synthesis of ammonia from the mixture of hydrogen and nitrogen. Measurements of the rate and quantity of nitrogen adsorbed by active iron catalyst led to a conclusion many years ago that the slow step in the synthesis of ammonia was the chemisorption of nitrogen by the surface of the catalyst.[24] This result was generally accepted until a few years ago when Horiuti[25,26] and his colleagues came to the conclusion that the slow step could not possibly be the chemisorption of nitrogen on iron. Their conclusion was based on their theory of the "stoichiometric number" of the individual steps in a catalytic reaction. Space does not permit a review of the theory of this stoichiometric number in the present article. It will perhaps suffice to point out that according to their ideas the stoichiometric number of the adsorption of nitrogen on iron would have to be unity if the slow step were to be the chemisorption of nitrogen. By a combination of kinetic measurements of the rate of synthesis of ammonia over iron catalysts close to equilibrium, and the rate of isotopic exchange of N^{15} between gaseous nitrogen and ammonia, the Japanese were able to calculate a value for this stoichiometric number; they reported a value of two rather than a value of one and concluded that the slow step in the synthesis had to be something other than the chemisorption of nitrogen. This would present a major disturbance in attempts to connect chemisorption and catalysis, if it were experimentally verified by other workers in the field. It so happens, however, that repetition of the work by Van Heerdin, Zwietering[27] and their colleagues in Holland has led to results that contradict those obtained by the Japanese. These workers find a stoichiometric number of one and therefore conclude that the slow step in the synthesis may well be the chemisorption of nitrogen. Furthermore, by additional measurements of the rate of nitrogen adsorption during actual synthesis, they show[28] that at and above 250° excellent agreement is obtained between the rate of nitrogen chemisorption and the rate of synthesis that occurs over an iron catalyst.

This classic and long extended study of the chemisorption of nitrogen on iron accordingly serves as an excellent example of the way in which chemisorption can be related to the mechanism of a catalytic reaction.

A consideration of chemisorption would be incomplete without recording some of the spectacular advances that are being made in obtaining detailed information as to the nature of the chemisorption complex. Foremost among these new tools is the use of infrared absorption.[29] Pliskin and Eischens[30] have shown that it is possible to study by infrared the detailed nature of the binding of molecules such as carbon monoxide, acetylene, ethylene and hexene on finely divided metals supported on a finely divided silica. The method seems capable of giving information that could be correlated with the catalytic activity of the particular metals involved and in some cases can apparently even be used to obtain information as to the nature of the intermediate that is formed in catalytic reactions.[31] This spectacular development of a method for measuring the infrared absorption spectra of a few chemisorbed molecules appears capable of giving us an insight into the details of chemisorption that, a few years ago, we would never have expected to be possible.

A second important set of tools for the study of the nature of the chemisorption of molecules on surfaces is the use of magnetic and conductivity measurements. Selwood[32–35] has shown that when gases such as hydrogen, ethylene, acetylene and benzene are chemisorbed on the surface of metallic nickel, the specific magnetization of the finely divided nickel is lowered in such a way as to indicate a decrease in the number of d-band vacancies in the nickel as a result of the adsorption. Specifically, he concludes that hydrogen on nickel appears to contribute one electron per hydrogen atom to the nickel crystal. Ethylene and ethane contribute, respectively, about two and about four electrons per chemisorbed molecule. Selwood, at first, suggested that these magnetic results were consistent with the observations of Suhrmann[36] to the effect that the adsorbed hydrogen increased the conductivity of nickel. More recent measurements,[37] however, have indicated that the first exposure of the fresh nickel surface to hydrogen results in a decrease and not an increase in conductivity. When sufficient hydrogen has been taken up, the trend apparently reverses itself and begins to yield an increase in conductivity. The latest conclusion[36] as to hydrogen chemisorbed on metallic surfaces such as nickel is that it forms primarily a covalent bond with the nickel. Slight polarization of

(24) P. H. Emmett and S. Brunauer, *J. Am. Chem. Soc.*, **56**, 35 (1934).

(25) S. Enomoto and J. Horiuti, *J. Res. Inst. Cat., Hokkaido Univ.*, **2**, 87 (1955); *Proc. Japan Acad.*, **28**, 499 (1955).

(26) S. Enomoto, J. Horiuti and Y. Kobayashi, *J. Research Inst. Cat.*, **3**, 185 (1955).

(27) C. Bokhoven, M. J. Gorgels and P. Mars, private communication.

(28) J. J. F. Scholten and P. Zwietering, private communication.

(29) W. A. Pliskin and R. P. Eischens, *J. Chem. Phys.*, **24**, 482 (1956).

(30) R. O. French, M. E. Wadsworth, M. A. Cook and I. B. Cutler, This Journal, **58**, 805 (1954).

(31) R. P. Eischens, *Z. Elektrochem.*, **60**, 782 (1956).

(32) P. W. Selwood, *J. Am. Chem. Soc.*, **78**, 3893 (1956).

(33) P. W. Selwood, *ibid.*, **79**, 3346 (1957).

(34) P. W. Selwood, "Advances in Catalysis," Vol. IX, Academic Press, New York, N. Y., 1957, p. 93.

(35) P. W. Selwood, *J. Am. Chem. Soc.*, **79**, 4637 (1957).

(36) R. Suhrmann, "Advances in Catalysis," Vol. VII, Academic Press, New York, N. Y., 1955, p. 303.

(37) J. J. Broeder, L. L. van Reijn, W. M. H. Sachtler and G. C. A. Schuit, *Z. Elektrochem.*, **60**, 838 (1956).

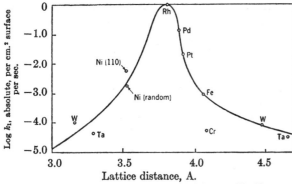

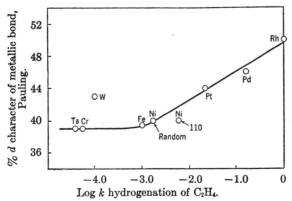

Fig. 1.—Activity at 0° of a series of thin metallic films per unit surface as a function of the distance between particular pairs of atoms in the lattice.[40]

Fig. 2.—Plot of the per cent. of d-character of a series of metals (by Pauling's theory) against the logarithm of the activity of the metals per unit area for hydrogenating ethylene.[43]

the bond occurs but the adsorbed atom is considered to be held in the surface primarily by covalent and not by ionic bonding. This interpretation is believed[36] to be consistent with both the magnetic data and the new conductivity data. There can be no doubt but that the combination of infrared measurements, magnetic measurements and conductivity measurements for gases on metals is going to give us a much more precise idea of catalytic action than we have been able to obtain in the past.

Chemisorption studies on semi-conductors have also received a great deal of attention. Some[38] have suggested that the word chemisorption of reacting gases on semi-conductors should really be called "ionosorption." This reflects the idea that on semi-conductors, most of the adsorbed particles are either positive or negative ions. There can be no question but that the chemisorption of gases on semi-conductors results in marked changes in conductivity. There may, however, still be considerable question as to whether the true reactants in catalysis over these semi-conductors are the completely ionized species that affect the conductivities or other types of chemisorbed gas including polarized molecules or molecules held by covalent bonds.

In summary, then, one can say that the chemisorption of gases on catalysts is receiving very intensive study at the present moment. A combina-

tion of infrared, magnetic and conductivity studies appears capable of giving us as complete a picture of the details of molecular interaction on solids as we now have for reaction between ions, radicals, atoms and molecules in the gas phase and in solution.

Catalysis.—For many years catalysis has been correctly classified as an art. Only within the last few years has evidence been developing to indicate that a firm scientific basis is being laid to explain catalytic phenomena. In the past, we have selected our catalysts by trying thousands of combinations of catalysts and promoters until we found one that worked. In the future, it seems quite probable that our job of selecting catalysts will be made much easier and that a knowledge of the fundamental properties of reacting gases and the electronic properties of the solid catalysts will go far toward narrowing our choice to a few promising solids. Some of the newer ideas and techniques that are helping to elucidate the mechanism of catalytic reactions and to throw light on the basic properties of catalysts will now be reviewed.

Dowden[39] has suggested that in view of the importance of the electronic factor catalysts might well be classified as conductors, semi-conductors and insulators. Conductors can be illustrated by metallic catalysts such as nickel, platinum and iron. Semi-conductors, by well known commercial catalysts such as zinc oxide, chromium oxide, molybdenum oxide and vanadium pentoxide. Insulator type catalysts may include such compounds as aluminum oxide and alumina–silica cracking catalysts. For the sake of illustrating the new approach catalytic hydrogenation over metallic catalysts will perhaps suffice.

During the period 1934–1948, Beeck[40-43] and his co-workers carried out their classical experiments on the study of the catalytic hydrogenation of ethylene over metallic films. They reached the conclusion that the activity of these metallic films for this reaction was apparently related to certain inter-atomic spacings in the metal. Their results for a number of metals, for example, were plotted according to the scheme shown in Fig. 1.[43] These results seemed to indicate that the spacing of atoms in rhodium was optimum for the ethylene hydrogenation reaction. In line with this, the activity of the rhodium per unit surface area was found to be some three orders of magnitude greater than that of nickel.

In the spring of 1950, Beeck[43] pointed out that these results could also be interpreted on the basis of the per cent. d-character of the various metals as defined by the theory of Pauling.[44] A plot of the data for the metallic films interpreted on this basis is shown in Fig. 2. Thus, Beeck and his co-workers suggested an electronic interpretation that suited

(38) K. Hauffe, "Advances in Catalysis," Vol. VII, Academic Press, New York, N. Y., 1955, p. 213.

(39) D. A. Dowden, *J. Chem. Soc.*, 242 (1950).
(40) O. Beeck, *Revs. Modern Phys.*, **17**, 61 (1945).
(41) O. Beeck, A. E. Smith and A. Wheeler, *Proc. Roy. Soc. (London)*, **A177**, 62 (1940).
(42) O. Beeck, "Advances in Catalysis," Vol. II, Academic Press, New York, N. Y., 1950, p. 151.
(43) O. Beeck, *Disc. Faraday Soc.*, **8**, 118 (1950).
(44) L. Pauling, *Proc. Roy. Soc. (London)*, **A196**, 343 (1949).

their data equally as well as the geometric interpretation.

About this same time Dowden and Reynolds[45] carried out a series of studies on the catalytic hydrogenation of styrene, benzene and a few other gases over nickel–copper and nickel–iron alloys. The results for styrene are illustrated in Fig. 3. Dowden[39] predicted that when enough copper was added to nickel to fill completely the d-band vacancies, one would probably destroy the catalytic activity of the nickel toward the hydrogenation of styrene. His prediction was beautifully confirmed by the experimental results illustrated in Fig. 3. The activity of a catalyst for the hydrogenation of styrene decreased steadily with an increase in the copper content and reached substantially a zero value when 38% copper was incorporated into the nickel. The addition of iron to nickel lowered the catalytic activity in spite of the fact that it increased the number of d-band vacancies. This led Dowden and Reynolds to suggest that some other electronic properties such perhaps as the density of electronic levels in the catalysts were also important.

Although the results of Dowden and Reynolds for the hydrogenation of styrene seemed to suggest a firm correlation between the electronic properties of the metallic catalysts and the activity toward this reaction, later experiments with other reactions have indicated that the problem is quite complicated. Specifically, Best and Russell[46] reported that catalysts containing 30–60% copper were many times more active than nickel for the catalytic hydrogenation of ethylene. The results seem to indicate definitely that the explanation for the alloy effect is not the same for all hydrogenation reactions.

Hall[47] recently completed a detailed study of the hydrogenation of ethylene and of C_6H_6[47b] over a series of nickel–copper alloy catalysts. His results like those of Best and Russell show that the addition of copper increases rather than decreases the activity toward the hydrogenation of ethylene until more than 80 atom % copper has been added (Fig. 4). Furthermore, he found that hydrogen added in the temperature range 100 to 300° acts as a promoter to the copper–nickel catalysts but as an inhibitor for pure nickel (Fig. 5). A theoretical explanation of the results is not yet forthcoming; it seems certain, however, that any satisfactory theory must explain the influence of the added copper and hydrogen on the proportionality factor A in the equation

$$\text{Rate} = Ae^{-E/RT}$$

as well as on the energy of activation E.

Very recently Kokes and the writer[48,49] have completed a study of Raney nickel catalysts and obtained results that throw additional light on the question of the action of alloying constituents on the activity of metallic catalysts. It has long been

(45) D. A. Dowden and P. W. Reynolds, *Disc. Faraday Soc.*, **8**, 172 (1950).

(46) R. J. Best and W. W. Russell, *J. Am. Chem. Soc.*, **76**, 838 (1954).

(47) (a) W. Keith Hall, Thesis, University of Pittsburg, 1956; (b) W. Keith Hall and P. H. Emmett, THIS JOURNAL, **62**, 816 (1958).

(48) R. J. Kokes and P. H. Emmett, to be published.

(49) R. J. Kokes and P. H. Emmett, to be published.

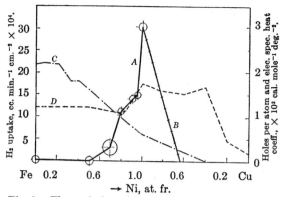

Fig. 3.—The variation with composition of the activity of a series of Fe–Ni (curve A) and Ni–Cu alloys (curve B) for the hydrogenation of styrene. The number of d-band vacancies per atom of Ni and the coefficient of the low-temperature electronic specific heat term are also shown (curves C and D, respectively).[45]

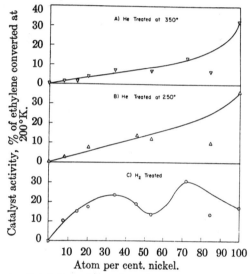

Fig. 4.—Catalytic activity of a series of nickel–copper alloys at 200°K. for the hydrogenation of ethylene.[47]

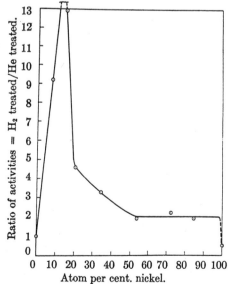

Fig. 5.—The promoting effect of pre-adsorbed hydrogen on the rate of hydrogenation of ethylene as a function of catalyst composition.[47]

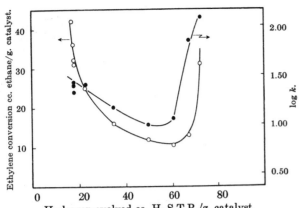

Fig. 6.—The activity of a Raney nickel catalyst for ethylene hydrogenation and for the ortho–parahydrogen conversion at −78°.[49]

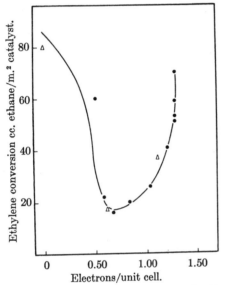

Fig. 7.—Activity per sq. m. of surface area for the hydrogenation of ethylene over a Raney nickel catalyst[49] and over a nickel–copper alloy[47] at −78° as a function of the number of electrons added by alloying components per unit cell. Each copper atom and each hydrogen atom is assumed to contribute one electron; each aluminum atom, three electrons. For the Ni–Cu alloys, the atom per cent. copper is 0, 15.2 and 27.6%, respectively, for the three points shown as triangles.[47] The solid line is for the Raney nickel catalyst.[49]

known that active Raney nickel catalysts contain a considerable amount of dissolved hydrogen. A few years ago Smith and his co-workers[50] pointed out that the removal of this dissolved hydrogen tended to destroy the activity of the nickel catalysts. In checking up on this relation for the ethylene hydrogenation, Kokes and the writer have now obtained the data shown in Fig. 6. The results indicate that as the dissolved hydrogen content of the original Raney catalysts is gradually removed, the activity falls. However, it reaches a minimum and again increases as a last portion of the dissolved gas is removed from the nickel catalyst. We were at first inclined to believe that the rise of activity during the removal of the last part

(50) H. A. Smith, A. J. Chadwell, Jr., and S. J. Kirslis, This Journal, **59**, 820 (1955).

of dissolved hydrogen might be the result of removing traces of oxygen poison from the surface of the nickel. However, a comparison with the results obtained by Hall for the addition of copper to nickel seems to indicate that the hydrogen is influencing the activity of the nickel because of some electronic factor. This is illustrated by the fact that a plot of catalytic activity per unit surface area for the nickel copper catalysts against the atom per cent. copper in the alloy as measured by Hall[47] yields the three points shown as triangles on the curve in Fig. 7. It is, of course, fully realized that when the metallic surface of the Raney catalysts has been measured by carbon monoxide chemisorption one probably will find that the curve for the Raney nickel catalyst per unit surface area of nickel will be shifted upwards from the present curve which is based on total surface area. Nevertheless, the shape of the curve for the Raney nickel catalyst as a function of hydrogen content will still remain essentially the same as the shape of the curve for the catalytic activities as a function of copper content. This suggests that electronic factors are involved in both instances.

The nature and amount of added hydrogen in the copper–nickel series studied by Hall is not yet known. However, for Raney catalysts a combination of magnetic measurements, differential thermal analysis and density measurements (by helium) as a function of the amount of hydrogen pumped out of the samples seem to afford strong evidence that the hydrogen promoter is present in the form of a substitutional solid solution with the nickel. Thus, the number of nickel vacancies as estimated from the density of the catalyst is very closely the number needed to account for the hydrogen content if one atom of hydrogen is present in each lattice vacancy.

In summary, one can say that catalytic hydrogenation appears to be related to electronic properties of the metallic catalysts; however, we are still far from reaching definite conclusions as to the exact way in which this electronic factor enters into the different types of catalytic hydrogenation reactions.

Space does not permit a review of the very interesting work that is now being carried out in the field of semi-conductors as catalysts. It can perhaps be summarized by saying the catalysts such as zinc oxide, chromium oxide, nickel oxide and cuprous oxide appear to exert their catalytic activity by transferring electrons to the reactant gases on the surface or accepting electrons from them.[51,52] Furthermore, both the semi-conductors and the metallic catalysts may have a considerable portion of their activities located in surface defects[53] and dislocations. Exact correlations are still few and far between but as methods become more exact for assaying the number and nature of these surface defects and dislocations, more quantitative correlations probably will follow.

One aspect of the electronic theory of catalysis

(51) W. E. Garner, "Advances in Catalysis," Vol. IX, Academic Press, New York, N. Y., 1957, p. 169.

(52) "Advances in Catalysis," ed. G. W. Frankenburg, Vol. VII, Academic Press, New York, N. Y., 1955.

(53) A. L. G. Rees, "Chemistry of the Defect Solid State," John Wiley and Sons, Inc., New York, N. Y., 1954.

merits a special mention. One might be inclined to say that the lattice defects and dislocations are merely specific forms of the "active points" postulated many years ago by Taylor.[54] In a measure this also could be said of the interfaces between promoter molecules and the catalytic substances. There is however one electronic concept that is very different from any involved in the early active point hypothesis. Volkenstein[55] has suggested that one variety of "active point" may actually be created on a homogeneous surface by the adsorption of a gas molecule or a gas atom. This idea has been utilized by Taylor and Thon[56] in explaining the kinetics of adsorption of gases on surfaces. They contend that the application of the Elovitch equation to such adsorption phenomena is an example of the way in which active adsorption sites can be created by the adsorption of the gaseous adsorbate and that the slowing down of the adsorption as a function of coverage during such an adsorption experiment can be best explained by the mutual self-destruction of these active points originally created by the adsorption of gases on the surface. Regardless of whether this picture of the creation of active sites by the adsorption of gases survives the test of time, there can be no question but that the analogous poisoning effect of various gases on surfaces may very logically and reasonably be interpreted on such a basis. As early as 1928, Brewer,[57] for example, pointed out that a single oxygen atom adsorbed on the surface of metal could be considered to poison the surface not by virtue of its occupying an active point by the influence that it exerted on the electronic properties of the metal for a number of molecular diameters around the adsorbed oxygen. Accordingly, the general idea of the importance of the electronic influence of gas molecules on solids is an accepted idea in catalysis, even though many are still skeptical of the evidence as to the creation of active sites by the chemisorption of gases on solids.

Two other research tools useful in the study of catalysts should be mentioned. The first of these is the use of isotopes in studying catalytic mechanisms. Many reactions are being studied with the help of the isotopes of hydrogen, nitrogen, oxygen and carbon. These can be illustrated by the extensive measurements that have been made with the help of carbon-14 in studying the mechanism of Fischer–Tropsch synthesis over iron and cobalt catalysts. By the addition of suspected intermediates in the form of radioactive compounds, it has been possible[58–62] to arrive at some fairly definite conclusions in regard to the various steps involved in the complicated synthesis of hydrocarbons over metallic catalysts. Up to the present time the radioactive compounds added in small amounts to carbon monoxide hydrogen feed-gas in such experiments include radioactive methanol, formaldehyde, ethyl alcohol, n-propyl alcohol, isopropyl alcohol, isobutyl alcohol, ethylene and carbon dioxide. Very recently to this list has been added experiments involving the use of radioactive ketene, $C^{14}H_2CO$.[63] These results combine to suggest that the carbon monoxide–hydrogen reaction forming higher hydrocarbons over metallic iron or cobalt catalysts probably proceeds through the formation of an oxygen surface complex similar to those that can be formed by the chemisorption of the alcohols or molecules such as ketene on the surface of the catalysts.

Attention should also be called to the tremendous importance of gas chromatography[64] as a new tool in studying catalytic reactions. As an analytical tool gas chromatography[65–67] makes possible a detailed analysis of products obtained in conventional catalyst test units with a speed and completeness far in excess of any heretofore obtainable. One particular type of set up emerging from this work and known as the micro-catalytic-chromatographic technique,[68] perhaps should be mentioned in closing. It has now been established that if a microquantity of reactant is added to a stream of carrying gas and passed through a small catalyst converter and directly onto a chromatographic column, detailed information can be obtained rapidly as to the quality and quantity of products formed by the particular catalyst with the particular added reactants. This technique promises to give valuable information in regard to the behavior of catalysts during the first fraction of a second of exposure to gaseous reactants as well as detailed information on the kinetics of catalytic reactions[69] and on the behavior of catalysts as judged by the use of radioactive reactants.

Space does not permit a complete enumeration of all the new tools that are being brought to bear in an endeavor to elucidate the way in which metallic, semi-conductor, or insulator catalysts exert their influence on reactants brought in contact with them. Not only the magnetic, conductivity and infrared measurements already mentioned but in addition nuclear magnetic resonance, contact potential, X-ray and electron microscope measurements will in the future be utilized or are at present being utilized in attempting to unravel the many facets of catalysis. One may venture the prediction that the next ten years or so will witness more progress in putting catalysis on a truly scientific basis than has been made in all work up to the present time.

Acknowledgements.—The writer here wishes to

(54) H. S. Taylor, *Proc. Roy. Soc. (London)*, **A108**, 105 (1925).

(55) F. F. Volkenstein, *J. Phys. Chem. U.S.S.R.*, **23**, 917 (1949).

(56) H. A. Taylor and N. Thon, *J. Am. Chem. Soc.*, **74**, 4169 (1952).

(57) A. K. Brewer, THIS JOURNAL, **32**, 1006 (1928).

(58) J. T. Kummer, H. H. Podgurski, W. B. Spencer and P. H. Emmett, *J. Am. Chem. Soc.*, **73**, 564 (1951).

(59) J. T. Kummer, T. W. De Witt and P. H. Emmett, *ibid.*, **70**, 3632 (1948).

(60) W. Keith Hall, R. J. Kokes and P. H. Emmett, *ibid.*, **79**, 2983 (1957).

(61) R. J. Kokes, W. Keith Hall and P. H. Emmett, *ibid.*, **79**, 2989 (1957).

(62) W. Keith Hall, R. J. Kokes and P. H. Emmett, to be published.

(63) G. Blyholder and P. H. Emmett, to be published.

(64) A. T. James and A. J. P. Martin, *Biochem. J.*, **50**, 679 (1952).

(65) A. I. M. Keulemans, "Gas Chromatography," Reinhold Publ. Corp., New York, N. Y., 1957.

(66) "Vapor Phase Chromatography," ed. D. H. Desty, Butterworths, London, 1957.

(67) "Gas Chromatography," D. H. Desty, Butterworths, London, 1958.

(68) R. J. Kokes, H. Tobin, Jr., and P. H. Emmett, *J. Am. Chem. Soc.*, **77**, 5860 (1955).

(69) H. H. Voge and A. I. M. Keulemans, to be published.

acknowledge the valuable assistance of his many co-workers and collaborators with whom during the past 35 years he has been studying the properties of catalysts and the nature of catalytic processes. He is glad also to acknowledge gratefully the support of his work by the Gulf Oil Company in the form of funds for the Multiple Petroleum Fel- lowship at the Mellon Institute; and direct and indirect support of his work at the Johns Hopkins University by W. R. Grace Company and the Davi- son Chemical Company. Finally, he wishes to thank the Kendall Company for the award, this year and for its continued support of the Kendall Award in colloid chemistry.

ADDENDUM 1972:

ADSORPTION AND CATALYSIS

P. H. Emmett
Portland State University
Portland, Oregon

In one sense it is relatively easy to update material covered in the earlier paper because there are essentially no corrections. The only items to be included therefore amplify or extend the contents of the earlier article without changing the general conclusions.

The comments here presented will be organized under the three main headings used in the initial paper, namely Physical Adsorption, Chemical Adsorption and Catalysis. Incidentally, this supplement does not represent a comprehensive coverage of everything that has been published on these subjects during the last thirteen years but represents rather the material that has come to the author's attention during the intervening years.

Physical Adsorption

Physical adsorption still finds its principal application to catalysis through its effective use in measuring surface areas and pore size by low temperature gas adsorption. Reviews and some modifications of the methods formerly used should perhaps be mentioned.

In surface area work the application of the B.E.T. procedure is still usually accepted without modification. Francoise Roquerol[1] compared areas obtained for a series of solids by using different adsorbates and concluded that the best all-around standard adsorbate is argon with a cross-sectional area per molecule of $16.8A^2$. The presumed advantage is that the amount of nitrogen required for a monolayer is more influenced by the presence of OH groups on the solid surface than is the adsorption of argon. It may be noted that 16.8 is 1.17 times the cross-sectional area, $14.4A^2$, calculated from the density of liquid argon[2].

When the B.E.T. theory was first proposed[4] it was recognized that areas obtained for materials having very small pores would be only approximate because each adsorbed molecule in a small diameter capillary would block off or cover a larger area than on a flat surface. This has been put on a quantitative basis for capillaries one or two molecules in diameter by Dollimore and Heal[5]. They pointed out that for capillaries of this kind the surface area might be as much as 3.66 times that calculated in the usual manner from the isotherms. Since we usually do not have detailed pore size distribution for pores as small as one or two molecular diameters in size we cannot hope to correct for the influence of such capillaries in any quantitative manner. Experimental data on zeolites has been used by Dubinin[6] to illustrate the uncertainties introduced by the presence of very small capillaries.

In discussing surface area work mention should be made of the "t-curve" concept introduced by the late Dr. de Boer.[7] He called attention to the fact that adsorption for nitrogen on a number of non-porous solids gave superimposable curves if plotted on the basis of the number of adsorbed layers as a function of relative pressure. If one then plotted the calculated thickness of the adsorbed layer for any solid against the thickness of the absorbed layer for some standard solid, a straight line should be obtained passing through the origin. On the other hand any capillary condensation or blocking of pores by adsorbed molecules would cause alterations in the slope of this "t-plot". These slope changes can be made to yield valuable information on pore size and shapes for a given solid adsorbent. Essentially the method uses the standard B.E.T. plots as a basis for calculating the layer thickness for the "t-plots".

Pore size measurements for solids as proposed by Barett, Joyner and Halenda[8] are still in use. Roberts modified the size calculation in such a manner as to facilitate computerized calculations[9]. Brunauer[10] suggests a method of calculation that did not make use of the assumption that the pores are cylindrical in cross section and in fact did not assume any shape factor for the solid being studies.

The use of pore size calculations in interpreting the kinetics of actual catalytic reactions has also received increasing attention. One of the more interesting suggestions[11] was that one should actually take into consideration not only the influence of pore size as calculated for an isothermal system, but the actual influence resulting from an apparent increase in rate due to internal temperature rise in the particles together with a decrease in rate due to pore diffusion limitations. The general correlations between size of pores and catalytic activity are still being worked on but have not yet reached a stage at which optimum pore size distributions can be specified for a given reaction.

Several examples have been found of adsorption isotherms that seem to defy the usual basis for differentiating between physical and chemical adsorption. For example Eischens[12] has discovered a sharp infra-red peak for the adsorption of nitrogen on finely divided supported nickel crystals. He interpreted the data as indicating a molecular chemisorption of the nitrogen. On the other hand, Hardeveld and Van Montfoort[13] in repeating the work have found similar infra-red peaks for nitrogen on nickel, on Pt and Pd. They concluded that physical adsorption of nitrogen is involved even though the calculated heat of adsorption as deduced from the adsorption isotherms is about 12 kcals per mole of nitrogen. They believe that particular types of sites on the nickel surface are able to produce such physical adsorption at room temperature. King[14] in reviewing the work concludes that the nitrogen is chemically adsorbed on nickel but physically adsorbed on Pt and Pd in producing the infra-red absorption peaks. A similar question might well be raised relative to the adsorption of CO by certain ion exchanged silica-alumina catalysts. Stone[15] found for example that each calcium ion exchanged into the big cages of Faugesite was capable of binding a carbon monoxide molecule at $25°C$ and pressures in the range 10 to 50 torr. Similar excessive adsorptions were found[16] for both nitrogen and carbon monoxide on an amorphous silica-alumina cracking catalyst that had been exchanged with calcium ions. For the amorphous catalysts an evacuation and dehydroxylation at $800°C$ produced equally high carbon monoxide (or nitrogen) adsorption on the exchanged and unexchanged gels. The heat of adsorption of the carbon monoxide reached a value of about 15 kcals on the amorphous catalyst and about 9 kcals for nitrogen. These values are 5 to ten times the heat of liquefaction whereas physical adsorption is defined as having heats of adsorption no greater than twice the heat of adsorption. The detailed nature of the forces giving rise to these extra high heats of what appears to be physical adsorption of carbon monoxide on the silica-alumina catalysts is not at present clear.

Chemical Adsorption

Earlier it was pointed out that the chemisorption of carbon monoxide and carbon dioxide at $-183°$ and $-78°C$ respectively on a doubly promoted ($Fe-K_2O-Al_2O_3$) iron catalyst enabled one to conclude that about 40% of the surface of the reduced catalyst was iron and 60% consisted of promoter ($K_2O-Al_2O_3$) molecules[17]. Recently exchange experiments[18] using $H_2{}^{18}O$ in hydrogen have confirmed the results obtained by the chemisorption measurements.

A type of chemisorption of hydrocarbons on silica-alumina catalysts is still believed to exist. However, the work of Hopkins[19] on the adsorption of radioactive isobutane on sieve catalysts has shown that the chemisorption is as large on an unexchanged sodium sieve as on one converted to an H-sieve though the activities of these two sieves for cracking hydrocarbons are perhaps several orders of magni-

tude different from each other. This and other observations such as the influence of water vapor[20] on cracking, on isomerization[21], and on chemisorption[22, 23] leaves some doubt that the chemisorption measured for butane[23], and heptane[20] by the radioactive method is the form involved in catalytic cracking.

Nitrogen chemisorption has received intensive study in its relationship to ammonia synthesis. It has been shown for example that at $400°C$ nitrogen adsorbs on the 111 face of iron but not on the 100 or 110 faces[24]. Furthermore exposure of a sample of iron to nitrogen seems to transform 100 and 110 faces into 111 faces. Isotope[25] and other measurements[26] seem to show that at $350°$, 50 to 100% of the adsorbed nitrogen is present as molecules whereas for promoted iron catalysis[27, 29] almost all the nitrogen is present as atoms. Finally, it has now been concluded that nitrogen chemisorption is the slow step in ammonia synthesis as deduced from some of the most recent measurements on the "stoichiometric number"–of the synthesis reaction[30].

Infra-red spectroscopy has been extended to the study of a large number of adsorbed molecules on various solids. Perhaps the most satisfying result has been the identification by Kokes and Dent[31] of the C_2H_5 adsorbed radical on ZnO as the intermediate in the hydrogenation of ethylene to ethane. This infra-red technique continues as one of the most valuable tools for giving us information about the state of the adsorbed species.

The influence of adsorbed species on the magnetic properties of adsorbents such as nickel has continued to be studied by magnetic techniques by Selwood[32] and by others[33]. Geus designated the surface hydrogen as "surface hydride" on the basis of evidence by Germer[34] that the addition of hydrogen to the 110 face of Ni will, according to LEED (low energy electron diffraction) experiments cause a rearrangement of the surface atoms of Ni to a form having twice the spacing of Ni atoms in the normal 110 face. Perhaps the biggest innovation in these last 13 years has been the introduction of a host of high vacuum techniques for measuring the purity and adsorptive properties of various faces of catalysts by using suitable techniques involving vacua of 10^{-10} mms or better. The nature of the adsorbed species, the energy levels of the adsorbed molecules and the contours of adsorption on a given surface have all been measured and promise to tell us many things about chemisorption that we have not known in the past.[35]

Catalysis

Dowden[36] raised the question originally as to the importance of d-band vacancies in metals on their effectiveness as hydrogenating catalysts. His results as well as later measurements[37] on the hydrogenation of styrene to ethyl benzene indeed seemed to show that decreasing the d-band vacancies in nickel catalysts by alloying the nickel with copper would decrease the activity to zero as soon as some 40 atom percent copper was added. However, later work[38] on other catalytic hydrogenations raised grave doubts as to the generality of this conclusion in regards to hydrogenation. Indeed

Sachtler[41, 42, 43] and his coworkers have advanced evidence to indicate that the surface of copper-nickel alloys prepared at low temperatures as thin films may indeed consist primarily of two compositions, one alloy containing about 80% Cu and the other about 3% copper. There are still questions about the formation of these alloys as a function of the temperature and method of preparation. However, attention is now focused on the fact that until techniques become available for giving an accurate analysis of the surface layer of such alloys little in the way of final conclusions can be drawn as to the importance of this electronic factor in hydrogenation over metals. The application of the Auger technique[35] in high vacuum work now gives promise of being able to furnish the necessary analysis of the surface composition of metal alloys being studied.

The question of whether or not dissolved hydrogen in Raney nickel catalysts alters the properties of the catalyst for hydrogenation has received renewed attention. It has been shown beyond question that a Raney nickel catalyst containing a quantity of unreacted aluminum is both capable of evolving hydrogen[44] (up to perhaps 100 cc per gram of Ni) on being heated to 400°C and of producing enhanced rates of hydrogenation[45]. This was initially reported as due to the presence of dissolved hydrogen in the freshly prepared Raney nickel. However, workers in Holland[46] have shown conclusively that the hydrogen evolved on heating a Raney nickel catalyst to 400° is due to the reaction of residual aluminum with water vapor held by the catalyst. The hydrogen evolved is equivalent to the difference between the initial and final aluminum contents of the Raney catalyst. Presumably, therefore, the evolved hydrogen with its attendant enhanced activity is due to the residual aluminum reacting with water. Some puzzling aspects of the behavior of Raney nickel still remain[47], though in general the behavior and properties of the catalyst are becoming better understood.

Much additional tracer work using ^{14}C compounds has been done since the previous paper was published. One experiment changes the complexion of the previous conclusions. It was pointed out that tracer experiments on Fisher-Tropsch synthesis indicated that radioactive ketene $^{14}CH_2CO$ behaved like radioactive ethyl alcohol[48] in forming a surface complex that serves as an intermediate in the synthesis of hydrocarbons from CO and H_2. Later work[49] showed that $CH_2{}^{14}CO$ did not build in to form higher hydrocarbons. Clearly the ketene molecule decomposes in contact with the catalyst to evolve CO. The CH_2 group but not the entire ketene molecule therefore appears as an intermediate in the synthesis.

In conclusion it should be pointed out that the number of basic research tools being focused on the mechanism of catalysis is increasing steadily. Actually experiments are now being carried out on catalytic reactions taking place on a few sq. cm. surface of particular faces of single crystals. The activities are being measured both at low pressure (about 10^{-7} mms) and at atmospheric pressure on samples prepared by standard ultra high vacuum techniques. One of the more startling results is the contrast between flat surfaces and "terraced surfaces"[35] as described by Somorjai and his coworkers[35]. By these techniques and with the help of molecular beam, Auger spectroscopy and other newly developed tools the scientific evaluation of the fundamentals of catalysis may be expected to increase at a rapid rate in the coming decades.

References

1. Rouquerol, F., Nuclear Research Center, Saclay, Report CEA R 2947, 1966
2. Brunauer, S. and Emmett, P. H., *J. Am. Chem. Soc.*, **59**, 2682 (1937)
4. Brunauer, S., Emmett, P. H. and Teller, E., *J. Am. Chem. Soc.*, **60,** 309 (1938)
5. Dollimore, D. and Heal, G. R., *Nature*, **208**, 1092 (1965)
6. Dubinin, M. M., *Vesti Akad. Khim Navak*, **1**, 6-29 (1968) Russian
7. de Boer, J. H., Linsen, B. G., and Osinga, J. T., *J. Catalysis*, **4**, 643 (1965)
8. Barrett, E. P., Joyner, L. and Halenda, P. C., *J. Amer. Chem. Soc.*, **73**, 373 (1951)
9. Roberts, B. F., *J. of Colloid and Interface Science*, **23**, 266 (1967)
10. Brunauer, S., Mikhail, R. S., and Bodor, E. E., *J of Colloid Interface Science*, **25**, 358 (1967)
11. Carberry, J. J., *A. I. Ch. Eng.-Journal*, **7**, 350 (1961)
12. Eischens, R. P. and Pliskin, W. A., *"Advance in Catalysis"*, **10**, 1-50, Academic Press, New York, 1958
13. Hardeveld, R. van, and Montfoort, A., van, *Surface Science*, **4**, 396-430 (1966)
14. King, D. H., *Surface Science*, **9**, 875 (1968)
15. Stone and Vickerman, *J. C. Trans. Far. Soc.*, **67**, 316 (1971)
16. Huang Yan-Ying and Emmett, P. H.,—to be published
17. Emmett, P. H. and Brunauer, S., *J. Am. Chem. Soc.*, **59**, 310 (1937)
18. Solbakken, A., Solbakken, W. and Emmett, P. H., *J. of Catalysis*, **15**, 90-8 (1969)
19. Hopkins, P. D. and Stoffer, R. L., *J. Phys. Chem.* **72**, 3345-7 (1968)
20. Matsushita Kun-ichi, and Emmett, P. H., *J. of Catalysis*, **13**, 128-140 (1969)
21. Hindin, S. G., Oblad, A. G. and Mills, G. A., *J. Am. Chem. Soc.*, **77**, 535 (1955)
22. Larson, J. G. and Hall, W. K., *J. Am. Chem. Soc.*, **85**, 3570 (1963)
23. MacIver, D. S., Emmett, P. H., and Frank, H. S., *J. Phys. Chem.* **62**, 935 (1963)
24. Brill, R., Richter, E. L., and Ruch, E., *Angew. Chem.* **6**, 882 (1967)
25. Morikawa, Y. and Ozaki, A., *J. of Catalysis*, **12**, 145-9 (1968)
26. Huang, Yin-Yang, and Emmett, P. H., *J. of Catalysis*, **24**, 101-105 (1972)
27. Morikawa, Y. and Ozaki, A., *J. of Catalysis*, **23**, 97-104 (1971)
28. Takezawa, N. and Emmett, P. H., *J. of Catalysis*, **11**, 131-34 (1968)
30. Tanaka, Yamamoto and Matsuyama, *Proc. 3rd Int. Cong. Catalysts*, Amsterdam 1965, 676.
31. Dent, A., and Kokes, R. J., *J. Am. Chem. Soc.*, **92**, 6709-18; 6718-23 (1970)
32. Selwood, P. H., *"Adsorption and Collective Paramagnetism"* Academic Press
33. Geus, J. W., Nobel, A.P.P. and Zwietering, P., *J. of Catalysis*, **1**, 8-21 (1966)
34. Germer, L. H. and MacRae, A. U., *Proc. Nat. Acad, Sciences*, **48**, 477 (1962)

35. Somorjai, G. A., *"Principles of Surface Chemistry"*, Prentice-Hall, 1972

36. Dowden, D. A., *Research,* **1**, 239 (1948)

37. Stoddart, C. T. H. and Emmett, P. H.,—to be published

38. Best, R. J. and Russell, W. W., *J. Am. Chem. Soc.,* **76**, 838 (1954)

39. Hall, W. K. and Emmett, P. H., *J. Phys. Chem.,* **63**, 1102 (1959)

40. Campbell, J. S. and Emmett, P. H., *J. of Catalysis,* **7**, 252 (1967)

41. Sachtler, W. M. H., and Jongepier, R., *J. of Catalysis,* **4**, 665 (1965)

42. Sachtler, W. M. H. and Dorgelo, G. J. H., *J. of Catalysis,* **4**, 654 (1965)

43. Van der Plank, P. and Sachtler, *J. of Catalysis,* **12**, 35 (1968)

44. Kokes, R. J. and Emmett, P. H., *J. Am. Chem. Soc.,* **82**, 4497 (1960)

45. Yao, Hsien-Cheng, and Emmett, P. H., *J. Am. Chem. Soc.,* **84**, 1086 (1962)

46. Mars, P., Scholten, J. J. F., and Zwietering, P., *2nd Int. Catalysis Cong.,* **1**, Paris, 1960, 1245-63

47. Sevenster, P. and Emmett, P. H.,—to be published

48. Blyholder, G. and Emmett, P. H., *J. Phys. Chem.,* **63**, 962 (1959)

49. Blyholder, G., and Emmett, P. H., *J. Phys. Chem.,* **64**, 470 (1960)

1959

FLOYD E. BARTELL

UNIVERSITY OF MICHIGAN
ANN ARBOR, MICH.

. . . pioneer teacher and a founder of colloid chemistry in America, for his numerous meritorious researches in fundamental and applied aspects of colloid and surface chemistry.

The manuscript of the following Kendall Award Address of the late Prof. Bartell was kindly made available by his son Laurence S., professor of chemistry at the University of Michigan.

Wetting of Solids by Liquids

F. E. Bartell
University of Michigan
Ann Arbor, Michigan

Wetting of solids by liquids while one of the most important and most commonly occurring processes is a phenomenon for which there exists but little reliable quantitative data. There is question as to what methods are suitable for measuring the energy changes representing the "degree of wetting". One of the most generally used methods is based upon measuring the contact angle formed between the liquid and solid and by applying the Young equation[1] to such data. At present there exists active controversy as to the validity of the Young equation. There also is question as to the reliability of some of the contact angle methods which have been used in obtaining data for the calculations

In this paper are presented data which have been collected from researches in our laboratories covering a period of over 30 years. All the adhesion tension or "degree of wetting" data are consistent and for a given solid-liquid system the values obtained by different methods are very nearly the same. This indicates that the methods used for determining the magnitude of contact angles are reliable and that the use of the Young equation is justified.[1,2,3]

Wetting is said to have occurred when a solid and a liquid come into contact in any manner so as to form a solid-liquid interface. The energy change can be represented by the well known Young equation:

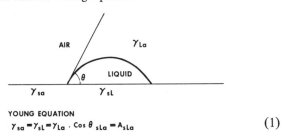

YOUNG EQUATION

$$\gamma_{sa} = \gamma_{sL} = \gamma_{La} \cdot Cos\, \theta_{sLa} = A_{sLa} \tag{1}$$

γ_{sa} = SURFACE TENSION, OR FREE SURFACE ENERGY, SOLID-AIR

γ_{sL} = INTERFACIAL TENSION, OR INTERFACIAL FREE SURFACE ENERGY SOLID-LIQUID

γ_{La} = SURFACE TENSION OR FREE SURFACE ENERGY, LIQUID-AIR

θ_{sLa} = CONTACT ANGLE, SOLID-LIQUID-AIR

$A_{sLa} = (\gamma_{sa} - \gamma_{sL})$ represents the change in energy which must occur when unit area of solid-air interface is replaced by unit area of solid-liquid interface. The magnitude of this value is referred to in this paper as the degree of wetting of the solid by the liquid. It is determined by γ_{La}. $Cos\, \theta_{sLa}$ and has been called the adhesion tension A_{sLa} for the system.

There is a great difference in the wetting tendencies of soft, low free surface energy solids and hard, high free surface energy solids. The freedom from contamination by adsorption shown by low free surface energy substances such as paraffin explains why investigators in different laboratories have been in such good agreement for the value of contact angles formed by water on paraffin. This constitutes one of the few cases in which there has been fairly good agreement in contact angle data. There is evidence that when values for paraffin have differed it has been because of the presence of impurities in the paraffin[4] or of surface roughness.[5] Investigators whose contact angle measurements have been limited largely to soft solids[6] have avoided many of the complications which would have been encountered with hard solids.

Hard, high free surface energy solids do not remain uncontaminated to the same extent as low free surface energy solids. Their surfaces are readily altered by adsorption. Fresh and clean surfaces of such high free surface energy solids as silver and gold[7,8] can be prepared by deposition of vapor of these materials in vacuum, but they can be kept uncontaminated only by keeping them in a practically perfect vacuum. It has been stated that within one hour in a "vacuum" as low as 10^{-9} mm Hg adsorption of gases can occur sufficient to form a complete monolayer on a solid surface.[8]

When a solid is used as a substrate for two immiscible liquids in contact upon its surface a modified Young equation can be used to indicate the relationship of the three boundary tensions thus:

$$Cos\, \theta_{sL_1L_2} = \frac{\gamma_{sL_1} - \gamma_{sL_2}}{\gamma_{L_1L_2}} \tag{2}$$

Unfortunately it is not possible to determine individually

the tension γ_{sL_1} or γ_{sL_2} so one cannot relate directly the magnitude of the angle θ to the magnitudes of these tensions. However, if one substitutes for the solid substrate the liquid mercury, one can measure not only the contact angle formed but also the three interfacial tensions Hg-L$_1$, Hg-L$_2$ and L$_1$-L$_2$. In one series of experiemtns mercury as substrate was used with benzene and with water[3]. The interfacial tensions of each of these pairs of liquids were measured when the pairs were first brought together, as also the rates of change of tensions with time which occurred when the systems were permitted to stand. Fairly large changes in interfacial tension for each of the mercury interfaces continued for over an hour.

By assuming that similar values existed and similar changes occurred at the interfaces when contact angles were formed by water drops on mercury, initial contact angles and the rate of change in contact angle values could be calculated. The same contact angles could also be measured. For the system mentioned very satisfactory agreement was obtained between the calculated and the observed contact angle values. Not only was the Young equation shown to apply to this system but it was found that the hysteresis effects observed in this system can be logically explained by a consideration of the alteration with time of the interfacial tensions existing at the mercury-water and mercury-benzene interfaces, these alterations being caused by adsorption at the interfaces.

Methods for Determining Degree of Wetting

The types of experimental methods which have been most generally considered or used for the determination of degree of wetting of a solid by a liquid are as follows:

1. Determination of the angle of contact formed by a liquid against a solid.
2. Application of a modified Gibbs equation to adsorption isotherm data.
3. Determination of the heat evolved when a solid is wetted by a liquid.

Methods for Determining Contact Angle

The most commonly used methods for determining contact angles are:

1. **Controlled drop on plate method.** The most generally used method. Used when a smooth flat surface is available.
2. **Bubble method.** Has been largely used in ore flotation studies. Used when a smooth flat surface is available.
3. **Capillary tube method.** Has found but limited use. Used when solid in form of transparent tube is available. Meniscus is photographed and contact angle is measured directly.
4. **Vertical rod method.** Has been used mainly with solid fibers and plates.
5. **Tilting plate method.** Has been used mainly with solid fibers, filaments and plates.

6. **Pressure of displacement method.** Used with powdered or finly divided solids.
7. **Rate of penetration method.** Used with powdered or finely divided solids.

As ordinarily conceived the measurement of contact angles involves one of the simplest of techniques; however those who have attempted to obtain reliable data have soon realized that the determination of significant angle values involves most careful study and highly exacting techniques. As has been stated this is especially true if one is using a solid with relatively high free surface energy. To be certain of maintaining a pure surface one must work in an almost impossibly high vacuum. Even with such a system one could hardly avoid some type of contamination or surface alteration as measurements progress. Moreover vapor of the liquids used always are present. Presumably real advances will be made when entirely new methods and new approaches become available.

When solid material is in the form of finely divided or powdered substance no single plane surface is available of sufficient size for application of "drop on plate" or similar method. A pressure of displacement method for use with powder was developed in 1927 and has been quite extensively used in our laboratory. The powder is firmly and uniformly compressed in a cell thus forming an extensive capillary system[9]. The effective radius of the pores is determined by noting the pressure built up by an advancing zero contact angle forming liquid. Subsequently by using similarly compressed powders and measuring displacement pressures of a contact angle forming liquid one can from the pressure build up determine the magnitude of the contact angle formed by the liquid against the powdered material. There is evidence that this method gives relatively reliable results though they cannot be regarded as precise results. The method requires very careful control. It is time consuming, often requiring a few hours to make a single determination. An experienced operator can obtain good duplication of results. Some operators however, seem to have found it practically impossible to obtain acceptable data. On the whole, this method is to be avoided if other methods for obtaining the desired data are available.

A method based upon measuring the relative rates of advance of liquids through uniformly compressed powders was developed by Shoemaker and Snow[10]. The method is fairly rapid and is reliable to about the same degree as the pressure of displacement method. To carry it out one must use, in a vertical tube, carefully compressed powder to which is added a given volume of liquid of known viscosity, and surface tension. This liquid must form a zero angle contact with the solid. The time required for this liquid to advance a given distance through the powder is determined. Next, a similar volume of a contact angle forming liquid of predetermined viscosity, and surface tension is added to a similarly compressed powder system. The time required for this liquid to acvance the prescribed distance is determined. From the collected data contact angle values can be calculated.

A constant, K, is determined from the surface tensions and viscosities of the liquids used

$$K = \frac{\gamma' \eta''}{\gamma'' \eta'} \qquad (3)$$

and

$$\cos \theta = \frac{Kt'}{t''} \qquad (4)$$

γ represents surface tension, η viscosity and t the time for liquid to flow a prescribed distance.

Significance of Contact Angle Data

Even though one may have serious doubts as to the true significance of his measured contact angle values he is justified in the view that the data are of value and should be preserved provided the exact conditions under which the data were obtained are known. When contact angle data are obtained under carefully controlled conditions for various solid-liquid systems many different relationships of interest, and in some cases of importance, are observed. For example, observation was made "that the order of increasing adhesion tension for liquids against carbon black was exactly the reverse of the order of increasing adhesion tension for the same liquids against silica"[11]. It was observed further in 1934 from data for pure organic liquids against water on high free surface energy solids that the angles formed by the interface between water and each of the organic liquids on the surface of a given solid were approximately the same for a series of organic liquids. This led to the construction of a diagram, known as the K_{n3} diagram[12] for "correlation of interfacial free surfaces energies". This diagram while highly empirical has served as a valuable guide in solving various practical problems. In the course of time some exceptions to the rule of constancy of interfacial contact angles were found. Low interfacial tension liquids such as cyclohexanol and mixtures of certain organic liquids were found to give deviations from the rule[13]. Furthermore, it was early noted that the rule did not apply to soft solids of low surface energy[14].

Data obtained on silver and gold[15] are excellent examples of the constancy of the interfacial contact angles formed both by water drops in organic liquids and by organic liquid drops in water on the solids. Such data are shown in Table 1 and 2.

Table 1

Stable Solid-Organic Liquid-Water Drop Contact Angles

Organic liquid	Water angle on silver		Water angle on gold	
	θ^a	θ^r	θ^a	θ^r
Isoamyl alcohol	127°	72°	117°	61°
n-Butyl acetate	129	68	116	57
Benzene	129	58	116.5	45
a-Bromo-hapthalene	127	65	118	55
Heptane	127	58	116.5	45

Table 2

Stable Solid-Water-Organic Liquid Drop Contact Angles

Organic liquid	Water angle on silver		Water angle on gold	
	θ^a	θ^r	θ^a	θ^r
Isoamyl alcohol	126°	71°	117°	59°
n-Butyl acetate	129	68	115.5	57
Benzene	129	57	116	45
a-Bromo-naphthalene	126	65	116	55
Heptane	127	57	115.5	45

When one considers a series of organic liquids so different in properties as those listed in Tables 1 and 2 one wonders why the interfacial advancing contact angle $\theta^a{}_{sL_1 L_2}$ values, for each of the five liquids with a given solid should be practically the same, i.e. 127° to 129° for Ag and 116° to 118° for Au. The authors have not been able to offer a suitable explanation.

At about the time the above data were obtained similar work was done with the metals bismuth, antimony and cadmium by P. A. Cardwell, W. W. Riches[16] and the author. It was found that for a given solid and stable water advancing interfacial contact angles were of practically the same value for the different organic liquids on that solid. The data obtained are shown in Table 3.

Table 3

Stable Solid-Water-Organic Liquid Drop Contact Angles

Organic Liquid	Bismuth		Antimony		Cadmium	
	θ^a	θ^r	θ^a	θ^r	θ^a	θ^r
Isoamyl alcohol	87°	32°	86°	34°	47.5°	21°
n-butyl-acetate	88	32	86	35.5	49	21
Benzene	86	34	87	33	49	22
α-Br-Naphthalene	88	34	86	34	48	22
Heptane	88	31	87	34	47	22.5

For the metals Bi and Sb the values for the stable interfacial advancing angles were all very nearly the same, i.e. 86° to 88°. For Cd, the values for the interfacial angles with the five different organic liquids and water were all practically the same, i.e. 47° to 49°.

For the three solids the "stable receding angles" had values quite different from the "stable advancing angles". But here again, as with the values for the advancing angles, the values for the receding angles for the different organic liquids on a given solid were of approximately the same value.

There is now very good evidence that in the majority of contact angle systems adsorbed layers exist which impart to the solid phase interfacial properties which are responsible for the magnitudes of the contact angles obtained and which may account for some of the unexpected and heretofore unexplained correlations of values obtained[17]. The altered surface of the solid may have properties very different from the pure or initial solid surface.

Modified Gibbs Equation Method

Free energy of immersion of powders can be determined by application of a modified Gibbs adsorption equation to data obtained from complete vapor adsorption isotherms. Bangham and Razouk[18] in 1937 showed that the Gibbs adsorption equation is applicable to the determination of free surface energy changes which occur on solids during adsorption of vapors upon them.

For non-porous solids Boyd and Livingston[19] and Harkins and coworkers[20,21] used the formula for the free energy lowering at relative pressure from zero to one:

$$-\phi_{S0/SL} = \gamma_{S0} - \gamma_{SL} = \frac{RT}{Ma_s} \int_{p/p_0 = 0}^{p/p_0 = 1} x/m \, d\ln p/p_0 + \gamma_{LV^0} \quad (5)$$

where $\phi_{S0/SL}$ = free energy lowering which occurs when unit surface solid-vacuum is replaced by unit surface of solid-liquid, γ represents any free surface energy or interfacial tension, subscripts S0, SL, and LV⁰ represent a solid-vacuum, a solid-liquid and a liquid-saturated vapor interface, respectively, M is the molecular weight of adsorbate, a_s is the specific surface area of solid, x/m is the weight in milligrams adsorbed per gram of adsorbent and p/p_0 is the relative pressure of the adsorbate. ($\gamma_{S0} - \gamma_{SL}$) represents free energy of immersion of solid in liquid.

By making appropriate changes in the method and theory, Dobay, Fu, and Bartell[22] altered the treatment to include porous solids, such as gels and compressed powders. The following formula was developed for the evaluation of the free surface energy lowering at $p/p_0 = 1$ on porous solids:

$$\gamma_{S0} - \gamma_{SL} = \frac{RT}{Ma_s} \int_{p/p_0 = 0}^{p/p_0 = 1} x/m \, d\ln p/p_0 \quad (6)$$

When free surface energy values are obtained with free flowing powders and the data treated using equation (5), the calculated values may be 20 to 40% too high, since in the case of vapor adsorption by a mass of loose powders a definite amount of capillary condensation of vapor occurs in the spaces between the particles of powder. When the powders are compressed into plugs they remain rigid and porous and the results obtained using data from compressed plugs treated by means of equation (6) are reasonably correct. Energy of immersion, or degree of wetting, values obtained with compressed plugs of silica are presented in Table 4 where they can be compared to corresponding data ob-

Table 4
Wetting of Solids by Liquids

Liquids γ_w (water) = 72.1	1 γ_{wo}	2 $\gamma_{wa} - \gamma_{wo} = A_{ow}$	3 SiO₂ Tripoli Disp. Press. ○	4 SiO₂ Ottowa Sand Disp. Press ●	5 SiO₂ Silex Disp. Press. ◊	6 SiO₂ Quartz Trans. Cap. Tube ×	7 BaSO₄ Disp. Press. △	8 Al₂O₃ Disp. Press. ⊖	9 CaF₂ Disp. Press. C	10 CaCO₃ Disp. Press. □	11 SiO₂ Silica Gel. Ads. Isoth. ▼	12 SiO₂ Linde Fine SiO₂ Comp. Plugs Ads. Isoth. ▽
						Adhesion Tension, A_{SL}					Free Energy of Immersion	
1. Water			76.3*	76.3	75.1	76.3	76.4	76.7	76.5		129	139.5
2. Amyl alcohol	5.0	67.1	72.3		70.6	73.1						
3. Aniline	5.7	66.4	68.6									
4. Ethyl carbonate	12.6	59.5				68.7		65.8		63.5		
5. Butyl acetate	13.2	58.9	66.9	64.5		66.3	64.6	64.3	65.2			
6. Nitrobenzene	25.3	46.8	56.1	57.7		57.2						
7. Benzene	34.7	37.4	45.9	44.1	44.5	45.4	44.3	45.	50.1	45.7	58.6	49.4
8. Toluene	36.1	36.0	48.1	43.2		46.5			48.5	44.2		47.0
9. Chlorobenzene	37.9	34.2		40.2								
10. Acet. tetrabromide	38.2	33.9		42.8	42.4	43.3	40.5	43				
11. Ethyl benzene	38.4	33.7		41.2								
12. Bromo benzene	39.8	32.5		39.3								
13. Propyl benzene	40.0	32.1		40.								
14. α Cl naphthelene	40.2	31.9		39.0		41.2						
15. Bromoform	40.3	31.8		37.4		40.9						
16. Iodo benzene	41.3	30.8		39.3		39.1						
17. Butyl benzene	41.6	30.5		38.7								
18. αBr naphthelene	41.6	30.5	35.8	39.6	39.0	41.1	39.9	41	43	40.5	38.4 35.	35.6
19. Carbon tetrachloride	44.5	27.6	34.2	36.3		35.6						28.8
20. Heptane	49.4	22.7					33.1	33.4			26.5 28.8	27.9
21. Hexane	51.0	21.1		25.9		29.9						

tained by distinctly different methods. On the whole the agreement is good.

Heat of Wetting

The important heat-of-wetting work of Zettlemoyer and coworkers[23], Pierce and Smith[24], Jura[25] and others should lead to generally acceptable methods for the determination of degree of wetting from heat of wetting data.

At the Atlanta meeting of the American Chemical Society, April, 1930 Bartell reported that "the decrease in the free surface energy which occurs when a polar solid and each of a series of organic liquids are brought together is in the same direction and of the same relative magnitude as the decrease in the free surface energy which occurs when the same liquids are brought into contact with water".

This conclusion was based partially on the work of Bartell and Greager[26] on liquid absorption of finely divided silica and calcium fluoride, in which work they found that a linear relationship existed between liquid absorption and adhesion tension for liquids which wet the solid with zero angle of contact, but was quoted directly from Bartell and Hershberger[27] who had further found relationships between adhesion tension and stability of suspensions, yield values, mobility of pastes and settling volumes of suspensions. At the same session of the same Atlanta meeting Harkins reporting upon work of Harkins and Dahlstrom[28] on wetting of pigments stated that "The energy relations at the interface between solid oxides and organic liquids are similar to those between water and the same organic liquids." Subsequently Harkins materially extended his researches along this line. Bartell and Whitney[29] in 1932 showed in a table of their data for Ottawa sand silica that the work of adhesion of a series of organic liquids against silica, W_a, was practically the same as the work of adhesion of the same organic liquids against water, W_a'. That is $\frac{W_a}{W_a'} = 1.12$ average.

To test further the above mentioned relationships and to check upon the numerical values for the degree of wetting of given substances obtained by different methods the writer has recently brought together practically all of the data on energy of immersion, or adhesion tension, organic liquids against polar solids, A_{so}, that have been obtained in various researches in his laboratories over a period of many years, and has compared these values with energy of immersion or adhesion tension data for the same organic liquids against water, A_{wo}. These collected data are given in Table 4 and also are plotted, A_{so} vs A_{wo}, in Figure 1.

In columns 3, 4, 5, and 6 of Table 4 under adhesion tension, A_{sL}, are given values obtained by different investigators using different kinds of silica such as Tripoli (kieselguhr)[11], Ottawa sand[29], Silex (ground quartz)[13] and a transparent capillary tube of silica[30]. The variation in values represents about the variation to be expected when one considers the difference in physical properties of the materials, the variation in methods used by different operators, etc.

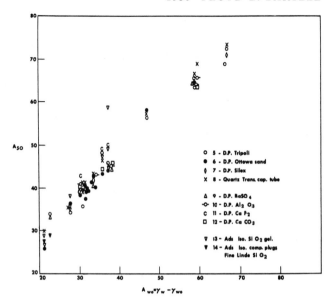

Figure 1. Correlation between solid-organic liquid and water-organic liquid adhesion tensions for various solids and organic liquids.

In columns 7, 8, 9 and 10 of Table 4 are given values obtained with the same series of organic liquids against barite[31], aluminum oxide[32], calcium fluoride[26], and calcium carbonate[33], each finely divided and subjected to the pressure of displacement method. The adhesion tension (or the energy of immersion values) show about the same agreement with each other and with silica as had been found with the different silica samples.

The data in column 11 of Table 4 were obtained with silica gel as solid using a vapor adsorption isotherm method and employing a modified Gibbs equation[34]. Column 12 gives values obtained by the same method using very fine Linde Silica highly compressed into plugs. The free energy of immersion data obtained for the different solid-liquid systems by the adsorption isotherm method are in reasonably good agreement with corresponding data obtained by the displacement pressure and other methods. The double values shown in column 11 for the liquids carbon tetrachloride and hexane are given because in each case two quite different samples of silica gel were used. We have no explanation for the apparently high value of 58.6 (apparently about 10 units too high) for benzene against silica gel. The free energy of immersion values for water shown in columns 11 and 12[35] are higher (129 and 139.5 resp.) than the corresponding adhesion tension values shown for the other columns. We believe these higher values are more nearly correct though we have no explanation for the lower values obtained in the other experiments.

Column 1 of Table 4 gives the interfacial tensions organic liquid against water and column 2, $\gamma_{wa} - \gamma_{wo}$, the adhesion tension of organic liquid against water. When the data in column 2 are compared with corresponding data in the columns to the right one notes that the energy changes which occur when water comes into contact with oxides and other polar solids are in the same order and of the same order of magnitude as those which occur when the same series of organic liquids are brought into contact with water.

The data plotted, A_{so} vs A_{wo}, in Figure 1 show, in spite of some scattering of points, that there is an approximately linear relationship between A_{so} and A_{wo} and that $A_{so} \approx A_{wo}$.

All the adhesion tension or "degree of wetting" data obtained by the different methods are consistent and for a given combination of materials show but small differences, this indicates:

(a) Methods used for the determination of contact angles are reliable.

(b) Use of the Young equation is justified.

(c) Application of a modified Gibbs equation to adsorption isotherm data gives results in good agreement with results from contact angle methods indicating that both methods are valid.

Bibliography

1. Thomas Young, *Phil. Mag.*, 165 (1805)
2. F. W. Fowkes and W. M. Sawyer, *J. Chem. Phys.*, **20**, 1650 (1952)
3. F. E. Bartell and C. W. Bjorklund, *J. Phys. Chem.*, **56**, 453 (1952)
4. B. R. Ray and F. E. Bartell, *J. Colloid Sci.*, **8**, 214 (1953)
5. F. E. Bartell and J. W. Shepard, *J. Phys. Chem.*, **57**, 211 (1953)
6. W. A. Zisman and co-workers, *J. Col. Sci.*, **5**, 514 (1950)
7. F. E. Bartell and P. H. Cardwell, *J. Phys. Chem.*, **64**, 494 (1942)
8. F. E. Bartell and J. T. Smith, *J. Phys. Chem.*, **57**, 165 (1953)
9. F. E. Bartell and H. J. Osterhof, *Ind. Eng. Chem.*, **19**, 1277 (1927)
10. M. L. Shoemaker and C. W. Snow, *J. Phys. Chem.*, **50**, 973 (1955)
11. F. E. Bartell and H. J. Osterhof, *J. Phys. Chem.*, **37**, 543 (1933)
12. F. E. Bartell, and L. S. Bartell, *J. Am. Chem. Soc.*, **56**, 2205 (1934)
13. F. C. Benner, PhD Thesis 1940, Univ. of Mich. (Microfilmed)
14. F. E. Bartell and H. H. Zuidema, *J. Am. Chem. Soc.*, **58**, 1449 (1936)
15. F. E. Bartell and P. H. Cardwell, *J. Am. Chem. Soc.*, **64**, 1530 (1942)
16. W. W. Riches, PhD Thesis, 1941, Univ. of Mich. (Microfilmed)
17. B. R. Ray and F. E. Bartell, *J. Phys. Chem.*, **57**, 49 (1953)
18. D. H. Bangham and R. I. Razouk, *Trans. Farad Soc.*, **33**, 1459 (1937)
19. G. E. Boyd and H. K. Livingston, *J. Am. Chem. Soc.*, **64**, 2383 (1942)
20. E. H. Loeser, W. D. Harkins and S. B. Twiss, *J. Am. Chem. Soc.*, **57**, 251, 591 (1953)
21. G. Jura and W. D. Harkins, *J. Am. Chem. Soc.*, **66**, 1356 (1944)
22. D. G. Dobay, Y. Fu and F. E. Bartell, *J. Am. Chem. Soc.*, **73**, 308 (1951)
23. F. H. Healey, J. J. Chessick, A. C. Zettlemoyer and G. J. Young, *J. Phys. Chem.*, **58**, 887 (1954)
24. C. Pierce and R. N. Smith, *J. Phys. Chem.*, **54**, 795 (1950); *J. Phys. & Colloid Chem.*, **54**, 354 (1950)
25. G. Jura and T. L. Hill, *J. Am. Chem. Soc.*, **74**, 1598 (1952)
26. F. E. Bartell and O. H. Greager, *Ind. and Eng. Chem.*, **21**, 1248 (1929)
27. F. E. Bartell and A. Hershberger, *Ind. and Eng. Chem.*, **22**, 1304 (1930); *J. Rheology*, **2**, 177 (1931)
28. W. D. Harkins and R. Dahlstrom, *Ind. and Eng. Chem.*, **22**, 897 (1930)
29. F. E. Bartell and C. E. Whitney, *J. Phys. Chem.*, **36**, 3115 (1932)
30. F. E. Bartell and E. J. Merrill, *J. Phys. Chem.*, **36**, 1178 (1932)
31. F. E. Bartell and H. Y. Jennings, *J. Phys. Chem.*, **38**, 495 (1934)
32. F. E. Bartell and C. W. Walton, Jr., *J. Phys. Chem.*, **38**, 503 (1934)
33. F. W. Albaugh, PhD Thesis, 1941, Univ. of Mich. (Microfilmed)
34. F. E. Bartell and J. E. Bower, *J. Colloid Sci.*, **7**, 80 (1952)
35. F. E. Bartell and J. J. VanVoorhis, *J. Phys. Chem.*, **61**, 1520 (1957)

1960

JOHN D. FERRY

UNIVERSITY OF WISCONSIN
MADISON, WISC.

. . . for his record of distinctive and unique accomplishment in studies of the dynamic mechanical properties of soft rubberlike solids as they are subjected to oscillating deformations.

The following Kendall Award Address is reproduced by permission from *The Journal of Chemical Education,* **38**, 110 (1961). The Addendum has been written by Prof. Ferry for this volume.

John D. Ferry
University of Wisconsin
Madison

Rheology in the World of Neglected Dimensions

Nearly half a century ago, Wolfgang Ostwald of the University of Leipzig published a book called "The World of Neglected Dimensions" (1). It was based on a lecture tour in this country and was designed to arouse interest in the novel phenomena of colloid chemistry; it fulfilled its goal, passing through many editions in both German and English. The theme expressed by the title was that colloidal systems are characterized by particle sizes in the no-man's-land between atomic dimensions and macroscopic dimensions—generally, in the range from 10 to 1000 Å.

In Ostwald's day, there were several ways of estimating the magnitudes of these mysterious dimensions, notably with the ultramicroscope and with ultrafilters. There was, however, no way of deducing particle dimensions from one of the most characteristic and unusual aspects of colloidal systems—their rheological behavior. Rheology is the science of flow, and it is especially concerned with anomalous viscous and elastic properties. Ostwald's book described some curious and spectacular rheological observations. Solutions of gelatin and certain soaps were then the favorite choices for illustrating such phenomena as enormous increases in viscosity with concentration; changes in viscosity with temperature; non-Newtonian flow, or so-called structural viscosity, where the apparent viscosity depends on the rate of flow; and flow elasticity or elastic recoil, in which a solution, after being stirred in one direction, will when the stirring stops reverse its direction and unwind.

Although the neglected dimensions and the peculiar rheology were both recognized as characteristic of colloids, it was necessary to wait a few decades before quantitative connections between them could be formulated. The two decisive steps were these, I believe: (a) demonstration in 1934 by Mark, Kuhn, and Meyer (2–4) that the phenomenon of rubberlike elasticity is due to constant wriggling motions and configurational changes of long flexible molecules, i.e., micro-Brownian motion; and (b) demonstration in 1949 by Kirkwood (5) that whereas external forces in phase with Brownian motion (velocity) represent energy dissipation, external forces in phase with Brownian displacement represent energy *storage*. Thus a system of large molecules, even in dilute solution, can store elastic energy while in a state of steady flow.

From these principles, deductions can now be made from rheological measurements yielding certain dimensions falling in the colloidal or macromolecular range. These dimensions may not be as clear-cut as a particle

diameter; for flexible macromolecules they involve root-mean-square averages (Fig. 1a), and they include average distances between network points in a cross-linked network (Fig. 1b), average distances between entanglement coupling loci in entanglement networks, etc. But these distances are all in Ostwald's old zone of 10–1000 Å.

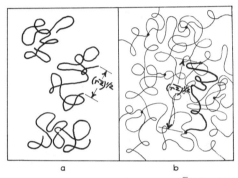

Figure 1. Average (root-mean-square) dimensions, $(\overline{r^2})^{1/2}$, in macromolecular systems. (a) End-to-end separation of an isolated flexible molecule; (b) vector between two junction points in a cross-linked network.

One relation between rheology and macromolecular dimensions is so well known that it is omitted from this discussion. From the viscosities of dilute solutions of flexible polymers, expressed as the intrinsic viscosity (6), deductions can be drawn about size and shape. We are concerned here with some less familiar relationships.

Storage of Energy in Steady-State Flow

It is easiest to introduce the concept of elastic energy storage in flow by discussing very dilute solutions. Consider a solution of long, thin rods. It is well known that in flow such solute particles are oriented, often producing birefringence (Fig. 2) (7). Now the less random distribution corresponds to a decrease in entropy (of course, entropy is also continuously being created by the flow and heat dissipation, but that is another matter) and hence to an increase in free energy which represents elastic stored energy. The same thing happens in a dilute solution of flexible coiled macromolecules. It is the interaction of the Brownian motion

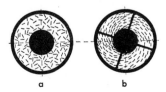

Figure 2. Orientation of rod-like particles between coaxial cylinders. (a) Random orientation at rest; (b) aligned orientation during flow. (From Reference 7).

Based on a Kendall Award address before the Division of Colloid Chemistry at the 137th Meeting of the ACS, Cleveland, Ohio, April, 1960.

with the hydrodynamic forces produced by the flowing solvent—first recognized by Kirkwood (5)—that gives the energy storage.

The theory has been worked out for the energy storage and dissipation for both rods (5) and coils (8). The results correspond mathematically to some rather simple mechanical models, shown together with the schematic hydrodynamics in Figure 3 (9). For

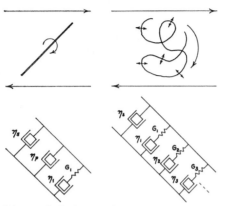

Figure 3. Schematic hydrodynamics for the behavior of rods and coils in a flowing solvent, and mechanical models corresponding to the energy storage and dissipation predicted by theory. (From Reference 9).

the rod, whose axis experiences random Brownian motion, there is a viscosity contribution from the solvent (η_s), and two from the solute, one of which acts as though in series with a spring; $\eta_1 = 3\eta_p$. During flow, the spring stores energy in proportion to the stress applied. For the flexible coil, there is a wide variety of Brownian motions involving configurational changes with displacements of segments both with respect to near neighbors and with respect to parts of the chain far away. As a result the equivalent mechanical model is more complicated, with a long series of springs that store energy. In the theory of Rouse (8), which leads to this result, the thermal motion of the coil is analyzed in terms of a set of normal modes (Fig. 4). These modes interact with the motions imposed by the flowing solvent.

Figure 4. Some normal modes of configurational rearrangement of a flexible macromolecule as in the theory of Rouse (8).

The stored energy in a flowing dilute solution of macromolecular rods or coils cannot be directly observed; neither can the elastic recoil, or unwinding, (which should in principle take place when the flow is stopped) because it is obscured by inertial effects. But the unwinding can easily be seen in a concentrated colloidal solution, as readily verifiable with a bottle of rubber cement or a bowl of turtle soup. A precise measurement can be visualized as in Figure 5, where a torque is applied to the inner cylinder. After a time, it settles down to a constant angular velocity of rotation, as shown by the linear portion of Figure 6; when

the torque is removed, it unwinds through an angle which is proportional to the recoverable elastic strain.

For a concentrated solution of flexible coils, a modification of the Rouse theory (10) gives:

$$\mathcal{E}_{st} = \mathfrak{T}^2 M / 5cRT \qquad (1)$$

$$\gamma_r = 2\mathfrak{T}M / 5cRT \qquad (2)$$

where $\mathcal{E}_{st}$ is the stored elastic energy per cc, γ_r the recoverable strain or elastic recoil, $\mathfrak{T}$ the shear stress applied during flow, M the molecular weight, c the concentration in g per cc, R the gas constant (units ergs deg^{-1} mole^{-1}), and T the absolute temperature. These equations provide a modern link between rheology and dimensions; from the observed elastic recoil, the molecular weight can be calculated. For a randomly coiled molecule as in Figure 1a, knowledge of M is equivalent to knowledge of average dimensions; usually the average end-to-end separation is proportional to $M^{1/2}$.

Equations (1) and (2) hold only for small stresses where the coiled molecules are negligibly distorted by the flow from the configurations they would assume in the solution at rest. They are also restricted to solutions in which all the molecules have the same molecular weight. More complicated expressions are available for blends of different molecular weights and molecular weight distributions (11, 12).

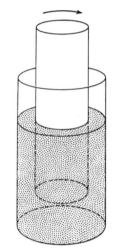

Figure 5. Shear of a macromolecular system between coaxial cylinders. (From Reference 12).

As a numerical example, consider a 10% solution of polystyrene which flows under a stress of 10^3 dynes/cm^2 and after cessation of flow undergoes a recoil of 20%. According to equation (2), the molecular weight is 1.25×10^6. For polystyrene in concentrated solution, the average end-to-end distance as in Figure 1a is expected (13) to be $7.35 \times 10^{-9} M^{1/2}$. Thus this dimension is 820 Å.

The Time Scale of Elastic Recoil and Related Rheological Effects

Another feature of Figure 6 which is worth scrutiny is the time required for steady-state flow to be established after the torque goes on and the time required

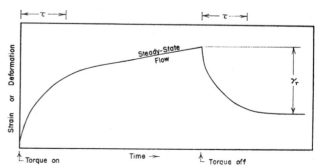

Figure 6. Behavior under constant shear stress: approach to steady-state flow; the steady state; elastic recovery following removal of torque.

for the recoil to be completed after the torque is removed. Strictly, each of these processes involves approaching an asymptotic limit, so the time cannot be defined in a simple manner, but qualitatively we can consider the time within which a major portion of the change is accomplished, as indicated by the interval τ in the figure. These two intervals are the same, and they are of the same magnitude as the so-called terminal relaxation time τ_1, which in the language of models is simply the ratio of the viscosity η_1 to the spring constant G_1 in Figure 3.

For concentrated coils, the terminal relaxation time is given by the Rouse theory as

$$\tau_1 = 6M\eta/\pi^2 cRT \tag{3}$$

where η is the steady-flow viscosity. This provides a second link between rheology and dimensions. For example, if a polymer with $M = 1.25 \times 10^6$ at a concentration of 15% has a viscosity of 400 poises, $\tau_1 = 0.8$ sec. Thus about 1 second is required for a major portion of elastic recoil to be accomplished, or probably several seconds for it to become complete within experimental precision.

The terminal relaxation time has still another interesting physical significance: it is approximately the ratio of the stored elastic energy to the energy dissipated as heat per second during steady-state flow.

If there is a distribution of molecular weights, both the approach to steady-state flow and the accomplishment of elastic recovery will have a more gradual course. In fact, the distribution of molecules of different sizes can in principle be calculated from the shapes of the curved regions of the plot in Figure 6. Even with this complication, the time scale for recovery can be used for order-of-magnitude estimations of molecular dimensions (11).

Networks of Coiled Macromolecules

In a vulcanized rubber, and related structures, flexible coiled macromolecules are entangled among each other and also cross-linked at widely separated points to form permanent junctions (Fig. 1b). The theory of rubberlike elasticity (2–4) provides a well-known relation between rheology and dimensions for such a system, often expressed in terms of the shear modulus, or stress/strain ratio at equilibrium, G:

$$G = \rho RT/M_c \tag{4}$$

where ρ is the density, and M_c is the average molecular weight between linkage points. Thus from elastic measurements M_c can be calculated, and from this the average distance between linkages as in Figure 1b. For a typical soft rubber, M_c may be 4000 and the average vector between cross-links about 50 Å, again in the colloidal domain.

All the dimensions thus far cited as being derivable from rheological measurements can be obtained from other experiments too, and in many cases the rheological relations would not normally be the primary source of information. But now we turn to a characteristic dimension which seems to be obtainable *only* from rheology, representing a phenomenon that is revealed only in rheology—the long-range coupling entanglements occurring in flexible macromolecules of high molecular weight.

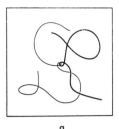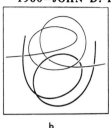

Figure 7. Concepts of long-range coupling entanglement: (a) by kinking; (b) by long-range looping. (From Reference 12).

Even in the absence of permanent cross-links as in Figure 1b, long flexible chains behave as if there were widely spaced points of rather strong coupling between adjacent macromolecules. This effect is over and above the local intertwining which must exist everywhere. The effects of local chain structure, insofar as they have been investigated, indicate that the effect is not due to kinking or knotting (Fig. 7a) but rather the mathematical expectation that two large loops will curve completely back on themselves (14) (Fig. 7b). The evidence for such long-range entanglement comes from several different rheological sources, and there are at least five ways of estimating the distance between entanglements.

The first involves the dependence of the steady flow viscosity on molecular weight, as pointed out some time ago by Bueche (15). As shown in Figure 8, the viscosity rises much more steeply above a critical value of the molecular weight corresponding (it is inferred) to an average number of two entanglements per molecule. The molecular weight between entanglements is of the order of 10,000 to 50,000.

The other estimates of entanglement spacings depend on various kinds of time-dependent rheological

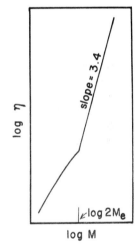

Figure 8. Dependence of viscosity on molecular weight (logarithmic).

calculations, illustrated in Figure 9. One involves the dynamic modulus, which is measured in small oscillating deformations; here G' is the stress in phase with the rate of strain divided by the strain. This quantity goes through an inflection with increasing frequency of oscillation as shown. Qualitatively, this means that at just the right frequency—high enough so that the entanglements do not slip within the period of oscillation, but not too high for rearrangements of the chain segments between the coupling points—the entanglement network has roughly the properties of a permanent network; the average molecular weight between entanglements, M_e, can be estimated from equation (4) with M_e substituted for M_c and G' for G. The other curves in Figure 9 refer to more complicated functions whose full explanation is beyond the scope of this review. The loss compliance J'' is the strain in sinusoidal deformations 90° out of phase with the stress, divided by the

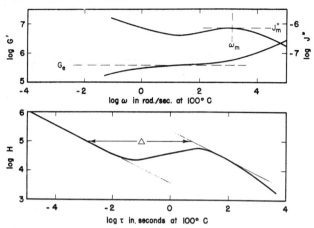

Figure 9. Sources of information on the spacing between entanglement coupling points (explanation in text).

stress. With changing frequency ω, it passes through a maximum, and according to a recent theory of R. S. Marvin (16) the spacing M_e can be calculated from either coordinate (J''_m or ω_m) of this maximum (12). The relaxation spectrum H is a distribution function of relaxation times which can be obtained from various rheological measurements. Its shape encompasses a plateau region from whose width (Δ in Fig. 9) the spacing M_e can again be calculated.

Some representative values of the number of chain atoms between entanglement coupling points (12) are given in Table 1. They range from about 200 to 1000, which for these flexible macromolecules correspond to vector separations of 70 Å to 160 Å, thus again in the range of neglected dimensions. The differences among values from different sources are due partly to approximations in the underlying theory and partly to the fact that they reflect different types of averages. Although the exact nature of the entanglement coupling phenomenon remains somewhat vague, these spacings represent significant dimensions which are derivable from rheological information alone.

Table I. Average Number of Chain Atoms Between Entanglement Coupling Points, from Various Rheological Measurements

Macromolecule	From η-M dependence	From G'	From J''_m	From ω_m	From Δ
Hevea rubber	...	500	140	200	...
Polyisobutylene	305	440	250	420	...
Polystyrene	360	660	200	...	190
Polyvinyl acetate	260	400	340	580	170
Polymethyl acrylate	...	...	200	380	...
Poly-n-octyl methacrylate	...	...	660	1160	1000

Rheology of Classical Colloidal Systems

Most of the above examples have been drawn from synthetic polymeric systems which did not exist in Ostwald's day. But there are fragmentary data in the literature indicating that similar calculations can be applied to some classical colloids such as soaps, and that in some of these rheology may be obliged to furnish the colloidal dimensions without much help from other kinds of measurements. For example, studies by

Bauer (17) on solutions of aluminum dilaurate in toluene reveal elastic recoil of the sort portrayed in Figure 6. Such solutions presumably contain long linear aggregates with some degree of flexibility (18). Application of equations (2) and (3) to the observed elastic recoil and (very roughly) to the terminal relaxation time yield molecular weights of the order of a million.

In aqueous solutions of colloidal electrolytes, similar viscoelastic phenomena are encountered but are still more difficult to explain (19). In dilute solutions of ammonium oleate, or of potassium oleate in the presence of neutral salt, there is spectacular elastic recoil, and transverse elastic waves can be propagated. In the experiments of Bungenberg de Jong (19, 20), spherical vessels are turned sharply through a small angle to excite waves which can be observed by the motion of suspended bubbles. From the wavelength, the dynamic shear modulus can be calculated by a complicated analysis of the elastic deformation with spherical geometry. It would be a good idea to study these systems by more modern rheological methods. What is the source of the viscoelasticity here? There seems no reason to expect long threadlike association complexes to be formed, simulating macromolecules, as in the aluminum dilaurate solutions in toluene. It has been postulated by Winsor (21) that there is an emulsion of different interpenetrating phases, representing different types of micelles—laminar and spherical. But it is not clear what the origin of the elastic deformation and energy storage can be in such a structure. Whatever the final explanation, the rheological phenomena will certainly be found to be related to structural features which fall in the range of Ostwald's neglected dimensions.

Literature Cited

(1) OSTWALD, W. "Die Welt der Vernachlässigten Dimensionen," T. Steinkopff, Dresden, 1914.
(2) GUTH, E., AND MARK, H., Monatsh., 65, 93 (1934).
(3) KUHN, W., Kolloid-Z., 68, 2 (1934).
(4) MEYER, K. H., VON SUSICH, G., AND VALKO, E., Kolloid-Z., 59, 208 (1932).
(5) KIRKWOOD, J. G., AND AUER, P. L., J. Chem. Phys., 19, 281 (1951).
(6) VAN HOLDE, K. E., AND ALBERTY, R. A., J. CHEM. EDUC., 26, 151 (1948).
(7) VON MURALT, A. L., AND EDSALL, J. T., J. Biol. Chem., 89, 315 (1930).
(8) ROUSE, P. E., J. Chem. Phys., 21, 1272 (1953).
(9) FERRY, J. D., Rev. Mod. Phys., 31, 130 (1959).
(10) FERRY, J. D., WILLIAMS, M. L., AND STERN, D. M., J. Phys. Chem., 58, 987 (1954).
(11) BUECHE, F., AND HARDING, S. W., J. Polymer Sci., 32, 177 (1958).
(12) FERRY, J. D., "Viscoelastic Properties of Polymers," John Wiley & Sons, New York, 1961, Chap. 10.
(13) FLORY, P. J., "Principles of Polymer Chemistry," Cornell University Press, 1953, p. 618.
(14) BUECHE, F., J. Polymer Sci., 25, 243 (1957).
(15) BUECHE, F., J. Chem. Phys., 20, 1959 (1952).
(16) MARVIN, R. S., in "Viscoelasticity-Phenomenological Aspects," J. T. BERGEN, Ed., Academic Press, New York, 1959, p. 27.
(17) BAUER, W. H., WEBER, N., AND WIBERLEY, S. E., J. Phys. Chem., 62, 106 (1958).
(18) LUDKE, W. O., WIBERLEY, S. E., GOLDENSON, J., AND BAUER, W. H., J. Phys. Chem., 59, 222 (1955).
(19) BUNGENBERG DE JONG, H. G., AND VAN DEN BERG, H. J., Proc. Kon. Ned. Akad. Wet., 51, 1211 (1948).
(20) BUNGENBERG DE JONG, H. G., op cit., 52, 457 (1949).
(21) WINSOR, P. A., J. Colloid Sci., 10, 94 (1955).

ADDENDUM 1972:

RHEOLOGY IN THE WORLD OF NEGLECTED DIMENSIONS PROGRESS FROM 1960 TO 1972

John D. Ferry
Department of Chemistry, University of Wisconsin
Madison, Wisconsin 53706

Since the presentation of a Kendall Award Address with the above title *(1)* in 1960, there has been substantial progress in understanding the relations between rheological properties of certain types of systems and the magnitudes of the dimensions, falling in the colloidal domain, which characterize their structure. In 1960, it was possible to estimate molecular sizes, cross-link spacings, and entanglement spacings in macromolecular systems by measuring magnitudes of energy storage in deformation or flow, or relaxation times obtained from the rates of viscoelastic processes such as elastic recoil. Both theory and experimental technique, as well as the variety of applications of such relationships, have advanced since then in numerous laboratories.

I. Rheology of Dilute Macromolecular Solutions

Perhaps the most striking recent application of a rheological measurement in dilute solution to determine a molecular dimension has been the observation of Zimm of elastic recoil in solutions of DNA at concentrations around 10^{-5} g./ml. *(2)*. From the magnitude of the retardation time (about 0.5 second in water, longer in glycerol-water mixtures) the molecular weight can be calculated, and if it is extremely large this may be the best method for molecular size determination. The energy storage is also related to molecular weight as explained in the original address on this subject *(1)*.

Even at such remarkably low concentrations, it is necessary to extrapolate data to infinite dilution to eliminate the effects of peripheral interactions of the giant molecules and obtain the behavior of isolated molecules from which their size can be calculated by theory. When the macromolecules are not quite so large, retardation or relaxation times and energy storage must be obtained from dynamic (sinusoidal) measurements and to make such measurements at very low concentration is exceedingly difficult. In 1961, with an apparatus developed by Birnboim *(3)*, the concentration range was pushed to below 1% at audiofrequencies, and extensive data were obtained *(4)* but could not quite be extrapolated to infinite dilution. In 1965, successful extrapolation was accomplished at kilocycle frequencies by Tanaka and Sakanishi *(5)* and in 1970 at audiofrequencies by Schrag *(6)* following another innovation by Birnboim *(7)*, in both cases from measurements at concentrations below 0.1%. Application of the new techniques to flexible linear macromolecules *(8)* in good and poor solvents quantitatively confirmed earlier theory *(9)* when the latter was

evaluated numerically by improved methods of calculation *(10)*. Relevant to the theme of molecular dimensions is the observation *(8b)* that the degree of hydrodynamic interaction between different parts of the molecule appears to be controlled by the average dimensions of the random coil, governed in turn by the thermodynamic interaction between polymer and solvent.

Recent results on branched polymers *(11)* have shown that relaxation times and energy storage depend on branch topology and length in a complicated manner which is not yet understood. It appears that rheological measurements in dilute solution may provide information to characterize the dimensions of branched polymers which could not be obtained in any other way.

II. Rheology of Entanglement Networks

In undiluted amorphous polymers of high molecular weight and in their concentrated solutions, the flexible macromolecules are extensively entangled and their long-range motions are dominated by an imperfectly understood phenomenon known as entanglement coupling; each molecule appears to be rather tightly coupled to others at widely separated points. In 1960, it was believed that the long-range motions were similar to those in very dilute solution except for being much slower, so that molecular weight could be related to energy storage and relaxation times by equations 1-3 of the original address *(1)*. Subsequent work by Tobolsky *(12)*, Nagasawa *(13)*, O'Reilly *(14)*, Odani *(15)*, Onogi *(16)*, and others showed these equations to be incorrect; the elastic recoil is almost independent of molecular weight if the latter is sufficiently large. It has been emphasized by Graessley *(17)* and Ziabicki *(18)* that the long-range molecular motions constrained by entanglements must be very different from those occurring in isolated macromolecules in very dilute solution. Although it is evident now that molecular sizes cannot be estimated from energy storage measurements in such systems, they can of course be related to the more traditional rheological measurement of steady-flow viscosity *(19)*. Combination of viscosity and elastic recoil for branched polymers in concentrated and undiluted systems can probably eventually furnish information about branch topologies and dimensions, though these relations are not yet understood.

The average spacing between entanglement coupling points is a dimension which can be obtained from several kinds of rheological measurements, and essentially from rheological measurements alone, as explained previously *(1)*. It now seems desirable to distinguish between those spacings based on a pseudo-equilibrium determination of modulus or compliance, and those based on viscosity or the viscoelastic time scale. A useful method of the first type involves integration over the loss component of the complex compliance *(20)*. Numerous entanglement spacings have been estimated in this manner *(21)*. However, the relation of this dimension to chemical structure is still uncertain.

III. Rheology of Cross-linked Networks Containing Entanglements

As previously described *(1)*, the average distance between cross-links in a loose network polymer is related to the equilibrium elastic modulus by the classical theory of rubberlike elasticity. In recent years, attention has been given to the effects of entanglements which may be present in such a permanently cross-linked network.

If the average number of cross-link points per original linear molecule is not too great, of the order of 10, for example, dangling branched structures may exist which are imperfectly attached to the network but extensively entangled with it *(22)*. Certain very slow relaxation mechanisms have been attributed to molecular motions which allow such untrapped entanglements to rearrange *(23)*. Other entanglements may be trapped between cross-links and contribute to energy storage in equilibrium deformations, to a degree which depends on the type and magnitude of deformation *(24)*. The topological relations between cross-links and trapped and untrapped entanglements are thus far largely a matter of conjecture, but may be clarified by making diverse rheological measurements and extending theory *(25)*.

IV. Expectations for the Future

Some of the topics mentioned in the original address have not received much recent attention. Rod-like macromolecules in dilute solution, either stiff or weakly bending; suspensions of hard or rubbery particles as found in latices; classical colloidal viscoelastic systems such as aluminum soaps and colloidal electrolytes—all of these provide promising opportunities for further investigation of the relations between rheological properties and structural features which fall within the range of Ostwald's neglected dimensions.

References

1. J. D. Ferry, *J. Chem. Educ.*, **38**, 110 (1961)
2. R. E. Chapman, Jr., L. C. Klotz, D. S. Thompson, and B. H. Zimm, *Macromol.*, **2**, 637 (1969); D. J. Massa, *Bull. Amer. Phys. Soc., Ser. II*, **17**, 317 (1972)
3. M. H. Birnboim and J. D. Ferry, *J. Appl. Phys.*, **32**, 2305 (1961)
4. (a) R. B. DeMallie, Jr., M. H. Birnboim, J. E. Frederick, N. W. Tschoegl, and J. D. Ferry, *J. Phys. Chem.*, **66**, 536 (1962); (b) N. W. Tschoegl and J. D. Ferry, *ibid.*, **68**, 867 (1964); (c) J. E. Frederick, N. W. Tschoegl, and J. D. Ferry, *ibid.*, **68**, 1974 (1964); (d) J. E. Frederick and J. D. Ferry, *ibid.*, **69**, 346 (1965)
5. (a) H. Tanaka, A. Sakanishi, M. Kaneko, and J. Furuichi, *J. Polymer Sci.*, **C15**, 317 (1966); (b) A. Sakanishi, *J. Chem. Phys.*, **48**, 3850 (1968); (c) A. Sakanishi and H. Tanaka, *Proc. 5th Intern. Cong. Rheol.*, **4**, 251 (1970)
6. (a) R. M. Johnson, J. L. Schrag, and J. D. Ferry, *Polymer J.*, **1**, 742 (1970); (b) J. L. Schrag and R. M. Johnson, *Rev. Sci. Instr.*, **42**, 224 (1971); (c) J. L. Schrag and D. J. Massa, *J. Polymer Sci.*, **A-2, 10**, 71 (1972)
7. M. H. Birnboim and L. J. Elyash, *Bull. Amer. Phys. Soc.*, **Ser. II, 11**, 165 (1966)
8. (a) K. Osaki, Y. Mitsuda, R. M. Johnson, J. L. Schrag, and J. D. Ferry, *Macromol.*, **5**, 17 (1972; (b) K. Osaki, J. L. Schrag, and J. D. Ferry, *Macromol.*, **5**, 144 (1972)
9. B. H. Zimm, *J. Chem. Phys.*, **24**, 269 (1956)
10. (a) A. S. Lodge and Y. J. Wu, in preparation; (b) K. Osaki, *Macromol.*, **5**, 141 (1972)
11. (a) Y. Mitsuda, K. Osaki, J. L. Schrag, and J. D. Ferry, *Polymer J.*, **4**, 24 (1973); (b) Y. Mitsuda, unpublished experiments
12. (a) A. V. Tobolsky, R. Schaffhauser, and R. Böhme, *Polymer Letters*, **2**, 103 (1964); (b) A. V. Tobolsky, J. J. Aklonis, and G. Akovali, *J. Chem. Phys.*, **42**, 723 (1965)
13. T. Fujimoto, N. Ozaki, and M. Nagasawa, *J. Polymer Sci.*, **A-2, 6**, 129 (1968)
14. J. M. O'Reilly and W. M. Prest, Jr., *Bull. Amer. Phys. Soc.*, **Ser. II, 13**, 371 (1968)
15. H. Odani, N. Nemoto, S. Kitamura, and M. Kurata, *Polymer J.*, **1**, 356 (1970)
16. S. Onogi, T. Masuda, and K. Kitagawa, *Macromol.*, **3**, 109 (1970)
17. W. W. Graessley, *J. Chem. Phys.*, **47**, 1942 (1967); **54**, 5143 (1971)
18. A. Ziabicki, *Pure Appl. Chem.*, **26**, 481 (1971)
19. G. C. Berry and T. GFox, *Adv. Polymer Sci.*, **5**, 261 (1967)
20. J. F. Sanders and J. D. Ferry, *Macromol.*, **2**, 440 (1969)
21. J. D. Ferry, *"Viscoelastic Properties of Polymers"*, 2nd Ed., Wiley, New York, 1970
22. N. R. Langley, *Macromol.*, **1**, 348 (1968)
23. (a) N. R. Langley and J. D. Ferry, *Macromol.*, **1**, 353 (1968); (b) C. P. Wong, J. Schrag, and J. D. Ferry, *Polymer J.*, **2**, 274 (1971)
24. (a) B. M. E. van der Hoff, *J. Macromol. Sci. (Chem.)*, **A1**, 747 (1967); (b) J. Janáček and J. D. Ferry, *J. Polymer Sci.*, **A-2, 10**, 345 (1972)
25. S. F. Edwards, *Proc. Phys. Soc.*, **91**, 513 (1967); *"Polymer Networks"*, edited by A. J. Chompff and S. Newman, Plenum Press, New York, 1970, 83

1961

STEPHEN BRUNAUER

PORTLAND CEMENT ASSOCIATION
SKOKIE, ILL.

. . . for his many outstanding contributions especially to the development of a reliable method for determining the specific surface areas of finely-divided solids, to the Brunauer-Emmett-Teller theory of the physical adsorption of gases, to experimental and theoretical knowledge of chemisorption and catalysis, to the determination of surface energies of inorganic substances, and to the elucidation of the hydration and structure of the calcium silicates, Portland cement, and concrete.

The following Kendall Award Address is reproduced by permission from *"Solid Surfaces and the Solid-Gas Interface"*, Advances in Chemistry Series, **33**; American Chemical Society, Washington D. C. 1961; p. 5.

Solid Surfaces and the Solid-Gas Interface

THE KENDALL AWARD ADDRESS

STEPHEN BRUNAUER

Basic Research Section,
Portland Cement Association, Skokie, Ill.

It is really heart-warming to see so many of my dear friends among you, and it is equally heart-warming to see so many unfamiliar faces, faces of those whom I have not had the privilege of meeting personally as yet. I never thought that the time would come when anyone would make a 20-minute address about me—let alone such an outstanding scientist and human being as Sir Hugh Taylor. I am profoundly grateful to him for honoring me with this wonderful address—it will always be one of the most treasured memories of my life—but I should like to assure him and assure you that I am keenly aware of my unworthiness of the honor bestowed upon me, and that I know full well that I am honored here because of his generosity and your generosity and not because of my own merits.

For all that, I confess that in spite of the knowledge of my inadequacy I am very happy to receive the Kendall Company Award for 1961. So I should like to express my sincere thanks to the Kendall Company for establishing the award and donating the funds for it, to the American Chemical Society for administering the award, to my friends who nominated me for the award, and finally to the Award Committee, whose members are unknown to me, but who selected me for this great honor.

I should like to say at this point also that practically all the contributions I have ever made were made in collaboration with others. So I have only partial share in all these contributions—and part of the honor I am now receiving should go to those with whom I collaborated. Nothing would please me more than to make my acknowledgment to each of them individually, but I am afraid this process would not be very interesting to most of you. I will mention, therefore, only four names now.

The first is the name of my first boss, the first chemist who befriended me, the first man with whom I collaborated, and the first man who made me interested in surface chemistry—Paul Emmett. I owe more to Paul Emmett than I can say here, and I wish to acknowledge it with deep appreciation.

Equal in importance in my scientific growth was Sir Hugh Taylor. Among other things, it was he who persuaded me to write the book, "The Adsorption of Gases and Vapors," and as I wrote each chapter I sent it to him for his comments. Hugh was a source of guidance for me through much of my scientific life, and a source of inspiration through all of it. For that and for his friendship I want to express my deep gratitude.

The two other names I have picked from the list of my coworkers, one from the first decade and one from the last. The first is Edward Teller, whose name is known to hundreds of millions, and the second is Lew Copeland, to whom I owe not only thanks for many things but also for organizing this wonderful symposium. To Edward Teller, to Lew Copeland, and to all my past and present coworkers I wish to express my profound gratitude.

Please, forgive me for the long introduction, but I would have felt bad if I had not had the opportunity to say these things. Coming to the subject of my talk, when Mr. Warren asked me last August for a title, I gave him, somewhat hastily, "Solid Surfaces and the Solid-Gas Interface." This may have created the impression in some of you that I intend to give you a bird's-eye view of these two broad fields in a single short lecture, but I do not intend to do that. I doubt that anyone could do that to an audience like this—I certainly could not. The second possibility was to give a bird's-eye view of the contributions that my coworkers and myself have made to these two fields, and that was my original idea. However, on second thought I discarded that, too. If I talk on any given subject for a few minutes, what can anyone get out of that, who does not know the subject, anyhow? So I have ended up by selecting only two topics, one from the field of the solid-gas interface, the other from the field of solid surfaces, and I have decided to discuss these in some detail. The first topic is the BET theory; the second is the surface of a particular solid.

The BET Theory

First, allow me to impart to you some of my present thoughts on the BET theory (3, 4) for whatever they are worth. [Reference (3) is a continuation and extension of the BET theory. Some authors refer to it as the BDDT theory; others call both papers together the BET theory.] Probably there are very few among you who have never made a BET plot, and many of you have made quite a few. I suppose, it will surprise no one among you if I say that I made the first BET plot. I did not know at that time that it was a BET plot, because that name did not exist yet. I tried to give a name to the theory that the three of us had developed, and I called it the multimolecular adsorption theory, which was probably not a very good name, but it was the best I could think of. Fortunately, somewhat later Professor Harkins invented the colorful name "BET theory," and that name has stuck.

The first BET plot was a plot of an adsorption isotherm that Emmett and I obtained a few years before, and it was the adsorption of nitrogen on something at liquid oxygen temperature. I had to make both liquid oxygen and liquid nitrogen myself from liquid air by rather crude means, and since liquid oxygen was easier to make, most of our isotherms were made at liquid oxygen temperature. This isotherm had eight points, ranging from 50- to 750-mm. pressure, a 15-fold range in pressure, and all of the points fell on a straight line—none of them deviated the breadth of a hair. I was astonished, and so was Paul Emmett when I showed him the plot. When I showed it to Edward Teller, he said that this cannot be true. No theory can fit that well. I plotted a lot more isotherms, and they all fitted just about as well. Figure 1 shows a groups of such isotherms. For example, see how accurately the line of curve 4 fits the eight points. Of course, this is because the data were obtained in the relative pressure range in which the BET equation fits, but at that time we did not know this. On one occasion, one of the plots showed a considerable curvature, so I tested the straight-edge I used, and found that it was curved, and not the BET plot. Occasionally, I found a point which fell off the line. Then I went back to the original data, recalculated the point, and found that my old calculation was in error—the point actually did fall on the line.

The fitting of the data with a curve is a necessary condition of the correctness

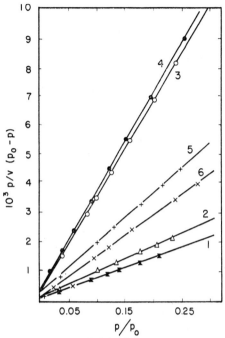

Figure 1. BET plots of nitrogen adsorption isotherms on six adsorbents at −183° C.

of a theory, but not a sufficient condition. This is well known now, but at the time we worked on the BET theory, it was not realized by surface chemists. In all prior theories of adsorption (1), with one exception, this single test of the theory was used to prove or disprove the theory. If the data fell on a straight line, the theory was proved. The one exception was the Polanyi theory, which also used a single criterion for its testing, but a different one. Since it offered no isotherm equation, the sole criterion was the correct prediction of the temperature dependence of the isotherm. We in the BET paper offered not one but four criteria. The first was the fit of the curve, an example of which is shown in Figure 1. The second was the correctness of temperature dependence, the same criterion which was used by Polanyi. Figure 2 shows how well the temperature dependence was reproduced by the theory. The parameters here were evaluated from one isotherm of each pair, and the other isotherm was calculated. If the theory had been wrong, we could have gotten isotherms which fell to the right or left of the experimental isotherms, but they fell on the right spot.

Parameter C. The two other criteria were the reasonableness of the two parameters evaluated from the straight lines, V_m and C. The C parameter includes the heat of adsorption. By assuming that the entropy term is of the order of unity, we calculated an average heat of adsorption for the first layer from C, and found that it was right in the ball park. If the model had been seriously in error, we could have gotten an average heat of adsorption which was ten times as large or one tenth as large as the known average heats of adsorption. Actually, the values were of the right order of magnitude, but they were always smaller than one would expect them to be, sometimes even smaller than 50% of the expected value.

One obvious explanation was pointed out by us at the start. The BET equation is always fitted in the region of monolayer coverage, let us say, in the range of $\theta = 0.5$ to 1.5 monolayers. The points fall off the curve at low coverages, indicating more adsorption and higher heats of adsorption. Thus, the C constant does not evaluate the average heat of adsorption for the entire first layer, but the average heat of adsorption for a part of the first adsorbed layer—that part which takes place on the energetically less active sites of the surface.

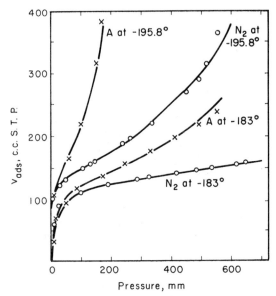

Figure 2. Adsorption isotherms of nitrogen and argon on iron

While this must certainly be a part of the explanation, examination of the heat of adsorption data in the literature indicates that this is not the whole explanation, nor even the most important part. However, a very interesting paper by Clampitt and German (*10*) goes a long way toward completing the explanation. These investigators have pointed out that the heats of vaporization of thin films of liquid are different from the heat of vaporization of the bulk liquid, and they have developed a theory for calculating the heats of vaporization of these thin films. Figure 3 shows their results for three liquids. The authors then proceeded to rederive the two-parameter BET equation by discarding the assumption that the heat of adsorption was constant for the second and higher adsorbed layers and was equal to E_L, the heat of vaporization of the bulk liquid. They assumed instead that the heat of adsorption for the second, third, etc., adsorbed layers was equal to the heat of vaporization of a film two, three, etc., layers in thickness. The beauty of their treatment was that this assumption did not add an extra parameter to the two, V_m and C; their final equation was formally identical with the two-parameter BET equation—only the meaning of C changed. The exponential term in C did not contain $E_1 - E_L$ only, but another term as well, which could be calculated from their theory.

Clampitt and German illustrated their results with a reinterpretation of certain data of Harkins and Boyd (*16*) for the adsorption of benzene on TiO_2 at 25° C.

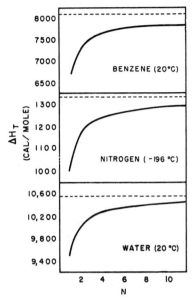

Figure 3.   Total heat of vaporization vs. number of adsorbed layers
above the first

Harkins and Boyd found by heat of emersion experiments that the value of $E_1 - E_L$ was 5.2 kcal. per mole, whereas the C constant of the BET plot gave a value of only 2.6 kcal. per mole. Clampitt and German calculated the third energy term in C, and showed that with their interpretation of C the value of $E_1 - E_L$ obtained from the BET plot was 4.3 kcal. per mole, which is much closer to the experimental value. If we combine the two explanations I have discussed, the discrepancies between experimental and theoretical values practically disappear.

Parameter V_m.   Next I should like to talk about the parameter V_m. The test of the correctness of V_m is, naturally, whether it gives the correct surface area for a given adsorbent or not. When we wrote the BET paper, the best that we could offer in support of the V_m values was their agreement with the Point B values (13). In addition, we pointed out the internal consistency of the results—i.e., different adsorbates gave approximately the same specific surface area for a given adsorbent. For example, for a silica gel sent to us by Lloyd Reyerson we used six different gases (argon, nitrogen, oxygen, carbon monoxide, carbon dioxide, and butane), and none of the surface areas obtained from the V_m values differed from the average by more than 10%. However, internal consistency is only a necessary condition of the correctness of V_m, but not a sufficient one. The question still remained whether the surface obtained from V_m was the true surface or not. By the time I wrote the book (1) a few years later, there was an imposing array of evidence that V_m gave the true surface. I shall recall to you the one I considered the most impressive proof at that time, and still do. For two carbon blacks, electron microscopic average particle size determinations were compared with average particle sizes calculated from BET surfaces. For Micronex, the electron microscope (12) gave an average diameter of 28 $\mu\mu$, nitrogen adsorption gave 28 $\mu\mu$, obtained by Smith, Thornhill, and Bray (25), and 31 $\mu\mu$, obtained by Emmett and DeWitt (14). For acetylene black, the electron microscope gave

51 $\mu\mu$, nitrogen adsorption gave 52 $\mu\mu$, obtained by Emmett and DeWitt, and 53 $\mu\mu$, obtained by Smith, Thornhill, and Bray. This is not merely internal consistency, but true accuracy. In the subsequent two decades, the number of such examples has multiplied.

Let us return now to Lloyd Reyerson's silica gel. Reyerson and Cameron (*24*), in 1935, published adsorption isotherms of bromine and iodine on silica gel, and obtained some beautiful and quaint isotherms of the kind that are now called Type III isotherms. The bromine isotherms are shown in Figure 4. Five years later, in the BDDT paper we showed that the simple, two-parameter BET equation can describe Reyerson's Type III isotherms even better than it can describe Type II isotherms. By making a single assumption that the value of the C parameter for bromine was 1, the four isotherms of Reyerson at four temperatures were accurately calculated. Thus, not only the course of the isotherms was described, but also their temperature dependence came out almost perfectly, as seen in Figure 4. The curves are theoretical; the points are experimental. I said that there was a single assumption, $C = 1$, but even this can hardly be called an assumption, because the isosteric heats of absorption calculated by Reyerson and Cameron strongly indicated that E_1 was very close to E_L. Now comes the thing that gave me one of the greatest thrills of my entire research career—and it still does quite a bit—the V_m value evaluated from the bromine data gave a specific surface for the silica gel that agreed within 5 or 6% with the surface obtained from Type II isotherms of the six other gases about which I talked before. As you see, these Type III isotherms do not have a Point B—there is nothing in the isotherm that indicates the completion of a monolayer. When $C = 1$, V is equal to V_m at the relative pressure of 0.5, but this point is an indiscriminate point on a continuously rising isotherm.

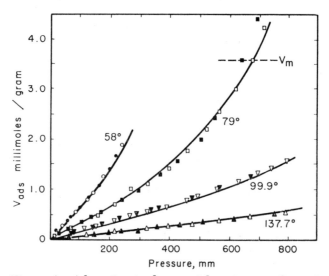

Figure 4. Adsorption isotherms of bromine on silica gel

By assuming a value of C which was considerably less than 1, the four Type III iodine isotherms of Reyerson and Cameron were very well fitted. Because the value of C was 1 for bromine and only a fraction of 1 for iodine, it was easy to see that it should be well over 1 for chlorine; so we predicted that chlorine would

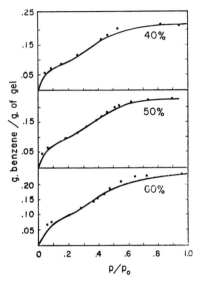

Figure 5. Adsorption isotherms of benzene on ferric oxide gel

give Type II and not Type III isotherms on silica gel. Whereupon Reyerson and Bemmels (23) went ahead and investigated the adsorption of chlorine on this same silica gel and did obtain Type II isotherms—and the isotherms agreed well with the BET theory.

Type IV Isotherms. I wish to say a few words now about the Type IV isotherms for which we advanced a theoretical explanation in the BDDT paper (3). Edward Teller and I independently derived the isotherm equations for this case, and he dictated his equation to me into the phone. It was a long one, as some of you know, and I got a different one, which was even longer. Because I could not get together with him for some time, I showed my derivation to Ed Deming, and he concluded that my derivation was correct. I am not saying this to boast that there was an occasion when I was right and Edward Teller was wrong, because I am quite sure that if one of us had been wrong, it would have been I. What happened was that we were both right. In the derivation, there is a parameter n, which represents the maximum number of layers that can fit into a capillary. Edward assumed that the n layers fitted the capillary exactly, whereas I assumed that the n layers did not fit the capillary exactly. In the paper you can find both equations.

These are four-parameter equations, and Lola Deming spent weeks at her desk calculating machine to fit the curves that Lambert and Clark obtained for the adsorption of benzene on ferric oxide gel (21). These isotherms are shown in Figure 5. I do not blame anyone for not using these equations, and actually very few have used them to date. Joyner and Emmett (20) were among the hardy souls who did. Clampitt and German published another paper very recently (11), which is well worth reading for all those who deal with Type IV isotherms.

Criticism of BET Theory

I should like to tell you only one more thing about the BET theory. Many years ago Cassie (9), and somewhat later Terrell Hill (18), derived the BET

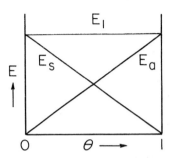

*Figure 6. Idealized variation
of heat of adsorption, E, with
coverage, θ*

E_s. *Surface condition*
E_a. *Adsorbate interaction contribution*

equation by statistical mechanics. In these derivations, the assumptions were
made that adsorption takes place on a uniform surface and that there is no
lateral interaction between the adsorbate molecules. Ever since that time—or perhaps even before that time—practically all writers of books and papers have repeated that the BET theory assumes a uniform surface and no interactions between
the adsorbate molecules. Back in 1952, there was a conference on solid surfaces,
the proceedings of which appeared in a book, edited by Gomer and Smith (*15*).
In this book, there is a paper by Hill (*19*), to which a discussion of mine is
attached (*2*), in which I pointed out that neither the BET theory nor the Langmuir
theory, on which it is based, makes such assumptions. Not many people have
read this discussion, so even last year at the Gordon Conference on Chemistry at
Interfaces I had to argue with several of our leading surface chemists on this
point, but I think I succeeded in convincing them.

For simplicity, we will talk about the Langmuir theory. This theory
assumes that the heat of adsorption is constant over the entire surface—
and so does the BET theory. Naturally, if one has a uniform surface and there
is no interaction between the adsorbate molecules, one gets a constant heat of
adsorption—this is a sufficient condition, but not a necessary one. One can get a
constant heat of adsorption in other ways, as well. This is illustrated in Figure 6,
taken from the Gomer-Smith book. Of course, this is an extremely idealized
example, but it will make the point clear. Most adsorbents have nonuniform
surfaces, and if you plot the differential heat of adsorption against fraction of
the surface covered, you get a diminishing function like E_s, because adsorption at
low coverages takes place on the energetically more active sites. Let us consider
next lateral interaction on a uniform surface. If you plot the energy of interaction
against the fraction of the surface covered, you get an increasing function like E_a,
because the more adsorbate you have on the surface, the more interaction energy
you will have. So these two effects are in the opposite directions, and if and
when they balance each other, you have a constant heat of adsorption. My good
friend Prof. Shereshefsky asked me how frequently I would expect such a balanc-
ing. My answer is that mathematically never or hardly ever, physicochemically
possibly quite frequently. If the sum of the two effects is a slightly decreasing
function, or a slightly increasing function, or if it goes up and down—but not too
strongly—about a mean, you have an approximately constant heat of adsorption,
and, physicochemically, that is all you need.

What I have been trying to do in this discussion is to show that the BET theory, and its continuation, the BDDT theory, can do a good deal more than merely give the specific surface areas of finely divided solids. It can give a description of the process of physical adsorption, and it can account for five different isotherm types—the account being, at worst, semiquantitative, at best, quantitative. I have also meant to imply not that all criticism of the theory is wrong, but that some of the criticism is wrong. I am well aware of the shortcomings of the theory and I not only hope but I am certain that the time will come when it will be superseded by a more complete and more accurate theory of physical adsorption. Right now we have a number of theories which can describe some of the aspects of physical adsorption more accurately than the BET theory.

Surface of Tobermorite

The second part of my talk deals with the surface of a particular solid, a calcium silicate hydrate, called tobermorite. The two main constituents of portland cements are two calcium silicates, which make up about 75% or more of a portland cement by weight, and both of these silicates produce tobermorite in their reaction with water. This tobermorite is the most important constituent of hydrated portland cement, concrete, and mortar. That is not the reason, however, for my talking about it—the reason is that it is a fascinating substance for a colloid chemist. I will discuss only two properties of the tobermorite surface: the surface area and the surface energy.

The tobermorite obtained in the hydration of tricalcium silicate (Ca_3SiO_2), β-dicalcium silicate (β-Ca_2SiO_4), portland cement, and concrete is a colloid, with a specific surface area of the order of 300 sq. meters per gram. To give an idea of how the elementary particles of tobermorite look, Figure 7 is an electron micrograph of a few particles (obtained by L. E. Copeland and Edith G. Schulz at the Portland Cement Association Research and Development Laboratories). These particles look like fibers, but if you watch them closely, you see that they are very thin sheets, rolled up as one would roll up a sheet of paper. At the lower end the sheets are partly unrolled. When one prepares tobermorite by the reaction of lime and silica, one usually obtains crumpled sheets, which are not rolled up. The electron microscopists tell us that the sheets are very thin, of the order of a single unit cell in thickness.

Tobermorite is a layer crystal, like many clay minerals—e.g., montmorillite and vermiculite. The unit cell of tobermorite is orthorhombic—i.e., a, b, and c are perpendicular to each other. The dimensions of the unit cell were determined by Heller and Taylor (17). Dimensions a and b lie within the layer, and c is perpendicular to the layer—it is the distance between two layers.

Let us calculate now the specific surface area of tobermorite, with the assumption that the sheets are one unit cell in thickness (5). Because the length and breadth of the sheet are very large compared to its thickness, just as for a piece of paper, it can be assumed that all the surface is on the two sides of the sheet, and the areas along the edges can be neglected. The surface contributed by a unit cell is 2 ab, and if there are N unit cells in the sheet, the surface area is 2 abN. If there are n molecules in the unit cell, and the weight of a molecule is m, the weight of a unit cell is nm, and the weight of the sheet is nmN. The specific surface area then is 2 ab/nm, and these are all known quantities. The result is 755 sq. meters per gram.

Let us assume next that the sheet is two unit cells thick. We can still neglect the edge areas, so the surface remains unchanged, but now the weight is doubled.

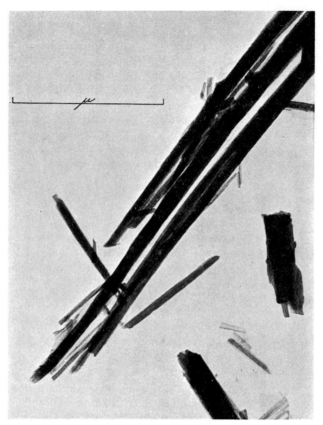

Figure 7. Electron micrograph of tobermorite obtained from hydration of β-dicalcium silicate

So we have a specific surface of 377 sq. meters per gram. Finally, let us assume that the sheet is three unit cells thick. Then the specific surface is 252 sq. meters per gram.

In a series of experiments (8), we determined the specific surface areas of 14 tobermorite preparations, made from different starting materials (Ca_3SiO_5 and β-Ca_2SiO_4), with different water to solid ratios (0.7 and 9.0), and using different methods of preparation. The BET method was used, with water vapor as the adsorbate. We use nitrogen, too, but it did not measure the total surface, as I will show. The largest specific surface we measured was 376 sq. meters per gram, the smallest was 245, and all the others fell between these limits. We concluded, therefore, that in one preparation practically all of the tobermorite sheets were two unit cells thick, in another practically all of them were three unit cells thick, and in all the rest the particles were either two or three unit cells thick.

I believe that this is a fine confirmation of the correctness of the V_m values obtained by water vapor adsorption. On the other hand, nitrogen adsorption gave surface areas that ranged from 20 to 100% of the surface areas obtained by water vapor adsorption. I said that nitrogen did not measure the total surface, and I give one evidence for it now and one later. Dr. Copeland determined the specific surface of a hydrated portland cement by small-angle x-ray scattering. The value he obtained checked the water adsorption value within 2 or 3%. On the other hand, nitrogen adsorption gave a surface about half as large. The other evidence will be given after I discuss the surface energy of tobermorite.

Surface energy is an abbreviation for surface total energy, and it can be defined as the difference between the total energy of a molecule of a given substance in the surface layer and the total energy of the same molecule inside the body of that substance. It is customary to express it in ergs per square centimeter. In the best determinations to date, the heat of solution method of Lipsett, Johnson, and Maass (22) has been used. A substance in different states of subdivision is dissolved in a suitable solvent, and the heat of solution is plotted as a function of the specific surface area. The slope of the curve is the surface energy. To be sure, what one obtains in this way is really the surface enthalpy, but the pv term in the surface region is small, so beginning with Gibbs and continuing with Harkins most investigators have made no distinction between surface total energies and surface enthalpies.

Some years ago, Kantro, Weise, and I set out to determine the surface energy of tobermorite. We came to grief, because the practice is usually not as simple as the principle. We decided, therefore, to work on some simpler but closely related substances, and we determined the surface energies of calcium oxide, calcium hydroxide, amorphous silica, and hydrous amorphous silica. Even these determinations did not turn out to be simple. We published these in two papers in the *Canadian Journal of Chemistry* in 1956 (6, 7). After this work, we felt strong enough to tackle tobermorite, and we published the surface energy of tobermorite in the same journal three years later (8).

I told you about the variability of the tobermorite specific surface, and in Figure 8 I show you the variation of the heat of solution with specific surface. The 14 preparations I mentioned before are represented by the 14 points on this curve, and the points randomly scatter around the least squares straight line. They do

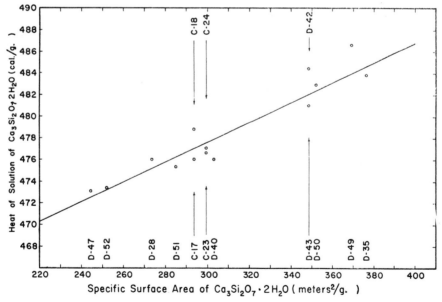

Figure 8. Variation of heat solution of $Ca_3Si_2O_7.2H_2O$ with specific surface area for 1.0-gram samples

not fall beautifully on a straight line, but if you knew the difficulties encountered in this system, you would not be the least bit surprised. The slope of this line gives a surface energy of 386 ± 20 ergs per sq. cm.

If I had been asked before we started to determine the surface energy of tobermorite, to make an educated guess of what the result would be, I would have argued this way. Speaking with some license, the surface of this calcium silicate hydrate is a sort of "chemical mixture" between the surface of calcium hydroxide and that of hydrous silica. Add to this the fact that the calcium hydroxide we used in the surface energy determination contained almost perfect crystals, the hydrous silica was amorphous, and the calcium silicate hydrate, tobermorite, was intermediate between these extremes, it was poorly crystallized. The surface energy of calcium hydroxide was 1180 ergs per sq. cm., that of hydrous silica 129 ergs per sq. cm., so I would have guessed that the surface energy of tobermorite would be the geometric mean of these two. It has turned out to be almost exactly that.

The surface areas of calcium hydroxide and hydrous amorphous silica were determined by means of nitrogen adsorption. The surface areas of tobermorite were determined by water vapor adsorption, and as you see, the three surface energies thus determined show a remarkable consistency. I said also that nitrogen adsorption did not give the entire surface area for tobermorite, and let me tell you now what I meant. Please look at the point D-28 in Figure 8. This preparation gave one of the smallest heats of solution, but it gave the largest area by nitrogen adsorption. On the other hand, preparation D-42, which gave a heat of solution more than 8 cal. per gram larger than D-28, had a nitrogen surface that was less than 40% of the nitrogen surface of D-28. If we would calculate the surface energy from these two preparations, we would get a large negative value, which is absurd. Actually, there was no correlation between heat of solution and nitrogen surface.

The probable explanation, I believe, is the following. The sheets of tobermorite roll into fibers, as in the electron micrograph (Figure 7). This rolling of the sheets plus the aggregation of the fibers create spaces into which nitrogen cannot penetrate but water can. This is not merely because of the difference between the sizes of the molecules, but also because the tobermorite surface is strongly hydrophylic.

I will say only two more things about this slide, and that will conclude my talk. I mentioned that we used a variety of ways to prepare the samples of Figure 8. The four points in the middle (C-17, 18, 23, and 24) were obtained by paste hydration. The calcium silicates were mixed with water for a few minutes, then the mixture was allowed to stand for about a year and a half. This type of hydration is very slow. The four points at the left end (D-47, 52, 28, and 51) were obtained by hydration in a small steel ball mill. This produces a much faster hydration—complete hydration was obtained in 6 weeks. My knowledge of solid state physics is negligible within experimental error, but I would think that the preparations that were banged around by steel balls for 6 weeks or longer should have more defects than the preparations that were allowed to stand peacefully. This would lead to higher energies for the ball-mill preparations, and higher heats of solution, but, as you see, the reverse is true. I do not mean to imply that there are no defects in our tobermorites, but I believe that these defects do not play a decisive role in these experiments.

The last remark is this. Please look at point D-49. This is the point that falls farthest away from the straight line—it is 2.6 cal. per gram too high. I am convinced that this is random error, but let us assume that it is not. Let us assume first that all of this difference is caused by some variation in the body structure, such as differences in defects, for example. The 2.6 cal. per gram amounts to only 0.58% of the heat of solution. Let us assume next that none of this difference is caused by variation in the body structure, but all is caused by variation in the surface structure. This amounts to only 7.6% of the surface energy of tobermorite. Please note that I am giving these figures for the preparation which showed the maximum deviation. If I made the same calculations for the average deviation, which is only 1.1 cal. per gram, the results would be less than half of the values I gave before.

The conclusion is that both the body structure and the surface structure of tobermorite are highly reproducible. Whether we use tricalcium silicate or β-dicalcium silicate as starting solids, whether we use a water to solid ratio of 0.7 or 9.0, whether we use paste hydration or ball-mill hydration or a third type which I have not discussed (which gave the six other points on the curve), we wind up with a tobermorite having very nearly the same body structure and surface structure.

Literature Cited

(1) Brunauer, S., "Adsorption of Gases and Vapors," Vol. I, "Physical Adsorption," pp. 53–149, Princeton University Press, Princeton, N. J., 1943.
(2) Brunauer, S., "Structure and Properties of Solid Surfaces," R. Gomer, C. S. Smith, eds., p. 395, University of Chicago Press, Chicago, 1953.
(3) Brunauer, S., Deming, L. S., Deming, W. E., Teller, E., *J. Am. Chem. Soc.* **62**, 1723 (1940).
(4) Brunauer, S., Emmett, P. H., Teller, E., *Ibid.,* **60**, 309 (1938).
(5) Brunauer, S., Kantro, D. L., Copeland, L. E., *Ibid.,* **80**, 761 (1958).
(6) Brunauer, S., Kantro, D. L., Weise, C. H., *Can. J. Chem.* **34**, 729 (1956).
(7) *Ibid.,* p. 1483.
(8) *Ibid.,* **37**, 714 (1959).
(9) Cassie, A. B. D., *Trans. Faraday Soc.* **41**, 450 (1945).
(10) Clampitt, B. H., German, D. E., *J. Phys. Chem.* **62**, 438 (1957).
(11) *Ibid.,* **64**, 284 (1960).
(12) Columbian Carbon Co., *Ind. Eng. Chem., News Ed.* **18**, 492 (1940).
(13) Emmett, P. H., Brunauer, S., *J. Am. Chem. Soc.* **59**, 1553 (1937).
(14) Emmett, P. H., DeWitt, T., *Ind. Eng. Chem., Anal. Ed.* **13**, 28 (1941).
(15) Gomer, R., Smith, C. S., eds., "Structure and Properties of Solid Surfaces," University of Chicago Press, Chicago, 1953.
(16) Harkins, W. D., Boyd, G. E., *J. Am. Chem. Soc.* **64**, 1195 (1942).
(17) Heller, L., Taylor, H. F. W., *J. Chem. Soc.* **1951**, 2397.
(18) Hill, T. L., *J. Chem. Phys.* **14**, 263 (1946).
(19) Hill T. L., "Structure and Properties of Solid Surfaces," R. Gomer, C. S. Smith, eds., p. 384, University of Chicago Press, Chicago, 1953.
(20) Joyner, L. G., Emmett, P. H., *J. Am. Chem. Soc.* **70**, 2359 (1948).
(21) Lambert, B., Clark, A. M., *Proc. Roy. Soc. (London)* **A122**, 497 (1929).
(22) Lipsett, S. G., Johnson, F. M. G., Maass, O., *J. Am. Chem. Soc.* **49**, 925 (1940).
(23) Reyerson, L. H., Bemmels, C., *J. Phys. Chem.* **46**, 31 (1942).
(24) Reyerson, L. H., Cameron, A. E., *Ibid.,* **39**, 181 (1935).
(25) Smith, W. R., Thornhill, F. S., Bray, R. I., *Ind. Eng. Chem.* **33**, 1303 (1941).

Received May 9, 1961.

1962

GEORGE SCATCHARD

MASSACHUSETTS INSTITUTE OF TECHNOLOGY
CAMBRIDGE, MASS.

. . . for his fundamental experimental and theoretical investigations of the physical chemistry of colloidal solutions and membranes, and particularly for the light they have shed on the basic processes and the molecular and ionic interactions in protein solutions and in ion exchangers.

Half a Century as a Part-time Colloid Chemist

George Scatchard
Massachusetts Institute of Technology
Cambridge, Massachusetts

March 26, 1962

Dr. Van Olphen, Dr. Williams, Dr. Stockmayer, Ladies and Gentlemen,

Let me start by expressing my gratitude to the Kendall Company who made this award possible, to the anonymous committee who selected me for the honor, and to all of you who have come to help me celebrate, especially those who are taking an active part. Some of you must have been surprised when you found that you were part of my symposium. I hope the surprise was a pleasant one for all of you.

I want to thank very especially Professor Stockmayer for his beautiful job of overpraising me. I ought not to have to talk after a speech like that, I should be allowed to sit back and dream how wonderful it would be if I were the man I had just heard about. When Stockmayer asked me what I was going to talk about so that he could discuss other aspects, I told him that he could talk about anything that is published. I planned to talk about unpublished work.

When one of my friends heard my title, he said, "So you did it with one hand tied behind your back." I replied, "Yes, but it took fifty years."

Let's talk about the part-time first. The best way to gain perspective in a field is to work part time in it and part time in a related field. Sure! I know that each of you has a way which makes your perspective better than those of the rest of us. But don't you think that my way is the second best? Then if you will only agree with me that research is incompatible with schizophrenia, you will agree that part time colloid chemistry automatically places one on the side of the unionists—Einstein and Svedberg for example—who held that a colloid particle is a macromolecule, like an ordinary micromolecule only more so, as opposed to the isolationists like Ostwald who claimed that the world of neglected dimensions was subject only to its own laws. My predecessor Dr. Brunauer assures me that there are no longer any isolationists. I think he is over-optimistic because his work is mostly with surfaces. There were certainly many around thirty-six years ago when I become a colloid chemist according to the definition of another predecessor,

the late Dr. Bartell, that is when I begin to teach a course in colloid and surface chemistry. Then I had to adopt a point of view. The surface chemistry was not too difficult. Lots of sense had been talked and relatively little nonsense. There was much to say about Gibbs, Langmuir, Harkins and Taylor. That same year Rideal's book appeared. The textbooks on colloid chemistry, on the other hand, seemed mostly cook-book or nonsense. I had to develop my own point of view. I tried to give the students enough background to enable them to separate the wheat from the chaff with some precision, for there was good work being done. You have just heard my best student, but I am not claiming the credit. We couldn't have spoiled him.

And now I'll reminisce with a series of short stories, each of which seems to me to have a point, if only how different the world was half or even a third of a century ago. In 1912, just fifty years ago, I spent five months touring Europe with no thought of any science. In the other seven months my science subjects were volumetric quantitative analysis, organic combustions and crystallography, which are very far from colloidal.

So I will go back a little further to my experience as a part-time practical colloid chemist. I started working in a drug store in 1904, a year before Einstein's first paper on the Brownian movement, and eleven years before the phrase "the world of neglected dimensions" was first published.

The drug store of those days was a do-it-yourself institution. The proprietary or "patent" medicines and cigarettes were about the only things pre-packaged. There were shelves and shelves of stock bottles with labels in Latin. We prepared the contents of many of them from the drawers of dried drugs below, also labelled in Latin.

I made ice cream, mechanics' hand soap, hot chocolate, opium suppositories, and a few other colloids. The ice cream contained only cream, sugar and vanilla extract. The sugar was delivered. I didn't go to the cow for the cream but I did go to the creamery and I made the vanilla extract from the dried beans. I slithered the ice from the front sidewalk to the backyard, broke it up with a mallet and turned the freezer by hand. But my colloid chemistry was only part time. I also swept and mopped the floor, waited on customers, delivered packages, bottled whiskey and kept books. A popular slogan in those days was "Let George do it," but that was a coincidence. There must have been half-a-million other Georges in the country, not counting the Pullman porters who had been baptized with another name.

My first research in colloid chemistry, which was also my first research in physical chemistry, was part of a minor in physiological chemistry forty-seven or forty-eight years ago. It concerned Bardach's test for proteins. Since I have found only one biochemist who knew what the test is before I told him, I will tell you too. Iodoform, which preceded mercurichrome as a household antiseptic, was characterized chiefly by its odor, but also by its bright yellow color and very regular hexagonal plates. Bardach found that if he tried to prepare iodoform in the presence of protein, he obtained mostly long needles instead of plates.

The first research from our group was to determine if all the available proteins gave the test; the second was to find if any other natural materials did; the third to see if these materials interfered with proteins giving the test; and the fourth was to find the minimum amount of protein which would give it. This should have completed the series by the standards of physiological chemical research of the time, but hormones were the fashion just then and iodine containing proteins were very exciting. It was in 1915 that Kendall finally succeeded in crystallizing thyroxine. The first of our group, the one who read Bardach's paper, had translated one phrase as "iodonitro compounds." So my research was to find out the composition of these yellow needles.

I also read Bardach and translated that phrase, "Iodstickstoff Verbindungen," as "nitrogen iodide compounds." Bardach performed his test by adding a little ethanol, a little iodine and a little ammonia to a dilute aqueous solution of the protein. The nitrogen iodide which he obtained as a by-product was black, apparently amorphous, perhaps colloidal, and Bardach would not commit himself to its being a single substance. When I stepped up from a test tube to a ten quart jar I used sodium carbonate in place of ammonia to get rid of this nitrogen iodide, which could not be left where a medical student or a janitor might touch it after it dried. I could only recover 98-99% iodoform from the needles, but I attributed this to inadequate technique. It was a beautiful example of a surfactant colloid, gelatin, changing the shape of crystals.

In 1916-17 I was research associate of Alexander Smith. Most of my work was on dry ammonium chloride. Brereton Baker had found that extremely dry ammonia and hydrogen chloride would not combine and that dry ammonium chloride would not dissociate. The density of the saturated vapor from dry ammonium chloride was nearly twice that from the slightly moist salt, but the vapor pressure was the same. You all see that there's a gimmick somewhere, but that's because you really believe the second law of thermodynamics. Not everyone was so sure in those days.

Baker seemed to be sitting pretty. If anyone failed to get results agreeing with Baker's, it was obvious that their materials were not sufficiently dry. My predecessor with Professor Smith fooled him, however. He found densities three or four times that from the moist salt. Publication was to await my extending his results, but I went to war before my apparatus was sufficiently dry, and I never got back to it. I don't know all the answer yet, but I am sure that much of it comes from the fact that the extra ammonium chloride was adsorbed on the thoroughly dried glass walls and was not in the vapor. If the dissociation is catalyzed by solid ammonium chloride as well as by water, everything can be explained.

I was doing more surface chemistry in a project to find material for a gas mask which would remove carbon monoxide from the air. It was fortunate that I did not accomplish enough to be given any credit. The material I worked on was developed into Hopcalite, which was named for Johns Hopkins and California. I still can't find a good way to insert any part of Columbia into that word, Hopcalite.

Going to War

My war work was under Grignard in a defensive gas warfare research laboratory. War gases are small molecules, though devilish ones. I remember only one experience related to colloid chemistry. Besides research we analysed any samples from German duds which the regular analytical laboratory thought might be interesting. They were shy about saying anything was not interesting until we had examined it. Sometimes they knew the materials were interesting. I was given one mixture of picryl chloride and diphosgene which was colored bright red, and told to find out what the red material was and why it was there. The analytical laboratory had identified it—though incorrectly—and had decided that some experimental shells had been sent to the front by mistake and that they were meant to be exploded over white chalk to show the spread of the "gas". The Germans were looked upon as devilish but superhumanly able scientists, just as some others are today.

When I said that I knew nothing of dye analysis I was given a huge French book of directions. I isolated ten milligrams of dye and followed directions. I got an answer with two confirmatory tests which satisfied the organic chemist in me, and I said that the dye came from an impurity in the phenol used to make the picryl chloride. A little later we learned that the British, who had much more material than I to work with, had used two different confirmatory tests and had reached the same conclusions.

After the war, in 1922, I dodged the opportunity to become a pioneer in a field of surface chemistry which later became quite well known. I did not know then that it involved surface chemistry, but I did know that it should be avoided. I will quote part of a letter to me from the Secretary of the Chamber of Commerce of the town where I grew up, "They are as confident as before that you could disclose none of the ingredients besides sulfuric acid and water. They still hold the $1000 offer in case you analyse the Electro-lyte, provided you prove the correctness of your analysis by making an Electro-lyte of the ingredients you find which will have a similar effect in battery tests as theirs."

In 1924 I published my first paper on colloid chemistry, a polemic on the effect of gelatin on transference numbers. Duncan MacInnes and I had decided to do something about the nonsense appearing in the Journal. This was my part of the crusade. My irritation at the wholly erroneous electro-chemistry was increased at seeing physical chemists discussing the reaction of acids and gelatin with so much less insight than the protein chemists and physiologists had shown.

In 1924 I also had my first experience with the theory of membranes in an informal consultation with a young physisian interested in research. He wanted to know the effect on osmotic pressure of a temperature difference between the two solutions. I persuaded him to simplify the problem by eliminating first the colloid and then the crystalloid solute, leaving only water, membrane and temperature difference. Then I thought I had flunked because all I could tell him was that it would depend upon the properties of the membrane. Now that the thermodynamics of irreversible processes has been invented I could tell him a little more upon what properties it depends, and would probably find that none of them are as easy to measure as the effect of temperature difference on pressure.

Last year I met this physician again and was surprised to learn that he was still deeply grateful, not for what I had told him, but because a physical chemist had been willing to treat a physician with a medical problem like a fellow scientist with a scientific problem. I was the only physical chemist in the Boston area, perhaps in the country, who would do this in 1924. How changed things are today!

Of course it was this same provincialism of many physical chemists, not only toward physicians, physiologists and physiological chemists, but toward all colloid chemists, that drove Ostwald and others to isolationism.

In 1925 Jim Conant came back from Europe sold on the idea that the real molecular weights of proteins are 200-600 and that what was studied in aqueous solutions were colloidal aggregates. This sat badly with Edwin Cohn, who was then determining minimum molecular weights from the amounts of some rarer elements or amino acids. So they agreed to settle it experimentally. The most precise measurements in the world were being made in T. W. Richards' laboratory where Cohn had worked with Richards' brother-in-law during the war, and Conant was Richards' son-in-law. So they went to Richards' laboratories. They started with the freezing point depression of zein in phenol. For a while it was one of those researches which gave one answer on Monday, Wednesday and Friday and another on Tuesday, Thursday and Saturday. I saw them both frequently. On the days when I saw Conant the molecular weight seemed small, when I saw Cohn it seemed large, when I saw them both together the answer was indefinite. Then they decided that the phenol must be splitting water out of the protein. I remembered how Bury had treated a similar problem. They added

calcium chloride and its lowest hydrate to give a constant water activity, and they found no measurable depression. They published a paper concluding that the molecular weight of zein in phenol is certainly greater than 10,000 and probably greater than 100,000, and showing no indication that there had ever been any disagreement between them.

I have already told you how I became a colloid chemist the next year.

I have tried to show you how chemistry, colloid chemistry and research have changed during the last fifty years, and now I want to talk about the fact that fundamentally it stays the same not only in different times but also in different fields. My favorite quotation from Kipling is, "and the things that you learn from the yellow and brown, they'll 'elp you a lot with the white." During the last war I was working part time at the Harvard Medical School on the fractionation of plasma proteins and part time at Columbia on the fractionation of uranium isotopes—almost all colloid or surface chemistry but part time in another sense. One week the same general problem came up in both places and also from one of my former students who was working on the de-icing of airplane wings.

I told you that my first thinking about membranes was concerned with physiological ones. Next came the membranes for osmotic pressure measurements, then those for gaseous diffusion, next ion-exchanger membranes, and then physiological membranes again, but now with the biophysicists—the experts.

Not only are the membranes much alike, but the thinking of the different kinds of scientists considering them is even more so. If you mention a membrane to a physicist, he thinks first of what I call a diaphragm—a membrane with very small holes but with thickness much less than the radii of the holes. This may be right for some physiological membranes, but it is certainly not for those usually studied or for synthetic membranes. When told this, the physicist jumps to right circular cylindrical pores with length very much greater than the diameter. You can persuade him to make the cross-section elliptical, to bend the pores a little and even to let the diameter vary along the length. It is impossible, however, to make most people see that most real membranes are like sand piles, brush piles or sponges in that the channels are not isolated, but that each splits into two or three, each part joins up with one or two more and then splits again, and that this cycle is repeated in about the diameter of a channel. Much of the difficulty comes from the fact that no one has succeeded in drawing a two-dimensional picture of a sponge-like structure such that both phases appear continuous.

This difficulty has led to serious differences among the physiologists. Five years ago when Professor Ussing was over from Denmark, Professor Solomon had the idea we could settle this by talking; that Professor Ussing should present his side of the case and Professor Chinard of Maryland should present the other side one morning and that I, with the lunch hour to digest their arguments, should settle the case in the afternoon. Thinking that this would be a

Harvard-M.I.T. party, I accepted and gave the title "Tertium quid," which nobody understood. Now that Harvard is translating its diplomas from Latin into English, I will translate this as "A Third something," which doesn't make it much easier to understand. Maybe my title helped increase the crowd. Some of the audience came from across the continent. Each of us left believing just what he believed when he came, but everyone had a grand time.

And now for a word about the work which Dr. Stockmayer could not mention because it is not yet published, and some of it not yet completed. There is a paper in the Hildebrand 80th birthday issue of the Journal of Physical Chemistry on the Gibbs Adsorption Isotherm, part of which was presented before this division eleven years ago and part of which is what seems to me an appropriately placed defense of Willard Gibbs against Edwin Guggenheim.

In the next few months I hope to know considerably more about gelatin, about the binding of small ions to albumin and about ion exchange resins. None of these will provide great breakthroughs but I hope that, like my best earlier work, they will broaden the basis of our understanding. I realized long ago that mine is not the kind of work which brings a Nobel prize. I am surprised that it brought a Kendall Award, but I am very happy about it, and I thank you all again.

1963

WILLIAM A. ZISMAN

UNITED STATES NAVAL RESEARCH LABORATORY
WASHINGTON, D. C.

. . . for his outstanding contributions to understanding the wetting of solids by liquids. These include the retraction method for the isolation of adsorbed monolayers, basic studies of contact angle phenomena, an explanation of the nature of autophobic liquids, the concept of a critical surface tension, and the relation of these properties to surface constitution. Together they have systemized present knowledge, and predicted new and useful effects.

The following Kendall Award Address is reproduced by permission from *"Contact Angle, Wettability and Adhesion"*, Advances in Chemistry Series, **43**; American Chemical Society, Washington, D. C. 1964, p. 1.

Relation of the Equilibrium Contact Angle to Liquid and Solid Constitution

W. A. ZISMAN

U. S. Naval Research Laboratory
Washington 25, D. C.

A review of the author's investigations of the equilibrium contact angles of pure liquids on low- and high-energy solid surfaces, both bare and covered with a condensed monomolecular adsorbed film, includes the critical surface tension of wetting and the effect of homology on spreading by pure liquids, the causes of nonspreading on high-energy surfaces, and the existence and properties of autophobic liquids and oleophobic monolayers. Constitutive relationships are summarized in a table of critical surface tensions of wetting. The theory and application of the retraction method of preparing adsorbed monolayers from solution and the conditions for mixed films are presented. Studies of the wetting behavior of solutions of various surfactants and the resultant explanation of the function of a wetting agent are generalized to include nonaqueous systems. Following estimates of the reversible work of adhesion of liquids to solids, the part played by wetting in obtaining optimum adhesion by adhesives is outlined, and a fundamental explanation is given of constitutive effects in the development of strong adhesive joints. Future areas of research on wetting and adhesion are indicated.

In his classic investigation of capillarity, Laplace [76] explained the adhesion of liquids to solids in terms of central fields of force between the volume elements of a continuous medium. This approach was illuminating about the origin of surface tension and energy and their relation to the internal pressure, and it resulted in the fundamental differential equation of capillarity which has been the basis of all methods of measuring liquid surface tension. For nearly a century this essentially mechanical treatment was elaborated by Gauss, Neumann, Poisson, Kelvin, Rayleigh, van der Waals, and many others; Bakker [3] has summarized their results.

Eventually, the Laplace treatment was abandoned because: in the resulting differential and integral relations among the surface tension, internal pressure of the liquid, the density, and its gradient, the parameters defining the interparticle field of force were not experimentally obtainable; the theory was unsuccessful in explaining common phenomena, some of which are the concern of this symposium; and as the molecular structure of liquids and solids became better understood, the central field of force approach became recognized as an oversimplification which had to be replaced by an electromagnetic and wave mechanics description of intermolecular fields of force. Even today knowledge of the force field in the vicinity of the molecules of a liquid is not precise enough for such calculations—except possibly in the case of the liquefied rare gases. The status of the statistical mechanical treatment of the subject and the mathematic problem to be solved is well indicated in the review of Hirshfelder, Curtiss, and Bird [60].

Over 150 years ago Thomas Young [104] proposed treating the contact angle of a liquid as the result of the mechanical equilibrium of a drop resting on a plane solid surface under the action of three surface tensions (Figure 1)—γ_{LV} at the interface of the liquid and vapor phases, γ_{SL} at the interface of the solid and the liquid, and γ_{SV} at the interface of the solid and vapor. Hence,

$$\gamma_{SV} - \gamma_{SL} = \gamma_{LV} \cos \theta \tag{1}$$

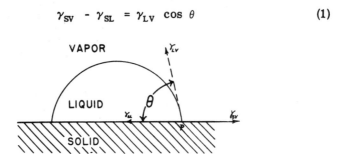

Figure 1. Contact angle of a sessile drop

The concept of the contact angle and its equilibrium was valuable because it gave a definition to the notion of wettability and indicated the surface parameters needing measurement. Today when we say that a liquid is nonspreading, we simply mean that $\theta \neq 0°$; and when the liquid wets the solid completely and spreads freely over the surface at a rate depending on the liquid viscosity and solid surface roughness, we say that $\theta = 0°$. A host of early experiments revealed that every liquid wets every solid to some extent—that is, $\theta \neq 180°$. Another way to express this point is that there is always some adhesion of any liquid to any solid. On a homogeneous solid surface, angle θ is independent of the volume of the liquid drop. Obviously, since the tendency for the liquid to spread increases as θ decreases, the contact angle is a useful inverse measure of spreadability or wettability.

Young's equation is deceptively simple; actually, there are present conceptual and experimental difficulties; and Equation 1 has been the source of many arguments. In the definition of γ_{SL} and γ_{SV}, neither of which we can conveniently and reliably measure, there is the difficulty that any tensile stresses existing in the surface of a solid would rarely be a system in equilibrium. Solids are rare whose surfaces are free of stresses which have penetrated from below the surface layer. Lester [77] has recently given a sophisticated treatment of Young's equation and has shown that it is correct so long as the drop of liquid rests on a solid which is not too deformable.

Another approach avoids specifying the field of intermolecular force between solid and liquid and instead resorts to thermodynamics. The first application of thermodynamics to capillarity appears to have been made by Thompson [101,102]; later came the classic and general treatment by J. Willard Gibbs [50]. Nearly 60 years had elapsed after Young's treatment before Dupré [31] introduced the reversible work of adhesion of liquid and solid, W_A, and its relation to γ_{SV} and γ_{SL} :

$$W_A = \gamma_{SV} + \gamma_{LV} - \gamma_{SL} \tag{2}$$

This equation is simply the thermodynamic expression of the fact that the reversible work of separating the liquid and solid phases must be equal to the change in the free energy of the system. Therefore, a correct derivation implies that the three terms on the right of Equation 2 are the nature of free energies per unit surface area of the solid-vapor, liquid-vapor, and solid-liquid interfaces, respectively.

As Sumner showed 25 years ago [99], the Young equation can also be derived thermodynamically for the ideal plane solid surface of Figure 1, provided that the system is treated as one in thermal and mechanical equilibrium and the quantities γ_{SL}, γ_{SV}, and γ_{LV} are defined as follows:

$$\gamma_{SL} = \left(\frac{\partial F}{\partial A_{SL}} \right) T, \mu_i$$

$$\gamma_{SV} = \left(\frac{\partial F}{\partial A_{SV}} \right) T, \mu_i \tag{3}$$

$$\gamma_{LV} = \left(\frac{\partial F}{\partial A_{LV}} \right) T, \mu_i$$

where F is the Helmholtz free energy (or the work function) of the system, A_{SV} is the area of the solid-vapor interface, etc., T is the temperature, and μ_i is the chemical potential of each component in the phases present. Implicit in this treatment, and also in Young's derivation, is the assumption that the contact angle is independent of the volume of the drop and depends only on the temperature and the nature of the liquid, solid, and vapor phases in contact. Later investigators have given more general thermodynamic derivations of the Young and Dupré equations, most noteworthy being those by Shuttleworth and Bailey [98] and Johnson [64].

It was not until 1937 that Bangham and Razouk [4,5] called attention to the importance of not neglecting the adsorption of vapor on the

111

surface of the solid phase in deriving the equilibrium relations concerning the contact angle; and they were the first to derive the following widely used forms of the Young and Dupré equations. Here the more precise system of subscripts due to Boyd and Livingston [25] is used in order to distinguish between the solid-vacuum and solid-liquid interfaces. Thus, γ_{S° is the free energy at the solid-vacuum interface, γ_{SV° the corresponding term for the interface of the solid with the saturated vapor, and γ_{LV° that for the interface of the liquid with the saturated vapor.

$$\gamma_{SV^\circ} - \gamma_{SL} = \gamma_{LV^\circ} \cos \theta \tag{4a}$$

$$W_A = \gamma_{S^\circ} + \gamma_{LV^\circ} - \gamma_{SL} \tag{4b}$$

and hence,

$$W_A = \left(\gamma_{S^\circ} - \gamma_{SV^\circ}\right) + \gamma_{LV^\circ}(1 + \cos \theta) \tag{4c}$$

Most textbooks neglect the first term on the right side of Equation 4c. Obviously, the quantity

$$W_{A*} \equiv \gamma_{LV^\circ}(1 + \cos \theta) \tag{4d}$$

is the reversible work of adhesion of the liquid to the solid when coated with an adsorbed film of the saturated vapor. The first parenthetical term in Equation 4c is simply the free energy decrease on immersion of the solid in the saturated vapor phase; for it Bangham and Razouk derived the following expression when the vapor obeys the ideal gas law:

$$\gamma_{S^\circ} - \gamma_{SV^\circ} = RT \int_0^{p_o} \Gamma d(\ln p) \tag{5}$$

where p_o is the pressure of the saturated vapor, R the gas constant, T the absolute temperature, and Γ the Gibbs absorption excess per unit area of the vapor on the solid.

Two investigators of the spreading of insecticides on leaves, Cooper and Nuttall [29], were the originators of the well-known condition for the spreading of a liquid substance, b, on a solid or liquid substance, a:

For spreading $S > 0$

For nonspreading $S \leq 0$

where

$$S = \gamma_a - \left(\gamma_{ab} + \gamma_b\right) \tag{6}$$

or using the subscripts just introduced,

$$S = \gamma_{SV^\circ} - \left(\gamma_{SL} + \gamma_{LV^\circ}\right) \tag{7}$$

Harkins soon afterwards [55-58] developed fully the usefulness of their concept, named S the "initial spreading coefficient," and from it derived the two relations

$$S = W_A - W_C \qquad (8)$$

and

$$S = -\left(\frac{\partial F}{\partial A}\right)_{T, \, \mu_i} \qquad (9)$$

Here W_C is the reversible work of cohesion of the liquid; from the Dupré equation for a liquid-liquid interface it is simple to show that W_c is twice the liquid surface tension. Equations 8 and 9 are especially suggestive about the physical cause of spreading; however, like Equation 1, they are deceptively simple. As Harkins pointed out, an "initial value" of the spreading coefficient exists for the condition that spreading can initiate; a "final coefficient" exists for the conditions that once spreading has occurred the liquid can remain spread. It turns out that much experimental information is needed to determine the final spreading coefficient.

Assuming that no surface electrification is involved, the above group of equations are the basic thermodynamic relations for describing the equilibrium contact angle and wetting phenomena. In so far as details of molecular structure of the substances and surfaces play an important part, these purely thermodynamic equations would not be expected to suffice to permit us to describe the wetting, spreading, and adhesion of liquids on solids.

Despite many attempts, little was learned about the constitutive aspects of the wetting and spreading of liquids on solids until the past two decades. Practically every investigator was engulfed in the difficulties of obtaining reproducible and significant contact angles. The oldest experimental difficulty, and the source of many controversies, was the occurrence of large differences between the contact angle, θ_A, observed in advancing the liquid boundary over a dry clean surface and the value, θ_R, observed in receding the liquid boundary over the previously wetted surface. There was much concern until very recently about which contact angle was more significant, and if both were significant, what function of the two was useful.

Some clarification of this problem resulted after Wenzel [103] developed a relation between the macroscopic roughness of a solid surface and the contact angle. Wenzel discussed the roughness factor, r (defined as the ratio of the true area of the solid to the apparent area or envelope), and its relation to the apparent or measured contact angle, θ', between the liquid and the envelope to the surface of the solid and to the true contact angle, θ, between the liquid and the surface at the air-liquid-solid contact boundary. He derived the well known relation

$$r = \frac{\cos \theta'}{\cos \theta} \qquad (10)$$

from the Young equation and from the definition of r; hence, Equation 10 is essentially a thermodynamic requirement. This relation is important, because surfaces having r = 1.00 rarely are encountered; perhaps the nearest to such a smooth surface is that of freshly fire-polished glass or freshly cleaved mica; usually r is significantly greater than 1.0. Wenzel's equation has been derived more generally and applied to woven and other regular structures by Cassie and Baxter [28] and by Shuttleworth and Bailey [98].

Several general consequences of Wenzel's equation should influence all research on contact angles. First, when $\theta < 90°$, Equation 10 indicates $\theta' < \theta$. But most organic liquids exhibit contact angles of less than 90° on clean polished metals; hence, the effect of roughening the metals is to make the apparent contact angle, θ', between the drop and the envelope to the metal surface less than the true contact angle, θ. In other words, each liquid will appear to spread more when the metal is roughened. Secondly, when $\theta > 90°$, Equation 10 indicates $\theta' > \theta$. Since pure water makes a contact angle of 105° to 110° with a smooth paraffin surface, the effect of roughening the surface tends to make θ' greater than 110°; θ' values of 140° have been observed. Thirdly, the experimental problem of accurately measuring the true contact angle is made difficult by the surface roughness, and Wenzel's equation makes it possible to estimate the resulting error. When $\theta = 10°$, the difference, $\theta - \theta'$, between the real and apparent angles will be 5° if r = 1.02. When $\theta = 45°$, the same 5° difference between θ and θ' will occur when r = 1.1. When $\theta = 80°$, the 5° difference in θ and θ' will occur when r = 2.0. This means that, in order to measure small contact angles accurately, the surface used must be much smoother than when large contact angles are measured. Unfortunately, this requirement has rarely been given sufficient attention.

Langmuir's Observations

Langmuir's investigations [71] had a profound influence on all research concerned with surface properties of solids and liquids. In his research prior to 1916 on the adsorption of gases and solids under high vacuum conditions, he had found good evidence that the major changes in the surface properties of solids had occurred with the adsorption of a monomolecular layer. He also had reasoned from the early x-ray findings about the structure of solids that the forces causing adsorption originated from the uncompensated field emanating from the atoms in the surface and that usually this adsorptive field of force was the residual electrical field of the valence electrons belonging to the surface atoms. In view of these conclusions and the fact, well known to chemists, that the fields of force giving rise to secondary valences in a compound are so localized that, for reaction to occur, the contact of atoms was necessary, he stated in 1916 [72] that: (i) such short-range force fields are responsible for nearly all types of adsorption, and (ii) a solid or liquid surface should have its adsorbing properties completely altered when covered by one layer of foreign atoms or molecules.

Langmuir's conclusion that the forces between molecules and the adsorbing surface come into play only at the immediate area of contact made predicting surface interactions much simpler than trying to compute quantities of interest through a precise knowledge of the central field of force between all of the molecules in the solid or liquid. In effect, he had discovered a convenient approximate method for investigating the constitutional aspects of adsorption as well as other surface properties. Langmuir later concluded that the adsorptive properties of the surface were determined essentially by the nature and packing of the atoms or groups of atoms in the surface of a solid or liquid, and he often referred to this concept as "the principle of independent surface action." Many years later, because he realized that its theoretical foundation might receive firmer support through use of the more recent scientific developments, he derived in a less widely known paper a limited justification for this principle [70].

Langmuir reported [75] in his famous 1919 lecture to the Faraday Society that an adsorbed monolayer of an organic polar compound could radically change the frictional and wetting properties of solid surfaces. He also emphasized the need for developing experimental methods to study oil films of solids, especially their adhesion and the effects of the resulting oil film on the contact angle with water. A method was described for depositing a condensed monolayer of oleic acid from its position as a compressed film on the surface of water, so that it would adsorb on a clean glass solid as it was withdrawn edgewise through the floating film. Such a monolayer always made glass and many other clean solids act hydrophobic and also lowered the coefficient of friction to only 0.1. Smooth clean surfaces of stearic acid, paraffin wax, myristyl alcohol, and cetyl palmitate exhibited large hydrophobic contact angles, the value for the last three substances being 110°. Langmuir expressed some surprise because the contact angles with the various waxy surfaces were not independent of the nature of the underlying solid; he had expected that they would depend only on the nature of the outermost hydrocarbon groups and so would be similar.

In two later addresses [73,74] Langmuir added other highly significant observations about the effects of adsorbed films on the wettability of solids. A trimolecular stearate film (prepared by the Langmuir-Blodgett technique) exhibited a contact angle of 55.4° with a white mineral oil, 51.7° with ethyl myristate, 48.7° with carbon tetrachloride, 48° with benzene, 1.5° with n-hexane, and 50° with water. The contact angle of this oil with barium stearate multilayers varied little with the number of monolayers in the film—e.g., it was 52° on a monolayer, 55.4° on three layers, and 55.9° on seven layers.

Langmuir offered the following highly suggestive explanation of why the mineral oil rolled off these monolayers: "The probable explanation is that the molecules are so tightly packed into an area of about 20 sq. A. per molecule that only the CH_3 groups at the end of the molecules are exposed on the surface. The properties of CH_3 may well be so different from CH_2 that a liquid consisting mostly of CH_2 does not wet a surface consisting entirely of CH_3." He also reported that as a drop of cetyl alcohol moved over the clean dry surface of glass, a monolayer of alcohol was left which the liquid could not wet. The following interesting observations were added later: "Stearic acid, however, when melted on the glass surface forms drops which show a large contact angle against the glass (oleic acid on glass gives a practically zero contact angle)," and also: "Molten stearic acid on chromium draws up into drops and leaves on the metal a monolayer which has a contact angle for water ($\theta > 90°$) and about 30° for petrolatum." These were the earliest reliable observations of adsorbed films which were hydrophobic and oleophobic.

Retraction Method and Wetting Properties of Resulting Films

Following an observation in 1941 that dilute solutions of pure heptadecylamine in white mineral oil exhibited considerable contact angles on the glass walls of the containing flask, Pickett and I [23] found that a monolayer of heptadecylamine could be adsorbed from solution on a polished clean solid glass slide or flat metal surface and also that the resulting coated surface could be slowly removed with the plane held vertically (see Figure 2) so that the solution was not transported along with it; the same phenomenon resulted when the heptadecylamine was adsorbed from any of a variety of nonpolar solvents. Any of numerous types of paraffinic polar compounds could be adsorbed and isolated from solution in the same way as the heptadecylamine.

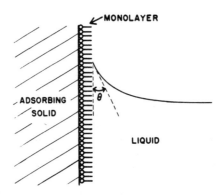

Figure 2. Retraction method

Later using a multiple dip method, Bigelow, Pickett, and I [23] adsorbed octadecylamine on a polished platinum foil dipper from a dilute solution in dicyclohexyl and proved that the average cross-sectional area per adsorbed molecule of the retracted film was 30 sq. A. and hence that the film was a condensed monolayer. In general, in order that such films could form, we found that the polar group had to be located at one extremity of the solute molecule with one or more methyl groups located at the opposite extremity and that the solute molecules must adsorb on the smooth solid surface with sufficient closeness of packing so that the outermost portion of the film is densely populated with methyl groups. A molecular configuration like a long rod or flat plate with a polar group attached to one end of the rod or the rim of the plate, and one or more methyl groups attached to the other end, satisfied these requirements. At my suggestion during World War II, Brockway and Karle examined retracted monolayers by electron diffraction [27,67] and found that the paraffinic polar compounds, octadecylamine and stearic acid, were oriented essentially along the normal to the solid surface with a random tilt of several degrees in the principal axis. These findings were confirmed and studied further by Bigelow and Brockway [21] and also by Menter and Tabor [83]. Further analysis of the data led Epstein [38] to suggest that the adsorbed polar molecules were clustered into two-dimensional, micelle-like, brush heaps.

In a subsequent study of the effect of temperature on the retraction and wetting process with Bigelow and Glass [22], we found a method by which the energy of adsorption of the polar molecule could be estimated and obtained values of from 10 to 14 kcal. per mole for stearyl derivatives; these obviously corresponded to a physical adsorption process. Furthermore, the molar energy of adsorption increased linearly with the number of carbon atoms in the polar molecule, and the energy increment per carbon atom was in reasonable accord with estimates of the energy of intermolecular cohesion per CH_2 group in adlineated (or crystalline) paraffinic compounds. An unexpected observation was that a wide variety of such polar paraffinic compounds could be adsorbed on smooth clean glass and metals by retraction from the molten compound; this made available an experimental method for preparing by retraction adsorbed condensed monolayers of the pure compound. Since this technique avoided any possibility that solvent molecules could remain trapped in the monolayer, it became a valuable method which was frequently used in subsequent comparisons of the properties of monolayers on solids.

Measurements of the wetting properties of various polished surfaces coated by retraction with condensed monolayers proved illuminating. Like Langmuir's observations on barium stearate monolayers and multilayers, these films were found to be both hydrophobic and oleophobic. A variety of organic liquids besides water and mineral oil were found unable to spread on such uncoated surfaces, and the contact angles exhibited were highly reproducible and independent of the nature of the solid substrate upon which a monolayer had been coated during the retraction process. Our results on the surface properties of these monolayers and the process by which they were produced obviously deserved attention because the retracted films were obtained under conditions of adsorption equilibrium at the solid-liquid interface; such films could and did occur in the arts and technology, whereas the Langmuir-Blodgett multilayers [71] are produced only by their one method; they do not occur at the solid-solution interface, and so despite their interesting properties they appear to be artifacts.

The large and reproducible contact angles observed on these retracted monolayers stimulated us to investigate if similar films could be adsorbed and retracted from solutions of various polar-nonpolar compounds in water [93]. In these and later experiments a simple platinum foil chimney was used to prevent inadvertently picking up by the Langmuir-Blodgett process any undissolved floating film-forming solute or contaminate [2,84,92]. It had been known for a long time that metals, glass, and minerals adsorbed a film when immersed in an aqueous solution of a polar adsorptive compound containing a hydrophobic group or radical such that upon removing the solid the aqueous solution would roll off the surface exhibiting a large contact angle. Our experiments confirmed our suspicion that usually the films retracted from aqueous solution are monomolecular and, when the solute concentration is not too low or the pH not inappropriate, are in the condensed state. Since water has such a high surface tension, it was not unexpected to find that by retraction from aqueous solutions one can isolate monolayers of hydrophobic polar compounds having the greatest variety in molecular structures [43,92].

Various experimental studies [2,78,92] revealed that the contact angle of any liquid on a condensed monolayer adsorbed on a polished solid surface was generally independent of whether it had been retracted from an aqueous or nonaqueous solution. For example, Table I summarizes the results obtained with Levine [78] on the wettability of a solid coated by using a variety of techniques to adsorb and retract a condensed monolayer of n-octadecylamine on platinum, stainless steel, and borosilicate glass. The identical packing of methyl groups in the condensed monolayer formed under each condition is demonstrated by the nearly constant value of the contact angle exhibited by methylene iodide (the maximum variation in the contact angle of 2° is the experimental uncertainty in our contact angle measurements).

Table I. Effect of Method of Film Preparation on
Wettability of Octadecylamine Monolayers

Films Prepared by	Methylene Iodide Contact Angle, °
Retraction from molten amine	69
Vapor phase adsorption	69
Retraction from hexadecane soln.	70
Retraction from dicyclohexyl soln.	68
Retraction from nitromethane soln.	69
Retraction from aqueous soln. of $C_{18}H_{37}NH_3Cl$	69

The interesting observation was made [2,92,93] on long-chain paraffinic films retracted from aqueous solution, that the contact angle of water on the resulting condensed film was 90° when the drop was advancing or receding! When the same compound was retracted from a nonaqueous solution in a nonpolar liquid, the advancing contact angle was 101° and the receding contact angle was 90° (see Table II). Water is a nearly unique liquid, in that it readily permeates between the long, adlineated, hydrocarbon chains of a close-packed monolayer of a fatty

Table II. Comparison of Hydrophobic Films of Amines

		Retracted Films Prepared from:					
		Molten Compound				Hexadecane Solution	Water Solution
n-Alkylamine	N^a	Thermal-Gradient Method[b]		Isothermal Method[b]		Oleophobic Method[b]	Hydro-phobic Method
		$\theta_A,°$	$\theta_R,°$	$\theta_A,°$	$\theta_R,°$	$\theta_A,°$	$\theta_R,°$	$\theta_A = \theta_R,°$
Butylamine	4	55	48	55	51	--	--	52[c]
Octylamine	8	81	67	74	69	73	68	69[c]
Dodecylamine	12	90	83	89	83	89	83	85[c]
Tetradecylamine	14	91	84	92	87	90	86	86[c]
Hexadecylamine	16	96	87	96	89	96	89	89[c]
Octadecylamine	18	102	89	102	91	101	90	90[b]

[a]Total number of carbon atoms per molecule.
[b]Measurements made at 20.0° ± 0.1°C.
[c]Measurements made at 25° ± 1°C.

acid, alcohol, or primary amine. Therefore, when the film was adsorbed from aqueous solution, it was saturated with water; hence the drop of water was moving over a water-saturated surface regardless of whether it was advancing or receding, and therefore θ_A and θ_R had to be equal. On one hand, because the monolayer adsorbed from the non-aqueous liquid was devoid of solvent molecules (including water), the drop of water was advancing over an anhydrous film and so the advancing contact angle had the higher value of 101°. On the other hand, the receding drop was moving over a surface which had become saturated with water abstracted by vapor transfer from the water drop during its prior advance, and it should have the same value as the water-soaked film retracted for the aqueous solution. The condition that the advancing and receding contact angles are different on condensed organic monolayers is unusual, and it occurs with a water drop simply because of the great permeability of condensed monolayers to molecularly dispersed water. When the liquid drop is not water, liquid molecules are usually too large to permeate into the condensed monolayer; hence $\theta_A = \theta_R$.

Especially significant about these early results was the wealth of reliable experimental evidence revealing the condition when contact angles were reproducible and interpretable in terms of the structure and composition of the surface phases. Ample justification existed for broadening the range of solid surfaces studied by means of the equilibrium contact angle.

Wetting of Low-Energy Solid Surfaces

In considering the wetting properties of solid surfaces, Fox and I found it helpful to coin a few convenient terms to identify the extremes of the specific surface free energies of solids [46]. As is well known, the specific surface free energies of liquids (excluding the liquid metals) are less than 100 ergs per sq. cm. at ordinary temperatures. But hard solids have surface free energies ranging from about 5000 to 500 ergs per sq. cm., the value being higher the greater the hardness and the higher the melting point. Examples are the ordinary metals, metal oxides, nitrides, and sulfides, silica, glass, ruby, and diamond. Soft organic solids have much lower melting points and the specific surface free energies are generally under 100 ergs per sq. cm. Examples are waxes, solid organic polymers, and in fact, most solid organic compounds. Solids having high specific surface free energies may be said to have "high-energy surfaces," and solids having low specific surface free energies have "low-energy surfaces." This terminology has since been widely adopted.

Because of the comparatively low specific surface free energies of organic and most inorganic liquids, one would expect them to spread freely on solids of high surface energy, since there would result a large decrease in the surface free energy of the system, and this is most often found to be true. But since the surface free energies of such liquids are comparable to those of low-energy solids, among these liquids should be found those exhibiting nonspreading on low-energy solids.

Our previous work on the retraction of monolayers from organic liquids and their oleophobic properties led us to propose that when any organic liquid would not spread on a high-energy surface, it did so because it contained a dissolved polar-nonpolar compound from which an oleophobic film had adsorbed on the surface. However, measurements on organic liquids from which polar-nonpolar adsorbable impurities had been carefully removed demonstrated that pure liquids like tri-o-cresyl phosphate and benzyl phenylundecanoate would not spread on high-energy surfaces. Because of these unexpected and at the time inexplicable results, a temporary detour was made to seek the missing information by studying the spreading of pure liquids on well defined low-energy surfaces.

Equilibrium contact angles of a variety of pure liquids were studied with Fox [46,47,48] and later Ellison [33,34] on surfaces of solid organic polymers free of contaminants, monomer, or plasticizers—examples are polytetrafluoroethylene, polyethylene, poly(vinyl chloride), poly(ethylene terephthalate), etc. Because of the large percentage error in low contact angles resulting from the surface roughness, great care was exercised in preparing the surfaces of these polymers in extremely clean and glossy-smooth condition. Each liquid used in observing contact angles on such surfaces was percolated slowly through columns packed with adsorbents to remove adsorptive contaminants. Using such solids and liquids it was found, in disagreement with past reports on contact angle phenomena, that these systems behaved reproducibly; furthermore, the advancing and receding contact angles were identical so long as the liquid drop was advancing or receding sufficiently slowly to be reasonably close to an equilibrium condition.

A comparison was made between the results of measurements of the contact angles of various volatile liquids on polytetrafluoroethylene when measured in air saturated with the liquid vapor and when measured in the open air. No significant differences were found in the contact angles of the n-alkanes until pentane or lower boiling homologs were used. Differences became significant in the series of dimethyl silicones only when homologs below the tetramer were used. This means that the adsorptivity of these vapors on polytetrafluoroethylene was so low at ordinary temperatures that the condensed vapor did not significantly affect the spreading of the liquid drop on the solid. Therefore, so long as attention was confined to measurements of the contact angles of the high boiling liquids on this and other low-energy surfaces, the measurements could be made in the open air rather than in a saturated atmosphere of the liquid vapor.

In general, a rectilinear relation was established empirically between the cosine of the contact angle, θ, and the surface tension, γ_{LV°, for each homologous series of organic liquids. Figure 3 illustrates the results with the n-alkanes on polytetrafluoroethylene [46]. The critical surface tension for wetting by each homologous series was defined by the intercept of the horizontal line $\cos \theta = 1$ with the extrapolated straight-line plot $\cos \theta$ vs. γ_{LV°, and it was denoted by γ_c. This intercept was found more valuable than the slope of the rectilinear graph for correlations between wettability and constitution. Even when $\cos \theta$ was plotted against γ_{LV° for a variety of nonhomologous liquids, the graphical points fell close to a straight line or collected around it in a narrow rectilinear band (see Figures 4 and 5). Certain low-energy surfaces, such as on polytetrafluoroethylene (Figures 6 and 7), exhibit curvature of this band for values of γ_{LV° above 50 dynes per cm. But

120

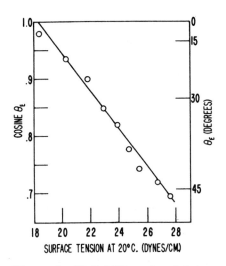

Figure 3. Wettability of polytetra-fluoroethylene by the n-alkanes [46]

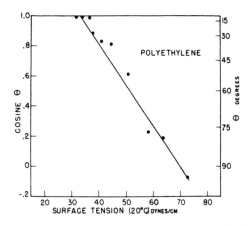

Figure 4. Wettability of polyethylene [48]

in those cases we found that the curvature results because weak hydrogen bonds form between the molecules of liquid and those in the solid surface. This is most likely to happen with liquids of high surface tension, because they are always hydrogen-bonding liquids. In general, the graph of $\cos \theta$ vs. γ_{LV° for any low-energy surface is always a straight line (or a narrow rectilinear band) as in Figure 4, unless the molecules in the solid surface form hydrogen bonds.

When rectilinear bands are obtained in this type of graph, the intercept of the lower limb of the band at $\cos \theta = 1$ is chosen as the critical surface tension, γ_c, of the solid. Although this intercept is less precisely defined than the critical surface tension of a homologous

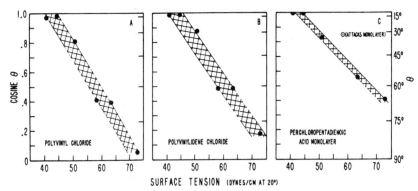

Figure 5.   Wettability by various liquids on surface of:

A. Poly(vinyl chloride)
B. Poly(vinylidene chloride)
C. Close-packed monolayer of perchloropentadienoic acid [33]

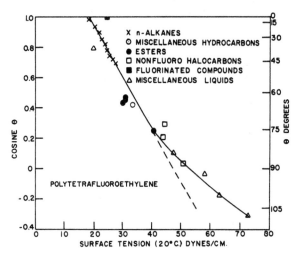

*Figure 6.   Wettability of polytetrafluoroethylene
by various liquids [47]*

series of liquids, nevertheless it is an even more useful parameter because it is a characteristic of the solid surface. It has proved to be a
useful empirical parameter whose relative values act as one would
expect of γ_{S°, the specific surface free energies of the solid. The
widespread occurrence of the rectilinear relationship between cos θ
and γ_{LV° in the rapidly growing body of reliable experimental data led
us to use γ_c to characterize and compare the wettabilities of a variety
of low-energy surfaces.

By comparing values of γ_c of structurally "homologous" or "analogous" solids, such as unbranched polyethylene and its various chlorinated

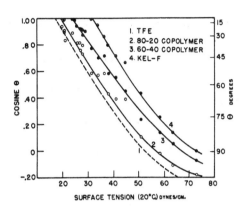

Figure 7. Wettability of copolymers of polytetrafluoroethylene and poly-chlorotrifluoroethylene [47]

or fluorinated analogs (see Table III) and by making the usually reasonable assumption that the surface composition of the solid polymer was the same as that of the horizontally oriented polymer molecule, it was possible to measure the effect of surface constitution on the wettability.

In the upper curve of Figure 8, γ_c values for polyethylene, poly-(vinyl chloride), and poly(vinylidene chloride) are plotted against the atom per cent replacement of hydrogen by chlorine. Although the introduction of the first chlorine atom in the monomer causes γ_c to rise from 31 to 39 dynes per cm., the addition of a second chlorine increases γ_c only to 40 dynes per cm. On comparing the upper and lower curves, striking differences are evident in the effect on γ_c of chlorine or fluorine replacement of hydrogen, both as to the direction of the change and the effect of progressive halogenation. Although polytetrachloroethylene does not exist, an organic coating with an outermost surface comprised of close-packed covalent chlorine atoms was prepared [33] by the retraction method to form a condensed, adsorbed, and oriented monolayer of perchloropentadienoic acid ($CCl_2 = CCl\text{-}CCl = CCl\text{-}COOH$) on the clean polished surface of glass or platinum. Not only is the graph of $\cos \theta$ vs. γ_{LV^o} for such a surface similar to those of the above-mentioned chlorinated polyethylenes, but the corresponding value of γ_c (43 dynes per cm.) is shifted in the appropriate direction—i.e., to higher values of γ_c. Extrapolation of the line defined by the experimental points for the two chlorinated polymers in Figure 8 to the value of γ_c for 100% hydrogen replacement indicates a value of 42 dynes per cm. Thus, the hypothetical polytetrachloroethylene surface should have a critical surface tension of wetting of 42 dynes per cm., which is only 1 dyne per cm. less than the experimental value found for the perchloropentadienoic acid monolayer. This shows how closely the wetting properties of the latter surface approximate those of a fully chlorinated polymeric solid surface.

Since the same results were obtained regardless of the nature of the polished solid substrate on which the perchloropentadienoic acid monolayer was adsorbed, it was evident that the wettability of these monolayer coated surfaces is determined by the nature and packing of

Table III. Critical Surface Tensions of Halogenated Polyethylenes [33]

Polymer	Structural Formula	Critical Surface Tension, Dynes/Cm.
Poly(vinylidene chloride)	H Cl H Cl H Cl -C-C -C-C -C-C H Cl H Cl H Cl	40
Poly(vinyl chloride)	H Cl H Cl H Cl -C-C -C-C -C-C H H H H H H	39
Polyethylene	H H H H H H -C-C -C-C -C-C H H H H H H	31
Poly(vinyl fluoride)	H F H F H F -C-C -C-C -C-C H H H H H H	28
Poly(vinylidene fluoride)	H F H F H F -C-C -C-C -C-C H F H F H F	25
Polytrifluoroethylene	F F F F F F -C-C -C-C -C-C H F H F H F	22
Polytetrafluoroethylene (Teflon)	F F F F F F -C-C -C-C -C-C F F F F F F	18

the outermost atom in exposed groups of atoms in the coating and not by the nature and arrangements of atoms in the solid substrate 10 to 20A. below the surface layer. This will exemplify the extreme localization of the fields of force of covalent bonded atoms responsible for the adhesion of liquids to organic solids.

In the upper curve of Figure 9 is plotted cos θ for each of the n-alkane liquids on a close-packed retracted monolayer of prim-octadecylamine on platinum or glass [94]. Comparable data for the surface of polyethylene cannot be shown here because the alkane liquids exhibit zero contact angles on this solid. In other words, the graph is so much above that of octadecylamine that it cannot be shown in this plot. The second curve gives the analogous results for polytetrafluoroethylene [46]. By structural analogy, one can reason that since the surface of close-packed -CH_3 groups is much less wettable than one of -CH_2- groups, a surface of -CF_3 groups should also be much less wettable than one of -CF_2- groups. This simple argument led Schulman and me [91] to study the wettability of close-packed adsorbed films of

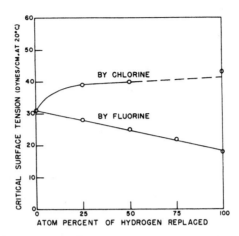

Figure 8. Effect of progressive halo-gen substitution on wettability of poly-ethylene-type surfaces [96]

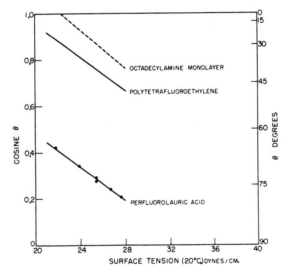

Figure 9. Comparison of effects of -CH₃, -CH₂-, and -CF₃ groups on wettability by n-alkanes [105]

perfluorodecanoic (or φ-decanoic) acid. This acid had only recently been prepared by the Simon's process of electrochemical fluorination. Soon afterward a study was made with Shafrin and Hare [53] of the homologous family of perfluoro fatty acids, and we then found that a condensed monolayer of perfluorolauric acid, $F_3C (CF_2)_{10} COOH$, was the most nonwettable surface ever reported; on it every liquid we studied was unable to spread. Figure 10 shows that very large contact angles

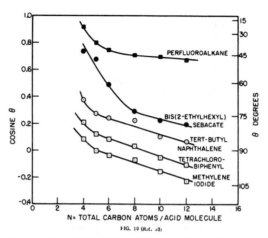

Figure 10. Wettability of condensed monolayers of perfluoroalkanoic acids by various
liquids [53]

are exhibited by common types of organic liquids upon retracted monolayers of any of the perfluoroalkanoic acids.

Since Berry [18] had succeeded in preparing the interesting and
related ψ-alkanoic acids

$$HF_2C(CF_2)_nCOOH$$

we investigated their behavior as condensed monolayers adsorbed by
retraction on polished platinum foil [32]. As expected, such coated surfaces also exhibited large contact angles with all liquids. Figure 11

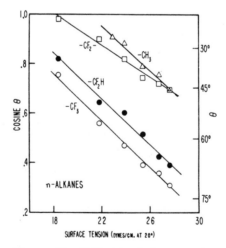

Figure 11. Effect of progressive
fluorination of ω-CH_3 group on wettability by n-alkanes [32]

compares the cos θ vs. γ_{LV} graphs for the n-alkanes on surfaces coated with close-packed $-CF_3$, $-CF_2H$, $-CF_2-$, and $-CH_3$ groups. Just as the hydrogen-donating liquids—water, glycerol, glycol, and formamide—all formed weak hydrogen bonds with the fluorine atoms in the surface of polytetrafluoroethylene, the same effect occurred in wetting surfaces covered with monolayers having outermost $-CF_3$ and $-CF_2H$ groups [32,53,91].

Our results on the wettability of surfaces covered with highly fluorine-substituted alkyl groups stimulated several research laboratories to apply these surface properties to polymeric coating materials for textile fibers and fabrics as a means of imparting to them nonstaining, oil-, and water-resistant properties. Such products are becoming very prominent today and the contact angle is used for product control and trade specifications [1].

Regularities in the cos θ vs. γ_{LV} graphs of various fluorine-rich surfaces are illustrated in Figure 12. From the cos θ = 1 intercepts

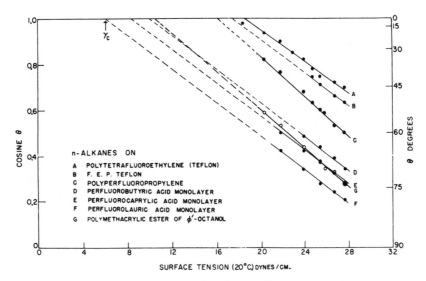

Figure 12. Contact angles for n-alkanes on various fluorinated surfaces [17,96]

on Figure 12, it is evident that γ_c has a value of about 18.5 dynes per cm. for the n-alkanes on the surface of Teflon. Values of about 17 and 15 dynes per cm. are obtained from curves B and C—i.e., the introduction of the perfluoromethyl group as a side chain in the polymer reduces γ_c, the reduction becoming greater the higher the surface concentration of exposed $-CF_3$ groups. An adsorbed, close-packed monolayer of a perfluoro fatty acid (curves D, E, and F) is an example of such a surface. The values of γ_c for such surfaces are, therefore, much lower than for surfaces comprised only of $-CF_2-$ groups. The closer the packing of the aliphatic chains of the adsorbed molecules, the closer the packing of the exposed terminal $-CF_3$ groups, and hence the lower γ_c. Thus, the value for a condensed monolayer of perfluorolauric acid (curve F) is only 6 dynes per cm., and this is the lowest value yet encountered; only the condensed inert gases could spread on such a surface. The value of 10.6 dynes per cm. for the polymethacrylic ester of perfluorooctanol (curve G) is the lowest encountered with any solid polymer [17].

Results to date of wettability studies on clean, smooth, plasticizer-free, polymeric solids of general interest are summarized in Table IV. Included in this table is the value for poly(vinyl alcohol) (γ_c = 37 dynes per cm.) reported by Ray, Anderson, and Scholz [88]; the same investigators also found a range in γ_c of 40 to 45 dynes per cm. for a series

Table IV. Critical Surface Tensions of Various Polymeric Solids [106]

Polymeric Solid	γ_c, Dynes/Cm. at 20°C.	Ref.
Polymethacrylic ester of ϕ'-octanol	10.6	[17]
Polyhexafluoropropylene	16.2	[16]
Polytetrafluoroethylene	18.5	[46]
Polytrifluoroethylene	22	[33]
Poly(vinylidene fluoride)	25	[33]
Poly(vinyl fluoride)	28	[33]
Polyethylene	31	[48]
Polytrifluorochloroethylene	31	[47]
Polystyrene	33	[34]
Poly(vinyl alcohol)	37	[88]
Poly(methyl methacrylate)	39	[61]
Poly(vinyl chloride)	39	[33]
Poly(vinylidene chloride)	40	[33]
Poly(ethylene terephthalate)	43	[34]
Poly(hexamethylene adipamide)	46	[34]

of hydroxyl-rich surfaces of the starch polymer type [90]. These values of γ_c are reasonably close to that of 43 dynes per cm. reported for the oxygen-rich surface of poly(ethylene terephthalate). Among the early reliable studies of contact angle vs. surface tension for smooth surfaces of various waxes, resins, and cellulose derivatives were those reported by Bartell and Zuidema [8]. If the cosines of their contact angles are plotted against γ_{LV°, good straight lines are obtained. The values of γ_c for their resin surfaces rich in exposed oxygen-containing groups fit in well with the data presented here on the relative wettability of oxygen-rich surfaces. Nylon, with its many exposed amide groups, has the highest γ_c value of the common plastics we have reported [34]. Since γ_c for all the polymers of Table IV are well below the surface tension of water (72.8 dynes per cm.), all are hydrophobic.

Effect of Constitution on Wetting of Low-Energy Surfaces

The widespread occurrence of the rectilinear relationship between cos θ and γ_{LV} in the large body of experimental data and the fact that these straight lines diverge away from the cos θ = 1 axis have made it possible to use γ_c to characterize the wettability of each low-energy surface. In Table V are presented the results of values of γ_c obtained from some comparative studies [96,105] of the wettability of a number of well-defined, low-energy, solid surfaces. In the first column is

Table V. Critical Surface Tensions of Low-Energy Surfaces [96]

Surface Constitution	γ_c, Dynes/Cm. at 20°	Ref.
A. Fluorocarbon Surfaces		
-CF$_3$	6	[53]
-CF$_2$H	15	[32]
-CF$_3$ and -CF$_2$-	17	[15, 16]
-CF$_2$-	18	[46]
-CH$_2$-CF$_3$	20	[95]
-CF$_2$-CFH-	22	[33]
-CF$_2$-CH$_2$-	25	[33]
-CFH-CH$_2$-	28	[33]
B. Hydrocarbon Surfaces		
-CH$_3$ (crystal)	22	[48]
-CH$_3$ (monolayer)	24	[94]
-CH$_2$-	31	[48]
-CH$_2$- and ⸺CH⸺	33	[43]
⸺CH⸺ (phenyl ring edge)	35	[43]
C. Chlorocarbon Surfaces		
-CClH-CH$_2$-	39	[33]
-CCl$_2$-CH$_2$-	40	[33]
=CCl$_2$	43	[33]
D. Nitrated Hydrocarbon Surfaces [45]		
-CH$_2$ONO$_2$ (crystal) [110]	40	[45]
-C(NO$_2$)$_3$ (monolayer)	42	[45]
-CH$_2$NHNO$_2$ (crystal)	44	[45]
-CH$_2$ONO$_2$ (crystal) [101]	45	[45]

given the constitution of the atoms or organic radicals in the solid surface arranged in the order of increasing values of γ_c. Literature references are given in the third column. Data have been grouped under the subheadings emphasizing the surface chemical constitution—i.e., fluorocarbons, hydrocarbons, chlorocarbons, and nitrated hydrocarbons.

Some important results included in Table V deserve a brief discussion. The surface of lowest energy ever found (and hence having the lowest γ_c) is that comprised of closest packed -CF$_3$ groups. The replacement of a single fluorine atom by a hydrogen atom in a terminal -CF$_3$ group more than doubles γ_c. A parallel and regular increase in γ_c has been observed with progressive replacement of fluorine by hydrogen atoms in the surfaces of bulk polymers. Data for polytetrafluoroethylene (-CF$_2$-CF$_2$-)$_n$, polytrifluoroethylene (-CF$_2$-CFH-)$_n$, poly(vinylidene fluoride) (-CF$_2$-CH$_2$-)$_n$, and poly(vinyl fluoride) (-CFH-CH$_2$-)$_n$ are listed in the order of increasing values of γ_c; however, this is also the order of decreasing fluorine content. As pointed out earlier, a plot of γ_c against the atom per cent replacement of hydrogen in the monomer by fluorine results in a straight line (Figure 8).

Among the hydrocarbons the lowest values of γ_c are found in surfaces comprising close-packed, oriented, methyl groups. The lowest value of 22 dynes per cm. results when the methyl groups are packed in the close-packed array found in the easiest cleavage plane of a single crystal of a higher paraffin such as n-hexatriacontane. The less closely packed arrangement found in a condensed adsorbed monolayer of a high molecular weight fatty acid is characterized by a γ_c value of about 24 dynes per cm. The great sensitivity of the contact angle (and hence of γ_c) to such subtle changes in the packing of the methyl groups comprising the surface of the solid is remarkable, and it has much significance in technological aspects of wetting and adhesion. The transition from a surface comprised of $-CH_3$ groups to one of $-CH_2-$ groups results in an increase in γ_c of some 10 dynes per cm.; this is to be compared with the increase of 12 dynes per cm. observed in going from a surface of $-CF_3$ to one of $-CF_2-$ groups. The presence of aromatic carbon atoms in the hydrocarbon surface also increases γ_c. Thus, the introduction of a significant proportion of phenyl groups in the surface in going from polyethylene to polystyrene raises γ_c from 31 to 33 dynes per cm. A further increase to 35 dynes per cm. results when the surface is composed solely of phenyl groups, edge on, as in the cleavage surface of naphthalene.

The results of many experiments summarized in Table V make it evident that the wettability of low-energy organic surfaces, or of high-energy surfaces coated by organic films is determined essentially by the nature and packing of the exposed surface atoms of the solid and is otherwise independent of the nature and arrangements of the underlying atoms and molecules. These findings exemplify the extreme localization of the attractive field of force around the solid surfaces of covalent-bonded atoms which are responsible for the adhesion of a great variety of liquids to solids; the field of force becomes unimportant at a distance of only a few atom diameters and hence there is little contribution to the adhesion by atoms not in the surface layers. However, when the constitution of the solid, or of the adsorbed monolayer, is such that either ions or large, uncompensated, permanent dipoles are located in the outermost portion of the surface monolayer, the residual field of force of the surface is much less localized.

Recent examples will be found in the unexpectedly strong wetting behavior of a solid coated with an adsorbed terminally fluorinated monolayer of a fatty acid or amine [95]. The outermost group of atoms (the $-CF_3$ group) has a strong dipole whose electrostatic field is not compensated by adjacent dipoles within the same molecule; hence, the external field of force is effective over much greater distances than that of nonpolar substances, so that the principle of localized action no longer holds. A solid coated with such a film is much more wettable by all liquids than is a coating made up of the fully fluorinated acid in which there is internal compensation of local dipoles. Subsequent similar studies with Shafrin [97] of a series of progressively fluorinated fatty acids synthesized by Brace [26] showed that only after the seven outermost carbon atoms were fully fluorinated, did the contact angles of various liquids on the surface approach those of a perfluoro fatty acid monolayer. In summary, our studies of wetting demonstrate that although there are understandable exceptions to Langmuir's "principle of independent surface action" it is usually true.

Wetting of High-Energy Surfaces

Many years ago Harkins and Feldman [57], extrapolating from measurements of the spreading coefficients of liquids on water and mercury [56,58], concluded that practically all liquids should spread on clean metals and other inorganic high-melting solids. But the research already summarized here has shown that when the liquid in contact with a high-energy surface is made up in whole or in part of polar-nonpolar molecules of certain types, there will be produced through adsorption at the solid-liquid interface a low-energy surface on which the liquid will not spread. When the adsorbed film comprises long-chain, un-branched, polar molecules, which are able to form a close-packed ar-ray with terminal $-CH_3$, $-CF_2H$, or $-CF_3$ groups, the resulting sur-faces permit spreading only by liquids having low surface tensions. When the adsorbed molecules are branched or cyclic structures [43], the resulting surfaces permit spreading by all liquids except those having high surface tensions.

It seemed possible to us by 1954 to explain the essential features of the wetting behavior of each organic liquid on high-energy surfaces, provided that information is available on the over-all configuration and packing of the adsorbed molecules of liquid. Therefore, the wetting behavior of over a hundred pure liquids comprising a great variety of organic and inorganic liquids was examined at 20°C. on surfaces of polished clean platinum, stainless steel, brass, fused silica, borosilicate glass, and synthetic sapphire (α-Al_2O_3). Some of the results obtained with Fox and Hare are summarized [44] in a condensed form in Tables VI and VII.

From a study of these data two problems were recognized, but their solution took several years of research. The first problem was defined by Hare and me [54] when we established that liquids such as 1-octanol, 2-octanol, 2-ethyl-1-hexanol, trichlorodiphenyl, and tri-o-cresyl phosphate exhibited appreciable contact angles on all these high-energy surfaces, no matter what extremes of purification were used. Examples of the observed equilibrium contact angles are given in Table VII. Our further investigation revealed that each liquid was nonspreading because the molecules adsorbed on the solid to form a film whose critical surface tension of wetting was less than the surface tension of the liquid itself. In short, each liquid was unable to spread upon its own adsorbed oriented-monolayer; hence, we named such sub-stances "autophobic liquids."

It followed logically that liquids which are not autophobic should have surface tensions which are less than the critical surface tensions of wetting of their adsorbed monolayers, and the data on γ_c vs. consti-tution agreed with this conclusion. For example, the polymethylsiloxane liquids spread on all high-energy surfaces because the surface tensions of 19 to 20 dynes per cm. [42] are always less than the critical surface tensions of their own adsorbed films. This follows because an adsorbed close-packed monolayer of such silicone molecules has an outermost surface of methyl groups which are not quite as closely packed as those in a single crystal of a paraffin. Since γ_c of close-packed stearic acid or octadecylamine is about 24 dynes per cm., the value of γ_c for the silicone monolayer must exceed 22; actually, it is 24 or more, depend-ent on packing [106]. Because γ_{LV^o} of this class of silicones is below

Table VI. Survey of Wettability by

Class of Liquid	Liquid Surface Tension, Dynes/Cm. at 20°C.
Open-chain aliphatic hydrocarbons	27 - 31
Open-chain methyl silicones	19 - 20
Open-chain aliphatic ethers	28 - 30
Open-chain aliphatic monoesters	27 - 29
Open-chain aliphatic diesters	28 - 34
Cyclic, saturated hydrocarbons	26 - 35
Aromatic-aliphatic hydrocarbons	28 - 38
Cyclic esters (dumbbell)	36 - 42
Cyclic esters (one ring)	30 - 35
Cyclic ethers	33 - 44
Phosphate esters (aromatic)	40 - 44
Phosphate esters (chlorinated aromatic)	44 - 46
Polychlorobiphenyls	42 - 46

20 dynes per cm., $\gamma_{LV} < \gamma_c$ and hence these silicones cannot be autophobic. A similar argument using the critical surface tension of polyethylene of 31 dynes per cm. and the fact that the surface tensions of liquid aliphatic hydrocarbons are always less than 30 dynes per cm., leads us at once to understand why such hydrocarbons cannot be autophobic.

The second problem encountered was to explain why all pure liquid esters spread completely upon the metals studied, yet as Tables VI and VII indicate, some spread on glass, silica, and α-Al$_2$O$_3$ and others did not. A careful investigation with Hare and Fox [44] finally revealed that the cause of these differences in spreadability is the hydrolysis of the ester immediately after the molecule has adsorbed upon hydrated surfaces such as those of glass, fused silica, and α-Al$_2$O$_3$. This is not surprising, since the polar group of the ester would be expected to adsorb on immediate contact with the solid surface unless prevented by steric hindrance, and since the surface molecules of the water of hydration and adsorption of the glass (being oriented) should be more effective in causing hydrolysis than bulk water. Hence, as the result of surface hydrolysis, two fragments of the ester result. The fragment which has a greater average lifetime of adsorption is more likely to remain, and eventually this molecular species coats the surface with a close-packed monolayer. Rapidly the surface becomes blocked or "poisoned" by the coating of the hydrolysis product and so the hydrolysis reaction ceases. Obviously, under these circumstances the volume concentration of hydrolyzed ester is too small to be measured by ordinary analytical methods. When the resulting adsorbed monolayer has a critical surface tension of wetting less than the surface tension

Higher Boiling Liquids [44]

Metals	Spreadability on Fused SiO_2	α-Al_2O_3
Spread	Spread	Spread
Spread	Spread	Spread
Spread	Spread	Spread
Spread	(No + yes)	(No + yes)
Spread	No	No
Spread	Spread	Spread
Spread	(No + yes)	(No + yes)
No	No	No
(No and yes)	No	No
No	(No and yes)	(No and yes)
No	No	No
No	No	No
No	No	No

of the ester, nonspreading behavior is observed—i.e., the ester is unable to spread upon the adsorbed film of its own hydrolysis product. Esters having a great variety of structures have been studied, and in every instance of nonspreading on glass, fused silica, and α-Al_2O_3, we have been able to give a similar explanation of the behavior.

As an example, consider the ability of bis(2-ethylhexyl) sebacate to spread freely on metals and its inability to spread on fused silica, glass, or α-Al_2O_3. On these hydrated nonmetallic surfaces the diester hydrolyzes to form 2-ethylhexanoic acid. The critical surface tension

Table VII. Some Autophobic Liquids and Their Contact Angles on High-Energy Surfaces [54]

Liquid	γ_{LV} at 20°C.	θ at 20°C., ° 18.8 stainless steel	Platinum	Fused silica	α-Al_2O_3
1-Octanol	27.8	35	42	42	43
2-Octanol	26.7	14	29	30	26
2-Ethyl-1-hexanol	26.7	< 5	20	26	19
2-Butyl-1-pentanol	26.1	–	7	20	7
n-Octanoic acid	29.2	34	42	32	43
2-Ethylhexanoic acid	27.8	< 5	11	17	12
Tri-o-cresyl phosphate	40.9	–	7	14	18
Tri-o-chlorophenyl phosphate	45.8	–	7	19	21

of wetting of a close-packed monolayer of 2-ethylhexanoic acid is about 28 dynes per cm. [44]. Since the surface tension of this diester is 31.1 dynes per cm. at 20°C., $\gamma_{LV} > \gamma_c$, and the diester cannot spread on the film of its hydrolyzed product. When adsorbed on metals at ordinary conditions of relative humidity, the diester is not in contact with hydrated water and so cannot hydrolyze; nevertheless, following de Boer's ideas [30], the molecule should be adsorbed lying on the surface as flat as possible to allow contact of the greatest number of polarizable atoms. The resulting adsorbed monolayer, because of the presence of the ester groups, must have a critical surface tension of wetting greater than that of polyethylene; in other words, γ_c is considerably greater than 31 dynes per cm. Hence, $\gamma_{LV} < \gamma_c$, and the diester must spread freely over its own adsorbed film and over the metal surface.

Once these two classes of nonspreading liquids were understood, it proved possible to explain the wetting and spreading properties of all of the many liquids investigated. To summarize the results of our investigations on the wetting of high-energy surfaces: Every liquid having a low specific surface free energy always spreads freely on specularly smooth, clean, high-energy surfaces at ordinary temperatures unless the film adsorbed by the solid converts it into a low-energy surface having a critical surface tension less than the surface tension of the liquid. Because of the highly localized nature of the forces between each solid surface and the molecules of a liquid and also between the molecules of each liquid, a monolayer of adsorbed molecules is always sufficient to give the high-energy surface the same wettability properties as the low-energy solid having the same surface constitution.

Theory of Retraction Method and Some Related Topics

After learning how to explain the spreading and wetting behavior of liquids upon high-energy solid surfaces, it soon was recognized [43,96,105] that the retraction method of preparing adsorbed monolayers is a logical consequence of the above-mentioned constitutive law of wetting. In essence, the retraction method is a process by which a clean, high-energy, plane solid is brought in contact with a liquid containing adsorbable polar-nonpolar molecules; adsorption occurs at the solid-liquid interface, and the surface is covered with a film which converts it to a lower energy surface. When the coated solid is held approximately vertically during withdrawal from the bulk liquid phase, the adhering layer of liquid exhibits an appreciable contact angle and then peels back or retracts, leaving the withdrawn, solid, low-energy surface dry but still coated with the monolayer. The monolayer-coated surface can be removed from the liquid unwetted only when the contact angle, θ, of the liquid with the solid is greater than zero; the permissible rate of removal can be faster the greater the value of θ. However, θ exceeds zero only for those liquids having surface tensions greater than the critical surface tension of the solid. Therefore, the necessary and sufficient condition for retraction is that $\gamma_{LV} > \gamma_c$. Adsorbed monolayers can thus be isolated from any liquid having a surface tension, γ_{LV}, greater than the γ_c value of the low-energy surface created by the initial adsorption process.

This explanation provides helpful guidance about the solvents one may use in the preparation of films by retraction from solution. The principal requirement is that γ_{LV} exceed γ_c; the greater the difference between γ_{LV} and γ_c, the larger the contact angle (according to the $\cos \theta$ vs. γ_{LV} relation) and the easier it is to perform the retraction operation to isolate the adsorbed film on the solid. In order that the adsorbed monolayer should emerge from the solution unwetted, it is usually advantageous to choose a solvent with a surface tension sufficiently in excess of γ_c of the adsorbed monolayer so that the solution exhibits a contact angle with the coated solid of 30° or more.

The earliest experiments with the retraction method [23] led to the conclusion that retraction was not possible when the polar-nonpolar solute had one or more cis double bonds—for example, oleic acid could not be isolated by retraction. Evidently, our difficulty arose from the fact that γ_c for an adsorbed film of oleic acid must be greater than for stearic acid; hence the solvent used for retraction of oleic acid must have a surface tension higher than that used with stearic acid. As the hexadecane used as the solvent in the early unsuccessful experiments with oleic acid had $\gamma_{LV} = 27.7$ dynes per cm., a solvent with a much higher surface tension is needed (a nonhydrocarbon would be preferable). It is now evident why on using liquids of high surface tensions such as water, ethylene glycol, and methylene iodide, it is possible to retract polar solutes as condensed monolayers having a wide variety of organic structures, including branched or cyclic types. Examples of $\cos \theta$ vs. γ_{LV} plots for such films [43] are given in Figure 13.

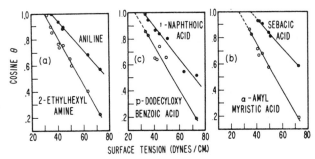

Figure 13. Wettability of various monolayers on platinum prepared by retraction from aqueous solution [43]

A method described earlier for preparing adsorbed monolayers on solids by retraction from the molten state of the pure compound [22] represents, in effect, an application of the autophobic property of the liquid; each of these compounds is autophobic or nonspreading at all temperatures in the portion of the liquidus range of the material in which the liquid surface tension exceeds γ_c of the adsorbed monolayer. A variant on the retraction technique [2] is the tilted plate method, in which the molten compound is placed along the inclined hotter edge. As the melt runs down the plane it cools and solidifies. A retracted monolayer forms on the surface near the hot edge. This method has advantages over the vertical plate retraction method when the contact angle is small (but not zero).

135

A temperature, τ, was found by us experimentally [22] above which it was no longer possible to use the retraction method to isolate the monolayer from the molten compound. The values of τ for each homologous series of aliphatic compounds were a rectilinear increasing function of the number of carbon atoms per molecule. For most pure homologous liquids not too near the critical temperature, the surface tension vs. temperature relations can be represented by a series of straight lines of negative slope. In contrast, the effect of raising the temperature of an adsorbed film by a few degrees is to decrease the surface density and lower adhesion very little; hence, γ_c increases only slightly per degree of rise. At a sufficiently high temperature, the condition is reached at which $\gamma_{LV} > \gamma_c$ and retraction is possible no longer. Hence, the temperature limit, τ, corresponds to the condition when γ_{LV} of the molten compound equals γ_c of the adsorbed film.

In the homologous series of fatty acids, only small variations were found in the value of γ_c characteristic of the acid monolayer at temperature τ. Whereas an increase in chain length in going from octanoic to octadecanoic acid resulted in a rise in τ from 23° to 106°, the corresponding change in γ_c was only to decrease from 28 to 25 dynes per cm. Since these values do not greatly exceed the value of γ_c characteristic of hydrocarbon surfaces in closest packing, they indicate that the adsorbed film is wetted by the melt at temperatures at which the adsorbed films are still relatively intact.

Formation of Mixed Films and Their Metastability

If in retracting a monolayer of stearic acid or octadecylamine from a solution in hexadecane, a mixed film of solute and solvent molecules is produced, it is not possible to detect the presence of the solvent molecules in the film by means of contact angle measurements using an appropriate liquid drop such as water, methylene iodide, etc. This result is a consequence of the fact we had already established that the wettability of an adsorbed monolayer is determined by the nature and packing of the outermost surface atoms or organic radicals. Bartell and Ruch [9,10], using an optical interference spectrometer, proved that octadecylamine films adsorbed and retracted from hexadecane solutions on polished chromium can comprise mixtures of solute and solvent. Levine and I [79] found shortly afterwards that such mixed films could also be formed by retraction on fire-polished glass surfaces and distinguished readily from films of solute by multiple-traverse boundary friction measurements. The mixed films prepared from solution in hexadecane of either stearic acid or octadecylamine proved to be metastable, for they did not occur if sufficient immersion time was allowed before retraction to approximate adsorption equilibrium. An immersion time of 24 hours was required for the adsorbed film to be free of hexadecane molecules. When the solvent used had a molecular structure such that it could not form adlineated films with the solute molecules, mixed films were not encountered. For example, either stearic acid or octadecylamine adsorbed and retracted from solutions in nitromethane, carbon tetrachloride, dicyclohexyl, or benzene showed no evidence of containing solvent molecules.

Recently Bewig and I have successfully used changes in the contact potential difference between metals for studying the formation of mixed films by retraction [20]. This method consists of measuring the change in the contact potential differences (between a clean metal electrode and a stabilized gold reference electrode) resulting from the adsorption of a condensed organic monolayer. The adsorbed monolayer studied was deposited on the clean metal electrode (platinum, nickel, or chromium) by retraction from a variety of pure solvents. Plots of contact potential difference, ΔV, vs. the length of time of immersion of the metal electrode in the solution readily distinguished between films free from or containing the solvent molecules. Our earlier results with the friction method [79] were verified. It was found that if the number of carbon atoms per molecule in the alkane solvent differed too much, mixed film formation became less prominent; it became insignificant when there was a difference of five carbon atoms per molecule. Solvents such as 1-phenyldodecane and methylene iodide did not form adlineated films with octadecylamine under any condition. Aliphatic solvents having methyl- branches, such as pristane and squalane, although less able to adlineate than unbranched alkanes, still formed mixed films with octadecylamine.

In summary, mixed films of solute and solvent can be prepared as retracted films; however, these are always transient or metastable systems, the probability of whose formation becomes greater the more fully does molecular adlineation of solute and solvent occur. If the concentration of polar solute is high enough, the film eventually becomes free of solvent as adsorption equilibrium is approached, and the retracted film is then free of solvent molecules. Where the solute and solvent molecules are so different in shape or size that the intermolecular cohesion between them through London dispersion forces becomes a minor factor, mixed films are never formed. Therefore, mixed films occur by the retraction process under special conditions; nonetheless, when they can be produced, a useful technique is available for studying the intermolecular interaction of solute and solvent molecules in the adsorbed state.

A by-product of the study of mixed films is better guidance in the selection of liquids suitable for reliable measurements of the wetting properties of adsorbed films. Whenever a liquid drop is placed upon a film, there is always the possibility that molecules of the drop will permeate into the film to form a mixed film system and so create the condition that the contact angle observed is dependent on other factors than the surface composition and packing of the original film. As pointed out earlier [2,92,93], this is precisely one of the limitations in the use of water for contact angle measurements in studying monolayers of paraffinic derivatives, and it gives rise to differences in the advancing and receding contact angles (see Table II). From the preceding discussion, it is evident that hexadecane or any other n-alkane liquid may also be able to permeate into and adlineate with the polar molecules in films of paraffinic polar compounds. For these reasons Levine and I concluded that it is generally preferable to use a nonlinear large molecule having a high surface tension as the reference liquid for many contact angle studies on adsorbed films and even on surfaces of plastics [78, 79]. Since methylene iodide has a surface tension of 50.8 dynes per cm. at 20°C., and a boiling point of 180°C., it will exhibit large contact angles with most organic surfaces and will not evaporate too rapidly. Furthermore, since the molecule is roughly spherical, it has little ability to adlineate to form mixed films. Its size is enough greater than that of the water molecule to limit greatly permeation into anything but loosely packed films. For these reasons we have used the equilibrium contact angle of methylene iodide very widely in our investigations of the properties of adsorbed films and other low-energy surfaces.

Wetting by Solutions - Aqueous and Nonaqueous

Present theories about the wetting ability of surface-active agents in aqueous systems either state or imply that spreading results because the wetting agent becomes selectively adsorbed on the surface so as to orient the hydrophilic group toward the aqueous solution [41,85]. Our studies on adsorption at solid-air and solid-liquid interfaces made it increasingly improbable that the major mechanism of wetting is the result of the ability of the nonpolar group of the wetting agent to adsorb by its hydrocarbon "tail" on the low-energy surfaces and so convert it into a high-energy surface. A more reasonable assumption is that spreading on low-energy surfaces is caused by the lowering of the surface tension of water. If the solute molecules do adsorb at the solid-liquid interface, it is one of the results of wetting—not the cause. Hence, changes in γ_{SL} must play a minor role in the presence of solutions of wetting agents. Therefore, we assumed that the ability of the aqueous solution to spread on such low-energy surfaces is determined by the value of the critical surface tension, γ_c, of the solid to be wetted and the amount of the wetting agent which must be dissolved in the water to depress the surface tension of water below γ_c.

These ideas were verified in experiments on two low-energy solids, polyethylene ($\gamma_c \approx 31$ dynes per cm.) and polytetrafluoroethylene ($\gamma_c \approx 18$ dynes per cm.), using a variety of well-defined aqueous solutions [14]. Pure water ($\gamma_{LV} = 71.9$ dynes per cm. at 25°C.) will not spread on either of these two surfaces. Fischer and Gans [39] have pointed out that conventional surface-active agents (derived from aliphatic or aromatic hydrocarbons) do not lower the surface of tension of water at 25°C. below between 26 and 27 dynes per cm. It follows from the definition of the critical surface tension that whenever any wetting agent lowers the liquid surface tension below 31 dynes per cm., the solution will spread on the surface of smooth clean polyethylene. However, aqueous solutions containing conventional wetting agents should not spread on polytetrafluoroethylene, since a wetting agent capable of depressing the surface tension below 18 dynes per cm. would be needed. Wetting agents used in our experiments included all three classes—i.e., anionic, cationic, and nonionic—and as many diversified hydrocarbon structures as feasible in order to show properties or general trends common to conventional surface-active agents.

Figure 14 shows the knee-shaped curve of solution surface tension vs. concentration characteristic of aqueous surface-active agents; the steeper the curve, the more efficient the wetting agent. It is generally assumed that the bend of the curve coincides with the critical micelle concentration (c.m.c.) of the respective compound in the aqueous medium. Since the discontinuities in the slopes of the individual curves of Figure 14 occur in the region of the c.m.c. values reported by various

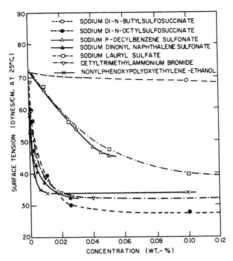

Figure 14. Surface tensions of conventional wetting agents in aqueous solution [13]

investigators, the above assumption can be made again for the agents discussed here.

Curves of surface tension vs. cos θ, obtained for wetting agents on polyethylene, are given in Figure 15, A and B. Analogous results using smooth polytetrafluoroethylene surfaces are given in Figure 16, A and B. At the critical surface tension (cos θ = 1) there is only a narrow

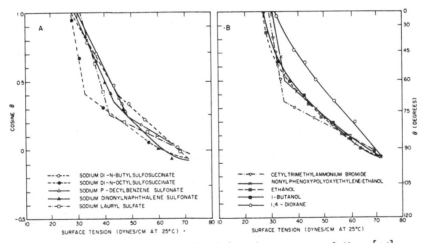

Figure 15. Wettability of polyethylene by aqueous solutions [13]

A. Anionic wetting agents
B. Other wetting agents

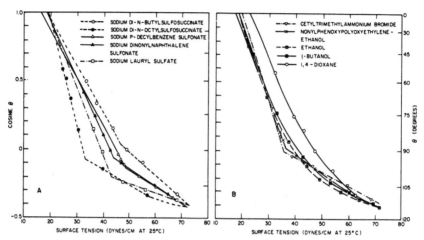

*Figure 16. Wettability of polytetrafluoroethylene
by aqueous solutions* [13]

A. Anionic wetting agents
B. Other wetting agents

spread of the converging curves for each organic surface. From these intercepts one can estimate γ_c values for polytetrafluoroethylene as between 16.5 and 19.5 dynes per cm. and for polyethylene as between 27.5 and 31.5 dynes per cm. These are in reasonable accord with the values reported in Table V, which are based on the wettability by pure organic liquids. Only solutions whose surface tensions were lower than 30 dynes per cm. spread freely on polyethylene. No aqueous solution of a conventional hydrocarbon derivative wetting agent had a surface tension low enough to spread freely on polytetrafluoroethylene. Hence, our observations agreed with the assumption that aqueous solutions will spread on a low-energy surface when the surface tension is less than the γ_c value of the solid.

Whereas the curves for the various wetting agents grouped in a narrow band intercepting the line cos θ = 1 at the value of γ_c, they did not converge to a point, as would be anticipated, at the value of γ_{LV} = 71.9 dynes per cm. for pure water but also exhibited a spread in θ of 2°, which is well within the experimental error in measuring θ. An unexpected and new phenomenon, however, was the abrupt change in slope in the middle zone of each of the curves of Figures 15 and 16. The data in the first two columns of Table VIII (calculated from Figures 14, 15, and 16) revealed that the discontinuities in slope for polytetrafluoroethylene occurred at approximately equal concentrations for a given wetting agent. Hence, the discontinuities in slope are determined by the constitution of the wetting agent and not by that of the solid. When the surface tension corresponding to each such discontinutiy was referred to the curve of surface tension vs. concentration in Figure 14, it was found that it corresponded to the concentration in the region of greatest curvature. Therefore, there was a close relation between the concentration at which discontinuity in the slope occurs in each wetting

Table VIII. Concentrations at Discontinuities in Slope for
Various Wetting Agents and Surfaces at 25°C. [13]

Compound	Concn., Moles/Liter		
	Polyethylene	Polytetra-fluoroethylene	C.m.c.
Na di-n-butyl sulfosuccinate	5.0×10^{-2}	4.2×10^{-2}	2×10^{-1}
Na di-n-octyl sulfosuccinate	6.8×10^{-4}	4.5×10^{-4}	6.8×10^{-4}
Na p-decyl benzene-sulfonate	1.2×10^{-3}	1.2×10^{-3}	3×10^{-3} at 50°
Na lauryl sulfate	3.0×10^{-3}	2.7×10^{-3}	8×10^{-3}
Cetyltrimethyl-ammonium bromide	4.3×10^{-4}	3.8×10^{-4}	1×10^{-3} at 60°

curve of Figures 15 and 16 and the c.m.c. of the wetting agent in the
same solution. The last column of Table VIII gives the best literature
values of the c.m.c. of each of these wetting agents at or near the temperature used. There is a fair correlation between the concentration
corresponding to the discontinuity in slope in the wetting curve and the
reported c.m.c.; also it appears that the c.m.c. is always slightly
higher.

The correctness of the preceding interpretation of the slope discontinuities in the curves of γ_{LV} vs. $\cos \theta$ was supported by experiments
on solutions of the following pure polar compounds which are not able
to form micelles: ethanol, 1-butanol, 1,4-dioxane, propylene carbonate,
diacetone alcohol, dipropylene glycol, 2-butanone, and tetrahydrofuran.
Surface tensions for these compounds agreed well with the available
literature values [87]. None of these polar solutes form micelles in
water and, as can be seen in Figures 15, B, and 16, B, no discontinuities
in the slopes of the curves of $\cos \theta$ vs. γ_{LV} were observed for either
solid surface. It was concluded, therefore, that no discontinuities in
slope occur without the formation of micelles.

Each intercept at $\cos \theta = 1$ fell within the expected range of γ_c
values (Figure 16, B), with the exception of the intercept for the

Table IX. Critical Surface Tensions by Aqueous Solutions
of Non-Micelle-Forming Compounds at 25°C. [13]

Solute	γ_{LV}, Dynes/Cm., Pure Solute	γ_c, Dynes/Cm.	
		Polyethylene	Polytetra-fluoroethylene
Ethanol	21.4	27.5	18.5
1-Butanol	23.7	27.5	18.5
2-Butanone	24.4	27.5	---
Tetrahydrofuran	27.4	29.5	---
Diacetone alcohol	30.2	29.0	17.0
1,4-Dioxane	32.4	31.5	22.5
Dipropylene glycol	33.1	30.0	18.0
Propylene carbonate	40.5	29.5	19.5

1,4-dioxane curve which occurred at the higher value of 22.5 dynes per cm. Wetting curves on polyethylene were almost coincident for 1,4-dioxane and propylene carbonate, for diacetone alcohol and 2-butanone, and for dipropylene glycol and tetrahydrofuran, and were very similar for ethanol and 1-butanol (Figure 15, B). Table IX shows the intercepts at cos θ = 1 for the compounds for polytetrafluoroethylene and poly-ethylene. It has been found in previous work [46,48] that the γ_c value of any one solid surface varies somewhat among the various homologous series of liquids.

Since discontinuities at the c.m.c. occur in curves of surface tensions vs. concentration and of either cos θ vs. concentration or cos θ vs. surface tension, the generalization is proposed that micelle-forming compounds will produce discontinuities in the slopes of other surface properties of surface active agents when plotted against either the concentration or the surface tension of the solution. For example, it may be that such will occur in plots of the interfacial tension vs. the concentration of the surface active solute in the aqueous phase.

Because of their hydrophobic-hydrophilic structures, each of the most effective wetting agents adsorbs to form a thin film on the free surface of the aqueous solution. At sufficiently high solute concentrations, this film will make the aqueous liquid appear to be a hydrocarbon liquid whose surface is comprised of the oriented, packed, hydrophobic groups characteristic of the organic structure of the wetting agent. Our experiments lead to the conclusion that bringing the low-energy surface of polytetrafluoroethylene (or polyethylene) into contact with this film-coated water did not change the orientation or packing of the film adsorbed at the water-air interface. Fowkes and Harkins [41] had reached essentially the same conclusion about paraffin when they found that the force-area curves for films of butyl alcohol, butyric acid, and butylamine adsorbed at the paraffin-water interface were nearly the same as those for adsorption at the air-water interface.

From these considerations one can predict that γ_{SL} will decrease as adsorption of the wetting agent increases. At high solute concentrations, γ_{SL} will become small because the interface will be between an organic low-energy solid and a hydrocarbon-like liquid. In other words, the much greater adhesion of the polar groups to the water and the much lower adhesion of the hydrocarbon groups to the low-energy solid surface will have caused the interface determining spreading and wetting to become the hydrocarbon-like outer surface of the film of adsorbed wetting agent. Hence, γ_{SL} will decrease with increased solute concentration and will become nearly constant at high concentrations as the adsorbed film approaches closest packing. Thus, the value of γ_c will vary with the nature of the wetting agent only in so far as there are differences in the nature and packing of the hydrocarbon groups of the wetting agent in the adsorbed state at the water-air interface. One would expect γ_c to be least for n-alkane long-chain derivatives; it will be greater on each surface for branched hydrocarbon derivatives; and it will be largest for aromatic polar compounds. Solutes like 1,4-dioxane or propylene carbonate (or the others in Table IX) adsorb a film at the water-air interface which has a higher free surface energy than one containing only hydrocarbon groups. Any discontinuity in the slope or first-order phase change of the force-area curve in the film adsorbed at the water-air interface will be accompanied by a discontinuity in the slope of the cos θ vs. γ_{LV} curve, in view of the conclusion that the contact angle is essentially that of a liquid having the same outermost hydrocarbon structure as the adsorbed solute. It is now evident why the discontinuity in the slope occurs at a concentration which is dependent only on the nature of the wetting agent and not on the nature of the solid surface.

The same experimental approach and conclusions should apply to the wetting of poly(vinyl chloride), polystyrene, and other low-energy solid surfaces. In previous studies we had shown that even a single, close-packed, adsorbed monolayer is sufficient to convert the wetting properties of a high-energy surface into those of a low-energy surface. One cannot predict that the same rule will hold for the wettability of any such film-coated surface by an aqueous solution. In predicting the spreading properties on low-energy surfaces of aqueous solutions of wetting agents, one cannot neglect γ_{SL} at solute concentrations so low that the liquid surface phase consists essentially of water; but at concentrations high enough for the surface of the water to approximate close packing of adsorbed solute molecules, γ_{SL} becomes constant and small in value, and the wetting behavior becomes predictable by the critical surface tension approach.

One might rate the effectiveness of any hydrocarbon-type wetting agent by the minimum concentration necessary at 25°C. to cause the solution to spread on smooth polyethylene (γ_c = 31 dynes per cm.). Another interesting measure of wetting efficiency is the minimum concentration of the agent necessary to wet the smooth clean surface of polytetrafluoroethylene (γ_c = 18 dynes per cm.); however, conventional wetting agents could not be so rated, since they do not decrease γ_{LV} below 26 to 27 dynes per cm. A sufficiently large depression of the surface tension of water necessitates use of a soluble surface-active agent whose hydrophobic group is an appropriately fluorinated hydrocarbon. Some of the properties of these rather "exotic" wetting agents have been reported by Scholberg, Guenthner, and Coon [89] and Klevens and Raison [68]. The low surface energies of films of such compounds have been discussed by us [53]. Results of an investigation of aqueous solutions of such agents for wetting low-energy surfaces have been reported with Bernett [14]. A measure of wetting efficiency of intermediate value is the minimum solute concentration necessary to spread an aqueous solution on the surface of a single crystal of hexatriacontane (γ_c = 22 dynes per cm.). Polystyrene (γ_c = 33 dynes per cm.) and poly(vinyl chloride) (γ_c = 34 dynes per cm.) might also be considered useful low-energy surfaces for relatively rating the wetting agents incapable of lowering γ_{LV} below 30 dynes per cm. The value of such a scheme for rating aqueous wetting agents is that it is quantitative and objective, requires small amounts of solution, has fundamental significance, and need not be confused by other properties such as detergency, capillarity, and surface roughness.

Of especial interest is that the same approach in defining wetting agents in aqueous solutions can also apply to nonaqueous solutions. Increasing interest has been shown in the behavior of surface-active agents at organic liquid-air interfaces. Many of the earliest such studies were made on compounds previously found effective as surface-active agents in aqueous solutions. Thus, McBain and Perry [81] showed that low concentrations of laurylsulfonic acids lowered slightly the surface tensions of a group of hydrocarbons. Jones and Saunders [65] later measured the surface tensions of a series of n-aliphatic acids in nitromethane, and Kaminski [66] observed the adsorption of lauric acid and lauryl alcohol at the mineral oil—air interface using the McBain microtome technique. In recent years it has been found that many silicones and fluorochemicals are remarkably surface active in organic liquids and so depress the surface tensions of organic liquids considerably. Banks [6,7] reported the formation of stable, insoluble films of polydimethylsiloxanes on oleic acid, olive oil, triacetin, or ethylene glycol. Almost simultaneously Ellison and I [35,36] demonstrated with an all-Teflon film balance the monomolecular nature of films of the linear polymethylsiloxanes and certain perfluorocarbon derivatives adsorbed as insoluble films on mineral oil, n-hexadecane, and tricresyl phosphate. Other silicones as well as various polyacrylates, polyalkylene ethers, organosilanes, and zein were also surface active on certain organic substrates. Scholberg, Guenthner, and Coon [89] found that certain fluorocarbon derivatives which were surface-active agents in water were also very effective in lowering the surface tensions of organic liquids. Blake, Ahlbrecht, and Bryce [24] reported the surface tension—reducing effects of a series of polyfluoroquaternary ammonium compounds in selected organic liquids. The preparation and physical properties of some partially fluorinated esters and ethers, which had been designed to have an organophobic-organophilic balance suitable for high surface activity, were described with O'Rear [107]; and with Ellison [37] and Jarvis [12,61,62,63] subsequent reports were made of a series of studies of the surface activity of these fluoro- compounds when dissolved in various organic liquids.

Estimates of Reversible Work of Adhesion

Bangham and Razouk's Equations 4c and 5 for the reversible work of adhesion and f_{SV^o} were used by Boyd and Livingston [25] and by Harkins et al. [11,59,80] to calculate W_A from experimental vapor adsorption isotherms for various liquids on a number of metallic and nonmetallic surfaces. Their results, summarized in Tables X and XI, show that in every instance f_{SV^o} is of the same magnitude as W_{A^*} and so cannot be neglected in the calculation of W_A. Each of these solids is a highly adsorptive, finely divided material, and excluding graphite, each is a hydrophilic high-energy surface.

Unfortunately, f_{SV^o} has not yet been measured for any well-defined, smooth, low-energy, solid surface. But one should not assume from the results in these tables that in dealing with low-energy solids f_{SV^o} will also be an important correction term in Equation 4c. On the contrary, there is good experimental evidence that whenever a liquid exhibits a large contact angle on a solid there is negligible adsorption of the vapor. Recent measurements with Bewig [19], using a special reference electrode for studying adsorption by contact potential changes, have demonstrated that negligible adsorption of most vapors occurs at ordinary temperatures on smooth clean surfaces of polytetrafluoroethylene. Extensive room-temperature adsorption measurements over

144

Table X. Literature Values of f_{SV° for Nonmetallic
High-Energy Surfaces[a] (Ergs/Sq. Cm. at 25°C.)

Solid	Liquid	$f_{SV^\circ} \equiv \gamma_{S^\circ} - \gamma_{SV^\circ}$	W_A
TiO_2	Water	300 (196[b])	370 (340[b])
	1-Propanol	114 (108[b])	138 (154[b])
	Benzene	85	114
	n-Heptane	58 (46[b])	78 (86[b])
SiO_2	Water	316	388
	1-Propanol	134	158
	Acetone	109	133
	Benzene	81	110
	n-Heptane	59	79
$BaSO_4$	Water	318	390
	1-Propanol	101	125
	n-Heptane	58	78
Fe_2O_3	n-Heptane[c]	54	94
SnO_2	Water	292 (220[b])	364 (364[b])
	1-Propanol	104 (117[b])	128 (163[b])
	Propyl acetate[b]	104	151
	n-Heptane[c]	54	94
Graphite	Water	64	136
	1-Propanol	95	118
	Benzene	76	96
	n-Heptane	57[c]	97[c]

[a]Data from [25] unless otherwise indicated.
[b][80].
[c][11].

Table XI. Literature Values of f_{SV° for Metallic
High-Energy Surfaces[a] (Ergs/Sq. Cm. at 25°C.)

Solid	Liquid	$f_{SV^\circ} \equiv \gamma_{S^\circ} - \gamma_{SV^\circ}$	W_A
Mercury	Water	101	174
	1-Propanol	108	132
	Acetone	86	110
	Benzene	119	148
	n-Octane	101	123
Copper	n-Heptane[b]	29	69
Silver	n-Heptane[b]	37	77
Lead	n-Heptane[b]	49	89
Iron	n-Heptane[b]	53	93
Tin	Water[c]	168	312
	n-Heptane[c]	50	90
	1-Propanol[c]	83	129

[a]Data from [25] unless otherwise indicated.
[b][59].
[c][80].

the entire vapor pressure range of p/p_o up to 1.0 by Martinet [82] and Graham [51] have also led to the conclusion that the vapor adsorption for the gaseous substances studied on this low-energy solid surface is but a small fraction of a monolayer. Hence, we have concluded [106] that f_{SV^o} is a small term in comparison with W_{A*} for each liquid and solid whenever γ_{LV^o} is greater than γ_c. The same conclusion should also apply to other low-energy solids at ordinary temperatures such as polyethylene, polystyrene, poly(vinyl chloride), etc.

Therefore, it was proposed that as regards any low-energy solid surface: liquids having γ_{LV^o} much greater than γ_c have W_A essentially equal to W_{A*}; as γ_{LV^o} closely approaches γ_c but exceeds it, $W_A - W_{A*}$ becomes more significant; and liquids having γ_{LV^o} less than or equal to γ_c may have appreciable values of $W_A - W_{A*}$. The same conclusions about the negligible value of f_{SV^o} apply to any high-energy surface which has been converted to a low-energy surface by the adsorption of a suitable condensed organic monolayer, with the one reservation that f_{SV^o} may become more significant if the molecule of the liquid is small enough to be able to penetrate readily through the condensed monolayer and so adsorb on the high-energy surface beneath.

In dealing with the adhesion of a liquid to a plane, nonporous, solid surface, according to Equation 4d, W_{A*} is given by $W_{A*} = \gamma_{LV^o} (1 + \cos \theta)$. However, experiments show that for any homologous series of liquid compounds and for all values of $\gamma_{LV^o} > \gamma_c$, the contact angle is related to γ_{LV^o} by the straight-line equation

$$\cos \theta = a - b\gamma_{LV^o} \tag{11}$$

Since γ_{LV^o} approaches γ_c as θ approaches zero, Equation 11 can be written

$$\cos \theta = 1 + b(\gamma_c - \gamma_{LV^o}) \tag{12}$$

Upon eliminating $\cos \theta$ between Equations 5 and 12, there results [106]:

$$W_{A*} = (2 + b\gamma_c) \, \gamma_{LV^o} - b\gamma_{LV^o}^2 \tag{13}$$

This is the equation of a parabola with the concave side toward the surface tension axis; it has a maximum value of W_{A*} occurring at

$$\gamma_{LV^o} = \frac{1}{b} + \frac{1}{2}\gamma_c \tag{14}$$

and the maximum value is given by

$$W_{A*} = \frac{1}{b} + \gamma_c + \frac{1}{4}b\gamma_c^2 \tag{15}$$

For example, consider smooth polyethylene for which $\gamma_c = 31$ dynes per cm. and $b = 0.026$ [48]; the maximum of W_{A*} occurs at $\gamma_{LV^o} = 54$ dynes per cm. and is about 76 ergs per sq. cm.

It is of interest to compute the value of W_{A*} for the many liquid-solid combinations reported in the past. Figures 17 to 20 show how

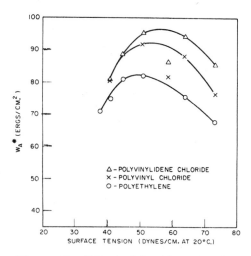

Figure 17. Effect of liquid surface tension on W_{A*} for chlorinated polyethylene [106]

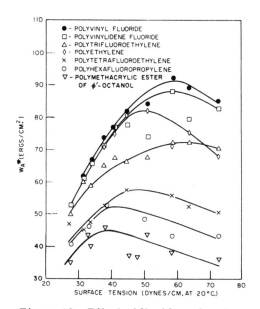

Figure 18. Effect of liquid surface tension on W_{A*} for fluorinated plastics [106]

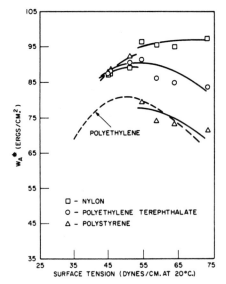

Figure 19. Effect of liquid surface tension on W_{A*} for some common plastics [106]

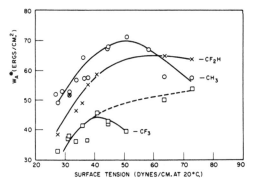

Figure 20. Effect of liquid surface tension on W_{A*} for a solid coated with a condensed monolayer [106]

W_{A*} varies as a function of γ_{LV^o} . In every case there results a parabolic curve with a maximum. However, in a few instances, such as in the case of polytrifluoroethylene (Figure 18), the decrease of W_{A*} after reaching the maximum value is greatly moderated by the effect of the hydrogen bonding action of the liquids of high surface tension—i.e., water, glycerol, and formamide, each of which is an effective hydrogen-donating compound. Similar effects are seen in the curve of Figure 20 for the close-packed monolayers terminated by -CF_2H groups and -CF_3 groups. In the latter case, if the data points for the hydrogen-donating liquids are excluded, the curve is seen to form a parabola with its maximum occurring at about 40 dynes per cm. It is also seen that the maximum values of W_{A*} for -CH_3, -CF_2H, and -CF_3 coated surfaces are approximately 70, 65, and 44 ergs per cm. sq., respectively. Inspection of Figures 17 to 20 also reveals that the extreme variation in W_{A*} among all the low-energy solid surfaces reported is only about threefold.

Relation of Wetting to Adhesives Art

Only in this century has the art of using adhesives been based on the properties and applications of liquid adhesives other than old glue formulations from fish and animal products or cements made from inorganic slurries or solutions. The subject of adhesives assumed a new stature with the advent of synthetic polymers having readily controlled and understood properties. However, there are many unanswered questions about the principles underlying the subject and accepted methods of application.

Figure 21 is a simplified diagram of a typical adhesive joint comprising two plane solid surfaces, identified usually as the "adherends," and a thin layer of liquid, the "adhesive." If the adhesive, while in the

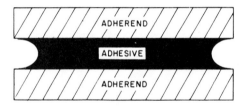

Figure 21. Idealized adhesive joint [106]

liquid state, has a zero contact angle, it will spread spontaneously over the solid to make intimate intermolecular contact with the surface of each adherend. Each adhesive is designed to increase in viscosity with the passage of time until it becomes solid; this may occur through freezing, evaporation of a volatile solvent, or polymerization. If the thermal expansion coefficients of adhesive and adherends are not too different, and if there is not too large a change in the density during solidification, internal stresses developed in the area during solidification of the adhesive will not be sufficient to shatter the joint or cause cracking. Obviously, a good joint must have sufficient strength to endure large externally applied stresses.

Tabor [100] has pointed out that according to Equation 4c, W_A for any system having $\theta = 0$ will be equal to or greater than twice the surface tension of the liquid; a simple calculation, assuming that the field of force emanating from the solid vanishes in about 3 A., and that the surface tension of the adhesive is 30 dynes per cm., results in an average tensile strength of the adhesive joint of 2000 kg. per sq. cm. This value is much greater than the tensile strength of common adhesives, for Kraus and Manson [69] obtained for polyethylene a tensile strength of 183 kg. per sq. cm. Therefore, the joint must break by cohesive rather than adhesive failure. Since the correction term, f_{SV^0}, in Equation 4c makes the adhesional energy even greater than $2\gamma'_{LV^0}$, it is apparent that when $\theta = 0$ the theoretical adhesive strength will always be much more than the observed tensile or shear strength of the adhesive material.

Real solid surfaces are neither flat nor free of pores and crevices, and no treatment of adhesion can disregard the realities of surface structure. Each adherend has a true surface area which is r times greater than the apparent or envelope area; hence the work of adhesion should be expected to be r times greater than that for the apparent surface area. However, the larger the contact angle the more difficult it becomes to make the liquid flow over the surface of each adherend to fill completely every crevice and pore in the surface. More often there are air pockets trapped in the hollows and crevices. Such difficulties with the formation of gas bubbles and pores are, of course, greatly amplified in dealing with viscous adhesives which solidify shortly after being applied to form the joint. Hence in practice the true value of W_{A*} lies somewhere between the value obtained from Equation 4c by using the apparent contact angle and r times that value. Where there are accessible pores, crevices, and capillaries in the surface of the adherend, the viscous liquid adhesive may penetrate to some extent and so increase adhesion, provided it does not harden too soon and an adequate supply of the liquid is available. However, in order to obtain the maximum adhesion, the adhesive should obviously be able to penetrate into each capillary and fill it.

An obvious approximation is to assume that the capillary rise equation

$$h = \frac{k\gamma_{LV^0} \cos \theta}{\rho R} \tag{16}$$

can be used (R is the equivalent radius of the capillary, $k = 2/981$, and ρ is the density of the liquid) and also that the liquid still wets the capillary wall according to the $\cos \theta$ vs. γ_{LV^0} relation of Equation 11. Eliminating $\cos \theta$ from the two equations results in the equation for the parabola [106]:

$$h = \frac{k\gamma_c}{\rho R} (b + \gamma_c + 1) - \frac{(bk)}{R\rho} \gamma_{LV^0}^2 \tag{17}$$

where h has a maximum when

$$\gamma_{LV^0} = \frac{1}{2} \left(\gamma_c + \frac{1}{b} \right) \tag{18}$$

For example, in smooth polyethylene the maximum capillary rise will occur when $\gamma_{LV^0} = 1/2 \ (31 + 38.4) = 34.7$ dynes per cm.; this value is only 3.7 dynes per cm. more than γ_c. Despite the unrealistic assumption made, this analysis reveals again that W_{A*} may go through a maximum as γ_{LV^0} increases, even when a porous interfacial surface is involved.

As the result of the intense localization of the residual field of force emanating from both the adherend and adhesive, when a liquid adhesive solidifies, the value of W_A for the joint should remain close to the value computed for the adhesive in the liquid state if stress concentrations do not develop in the process. Since the forces causing adhesion are effective to little more than the depth of one molecule in both surfaces, they will be unaffected by changes of state of the bulk phases so long as allowance is made for any resulting changes in the surface density or molecular orientation occurring at the joint interface [106]. The former can be estimated from the change of density on solidification, but the latter may be difficult to compute, since reorientation effects could originate through any crystallization process starting at or near the interface. It is also usual for internal stresses and stress concentrations to develop when the adhesive solidifies, the most common cause being the difference in the thermal expansion coefficients of adhesive and adherend. In some applications, care is taken to match the thermal expansion coefficients of adhesive and adherend; however, in others this matching process is not critical. Hence, the actual strength of the adhesive joint is usually considerably less than the theoretical (or thermodynamic) value because of the development of internal stress concentrations.

Mylonas [86] has shown that in a lap joint poor wetting of the adherend tends to produce a greater stress concentration at the free surface of the adhesive where failure is most likely to be initiated. As the contact angle, θ, becomes large, the maximum stress concentration increases and moves toward the edge where the adhesive makes contact with the adherend, the stress concentration factor increasing from about 1.2 when $\theta = 30°$, to about 2.5 when $\theta = 90°$ (Figure 22). Furthermore, Griffith [52] has shown that failure of the adhesive may occur at a relatively small applied stress if there are air bubbles, solid inclusions, or surface defects; it occurs because stress concentrations result which are much higher than the mean stress applied across the specimen. His conclusion is very important, because the most probable effect of poor wetting is the development of air pockets or voids at the adhesive-adherend interface. Even when $\theta = 0$, there may be gas pockets formed at the adhesive-adherend interface around which stress concentrations can build up, for if the adhesive is too viscous when

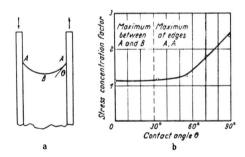

Figure 22. Variation of stress concentration maxima in lap joint adhesive with angle of contact [86]

applied, it may never penetrate the accessible surface pores before polymerizing. Of course, this situation is the more aggravated the larger the contact angle or the rougher the surface.

In adhesives technology it is common practice to roughen the surface of each adherend, or "give it tooth," and so obtain a stronger joint. This practice can be justified theoretically, with certain limitations, by the following considerations. If the gas pockets formed in the surface depressions of the adherend are all nearly in the same plane and are not far apart (as on the upper adherend of Figure 23) there may be crack propagation from one pocket to the next, and the joint may break as if it had a built-in "zipper." Therefore, if roughness must be accepted, the kind of roughness shown on the lower adherend would be preferable because crack propagation along a plane would be less probable.

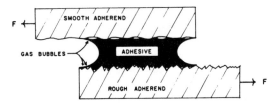

Figure 23. Effect of surface roughness on coplanarity of gas bubbles [106]

"Abhesives" are materials used in the form of films or coatings which are applied to one solid to prevent—or greatly decrease—the adhesion to another solid in intimate contact with it. Such materials are employed in molding, casting, or rolling operation; therefore, it is also common to refer to the film as the "parting," "mold-release," or "antistick" agent. Examples of materials commonly used for such purposes are the polymethylsiloxanes; the high molecular weight fatty acids, amines, amides, and alcohols; various types of highly fluorinated fatty acids and alcohols; and the fluorocarbon resins. Usually a condensed monolayer or thin film of the agent is sufficient to cause the optimum parting effect. Table XII gives the γ_c values of each of these films. Each abhesive coating will convert the solid into a low-energy surface, and any liquid placed on such a coated surface will exhibit an equilibrium contact angle which will be larger as $\gamma_{LV^o} - \gamma_c$ increases. When θ is large enough, such poor adhesion results on solidification of the liquid that the application of a modest external stress suffices for

Table XII. Critical Surface Tensions of Wetting of
Surfaces Coated with Abhesive Films [106]

Coating Material	γ_c, Dynes/Cm. at 20°C.
Polymethylsiloxane film	24
Fatty acid monolayer	24
Polytetrafluoroethylene film	18
Ψ'-Fatty acid monolayer	15
Polymethacrylic ester of Φ'-octanol	10.6
Perfluorolauric acid monolayer	6

parting or splitting of the joint along the adhesive-adherend interface. The excellent and easy parting action usually achieved is the result of the formation of a deliberately weakened joint; it should be explainable by any general theory of adhesives.

It was pointed out earlier that W_A exhibits a maximum variation among the various low-energy surfaces of about threefold in going from the least to the most adhesive liquids. Such a range is too small to explain the effectiveness and easy parting action of a good adhesive. However, this range of values of W_A is for flat, nonporous, smooth, solid surfaces. The roughness factor, r, of the uncoated surface of the mold could raise W_A by a factor of at least 1.5 to 3, and it could be much larger the less smooth the surface finish; however, if the material to be molded is (or becomes) viscous rapidly during mold injection or application of the liquid adhesive, poor wetting will cause voids or gas pockets to be produced at the interface between the material molded and the adhesive, and thereby the adhesion will be greatly decreased through stress concentrations by some unknown fraction, 1/g. Furthermore, if $\theta > 30°$, the resulting stress concentration factor of from 1.2 to 2.5 at the adhesive-plastic interface will contribute an additional decrease in the adhesion, and so the adhesional work per unit area of the apparent or envelope area at the molding interface will be $(r/gs)W_A$. Therefore, for both reasons the greater the value of θ, the greater the effectiveness of the release agent in weakening the joint formed.

The large values of θ encountered with the various types of organic liquids on such low-energy surfaces as those shown in Table XII are good evidence in favor of this explanation of abhesive action. As a general proposition, any low-energy surface will be more effective as a release agent (or abhesive) the lower its γ_c value [106]. It can also be concluded that the smoother the finish of the outer coating of the mold and the lower the surface tension and viscosity of the material being molded (or the lower its contact angle with the adhesive), the greater will be the external stress required to cause the desired parting action. In conclusion, a good adhesive joint can be made on any low-energy surface, including any abhesive surface, if the surface is not too rough or if the adhesive has $\gamma_{LV} < \gamma_c$ —i.e., if there is good wetting by the adhesive.

There has been much discussion about the necessity for using adhesives capable of forming chemical bonds with the adherends. But the preceding considerations concerning the relation of wetting to adhesion have made it evident that the energy involved in the physical adsorption to the adherend of molecules of adhesive is more than sufficient to form joints which are stronger than the cohesive strength of existing adhesives. Of course, chemical bonding may offer advantages other than that of increasing the joint strength—for example, greater heat, water, or chemical resistance may result. Under some circumstances a few monolayers of water adsorbed on the adherend can be tolerated because they are readily displaced by the adhesive or absorbed in the adherend or adhesive; but usually bulk water cannot be tolerated because of its low shear strength and its adverse effect on spreading of the adhesive. However, it is usually a great advantage, when selecting an adhesive for a given application, to be released from the very limiting requirement of finding a material capable of chemically bonding to the adherends.

In summary, for optimum or theoretical joint strength, it is essential to keep the contact angle between the liquid adhesive and the adherends as small as possible in order to obtain good spreading and to minimize the buildup of stress concentrations. Obviously, the interface of each adherend must be kept as smooth and free as possible of low-energy surface films and dust, in order to prevent forming voids, gas pockets, or occlusions. In applying liquid adhesives the viscosity should be as low as possible, in order to increase the extent of capillary flow into pores and crevices. Maximum spreading and capillarity will be obtained with adhesives having the highest surface tension compatible with obtaining a low contact angle. When conditions of complete wetting and freedom from the formation of gas pockets and occlusions prevail, the adhesion to either high-energy or low-energy surfaces will usually be ample, and generally failures of the joint will be in cohesion. If the surface of the adherend is to be roughened, it is better done in such a way as to keep any surface gas pockets formed from being coplanar or nearly so over a large region. When difficulties are encountered in finding a suitable adhesive for a specific application, or in seeking new adhesives, it should be helpful to try an adhesive having a γ_{LV} value (while in the liquid state), which is less than the γ_c value of the adherend surface, and, of course, to apply it under conditions that will minimize the tendency to occlude gas bubbles in the surface.

Future Research on Wetting and Adhesion

An exciting beginning has been made in explaining the wetting, spreading, and adhesive behavior of liquids on solids. We have established that the observation of the equilibrium contact angles of liquids on solids, under appropriately controlled and now well understood conditions, is a powerful method of investigating the surface properties of either solid surfaces or adsorbed monolayers. From the diversity of new surface chemical phenomena revealed by our investigations, it should be apparent why many past investigations of wetting and adhesion have foundered or gone astray. It is hoped that a solid foundation has been laid for a more understanding experimental and theoretical treatment of many other properties of solid-liquid interfaces. For example, the calorimetric measurement of heats of wetting to characterize either wetting properties or solid surface states will be much more revealing now that we understand the major effects arising from adsorbed films on the surfaces or from trace impurities or additives in the liquids.

Langmuir's "principle of independent surface action" has been found useful and often correct, and the conditions under which it should not be used are now more evident. Our experimental results on the effect of surface constitution on the spreading and adhesive properties of liquids on solid surfaces or on adsorbed films are convincing examples of the wide occurrence in solids and liquids of extremely localized surface forces. Our Table V of critical surface tensions, although empirical and approximate in nature, has proved a powerful tool for describing and predicting important properties of solid surfaces and adsorbed films; undoubtedly, future investigations and applications will be guided by such considerations.

With these new concepts we have been able to develop a simple general treatment of the wetting properties of solutions—both aqueous and nonaqueous. Undoubtedly, the principal function of a wetting agent is to lower the surface tension of the solution below the critical surface tension of wetting of the solid surface to be wetted. One of the most interesting by-products of our investigations has been the recognition of the major part played by incomplete wetting and by low-energy surfaces in the adhesives and related arts. When Griffith's important effect of stress concentrations around voids or gas pockets was coupled with our results on the effects of surface composition on wetting and spreading by the liquid adhesives, a general and useful theoretical treatment of adhesional joints and adhesives emerged.

Many applications in research and industry have already been made of the results of the investigations reviewed here; undoubtedly many more will be made. These include: studies of other physical and chemical properties of adsorbed retracted monolayers, a full exploration of the nonspreading properties of watch, clock, and fuse oils; the all-Teflon film balance for investigating the mechanical properties of surface films on organic liquids; the development of new and improved treatments for chemical-, oil- and water-repellent finishes for clothing, canvas, etc.; improved finishes for textile fibers and tire cords; an explanation and generalization of the many processes based on mold-release agents or abhesives, which are important in the plastic and paper industries; an explanation of the spreading properties of printing inks and the key to developing better inks as well as those capable of writing on greasy surfaces; both improved and new adhesives; improved polishes and cleaners for plastic floor coverings and tile; and a long-sought explanation of the relation of wetting properties of liquids to their effectiveness as lubricants.

Despite these many advances, certain aspects of the subject of this paper have been neglected and further investigation is much needed. No reliable data are available on the effect of temperature on the equilibrium contact angles of any systems; all of the measurements reported here were made at 20°C. and 50% relative humidity. However, several general effects of variations in the temperature and humidity on the contact angle can be outlined. It is well known that the surface tensions, γ_{LV°, of pure organic liquids decrease linearly with rising temperature until close to the critical temperatures. Hence, if the temperature is T, then

$$\gamma_{LV^\circ} = C_1 - C_2 T$$

But we have shown that $\cos \theta_E = a - b\gamma_{LV^\circ}$, when the surface composition of the solid is constant. Here C_1, C_2, a, and b are positive. Therefore,

$$\cos \theta_E = (a - bC_1) + bC_2 T$$

and $\cos \theta_E$ must increase (or θ_E decrease) with increasing temperature. In addition, the effect of raising the temperature will be to cause surface expansion and increased desorption of any physically adsorbed compounds. Temperature increase will therefore decrease the packing of such adsorbed films, and this will raise the critical surface tension of the system and thus cause θ_E to decrease with rising temperatures.

When the temperature becomes high, there may result chemical changes in the liquids, such as hydrolysis, oxidation, and pyrolysis, or there may develop surface-chemical changes in the solid due to oxidation, dehydration, or crystallographic rearrangement. The products of chemical reaction in the liquids may be highly adsorbable and may cause the formation of new low-energy surface films on which the liquids will not spread. This effect may overbalance the above-mentioned normal decrease in θ_E with rising temperature. The oxidation or dehydration of these inorganic solid surfaces may greatly alter the wetting behavior of the liquids. But if the chemical reactivity or adsorptivity of the liquid is little changed thereby, no large change in wettability is to be expected.

Decreasing the relative humidity at ordinary temperatures will have little effect on θ_E unless the atmosphere is made so dry as to dehydrate the solid surface (as in silica and α-alumina). But usually long exposure to very dry air will be required to change the contact angles significantly. As the relative humidity approaches 100%, increased condensation of water on the surfaces of both metals and metallic oxides will invite the hydrolysis *in situ* of some adsorbed liquids, and so the wetting properties of these surfaces will become more like those of such highly hydrated surfaces as glass, silica, and alumina.

Obviously, data are needed on the values of f_{SV^o} for a variety of low-energy solid surfaces; with these and the values of W_{A*} given here, one would no longer have to estimate W_A, and hence our knowledge of adhesion would become more precise.

The empirical nature of γ_c is obvious, and it would be helpful to replace γ_c by parameters having a sound basis in thermodynamic or statistical mechanical considerations. Recent efforts by Fowkes [40] to relate γ_c to the dispersion forces between molecules at the interface have been especially promising in leading to tractable equations. An interesting direct correlation has been recently pointed out to us by Gardon [49] between the value of γ_c of a solid polymer and the Hildebrand solubility parameter, δ, which is defined as the square root of the molar energy density—i.e., $\delta = \sqrt{E/\rho'}$. A simple consideration of the Young equation and the definition of γ_c indicates that when $\cos \theta = 1$,

$$\gamma_{SV^o} - \gamma_{SL} \doteq \gamma_c$$

Unfortunately, the effect of constitution on either γ_{SV^o} or γ_{SL} is still unknown, and neither quantity can be studied until a satisfactory experimental method for measuring it has been found. However, neither quantity appears to be a simple function of the constitution of the solid and liquid phases. A more precise theory of wetting of low-energy surfaces should at least include γ_{SL}. Probably the lateral spread in the data points of our graphs of $\cos \theta$ vs. γ_{LV} for a given solid surface is due to the variation of γ_{SL} among the liquids used.

Finally, our investigations have been concerned with the wetting of a solid by one liquid, and nothing has been said about the classic problem of the competitive wetting and adhesion of two immiscible liquids with respect to a given solid surface. However, it is obvious that the findings reported here for the simpler systems should prove helpful in learning more about the more complex.

Literature Cited

(1) Am. Assoc. Textile Chemists and Colorists, Piedmont Section, <u>Am. Dye-stuff Reptr</u>. **53**, 25 (Feb. 4, 1963).

(2) Baker, H. R., Shafrin, E. G., Zisman, W. A., <u>J. Phys. Chem</u>. **56**, 405 (1952).

(3) Bakker, G., "Kapillarität und Oberflächenspannung," in Wien-Harms "Handbuch der Experimental Physik," Vol. VI, Academischer Verlag, Leipzig, 1928.

(4) Bangham, D. H., <u>Trans. Faraday Soc</u>. **33**, 805 (1937).

(5) Bangham, D. H., Razouk, R. I., <u>Ibid</u>., **33**, 1459 (1937).

(6) Banks, W. H., <u>Nature</u> **174**, 365 (1954).

(7) Banks, W. H., <u>Proc. 2nd Intern. Congr. Surface Activity</u>, Vol. I, p. 16, Academic Press, New York, 1957.

(8) Bartell, F. E., Zuidema, H. H., <u>J. Am. Chem. Soc</u>. **58**, 1451 (1936).

(9) Bartell, L. S., Ruch, R. J., <u>J. Phys. Chem</u>. **60**, 1231 (1956).

(10) <u>Ibid</u>., **63**, 1045 (1959).

(11) Basford, P. R., Harkins, W. D., Twiss, S. B., <u>Ibid</u>., **58**, 307 (1954).

(12) Bernett, M. K., Jarvis, N. L., Zisman, W. A., <u>Ibid</u>., **66**, 328 (1962).

(13) Bernett, M. K., Zisman, W. A., <u>Ibid</u>., **63**, 1241 (1959).

(14) <u>Ibid</u>., p. 1911.

(15) <u>Ibid</u>., **64**, 1292 (1960).

(16) <u>Ibid</u>., **65**, 2266 (1961).

(17) <u>Ibid</u>., **66**, 1207 (1962).

(18) Berry, K. L., U.S. Patent **2,559,629** (July 10, 1951).

(19) Bewig, K., Zisman, W. A., <u>Advan. Chem. Ser</u>., No. **33**, 100 (1961).

(20) Bewig, K., Zisman, W. A., <u>J. Phys. Chem</u>. **67**, 130 (1963).

(21) Bigelow, W. C., Brockway, L. O., <u>J. Colloid Sci</u>. **11**, 60 (1956).

(22) Bigelow, W. C., Glass, E., Zisman, W. A., <u>Ibid</u>., **2**, 563 (1947).

(23) Bigelow, W. C., Pickett, D. L., Zisman, W. A., <u>Ibid</u>., **1**, 513 (1946).

(24) Blake, G. G., Ahlbrecht, A. H., Bryce, H. Y., "Quaternary Ammonium Compounds from Fluorinated Acids," Division of Industrial and Engineering Chemistry, 126th Meeting, A.C.S., New York, N. Y., September 1954.

(25) Boyd, G. E., Livingston, H. K., <u>J. Am. Chem. Soc</u>. **64**, 2383 (1942).

(26) Brace, N. O., <u>J. Org. Chem</u>. **27**, 4491 (1962).

(27) Brockway, L. O., Karle, J., <u>J. Colloid Sci</u>. **2**, 277 (1947).

(28) Cassie, A. B., Baxter, S., <u>Trans. Faraday Soc</u>. **40**, 546 (1944).

(29) Cooper, W. A., Nuttall, W. A., <u>J. Agr. Sci</u>. **7**, 219 (1915).

(30) De Boer, J. H., <u>Advan. Colloid Sci</u>. **3**, 27 (1950).

(31) Dupré, A., "Théorie Méchanique de la Chaleur," p. 369, Gauthier-Villars, Paris, 1869.

(32) Ellison, A. H., Fox, H. W., Zisman, W. A., <u>J. Phys. Chem</u>. **57**, 622 (1953).

(33) Ellison, A. H., Zisman, W. A., <u>Ibid</u>., **58**, 260 (1954).

(34) <u>Ibid</u>., p. 503.

(35) <u>Ibid</u>., **59**, 1233 (1955).

(36) <u>Ibid</u>., **60**, 416 (1956).

(37) <u>Ibid</u>., **63**, 1121 (1959).

(38) Epstein, H. T., <u>Ibid</u>., **54**, 1053 (1950).

(39) Fischer, E. K., Gans, D. M., <u>Ann. N. Y. Acad. Sci</u>. **46**, 371 (1946).

(40) Fowkes, F. M., <u>Advan. Chem. Ser</u>., No. **43**, 99 (1963).

(41) Fowkes, F. M., Harkins, W. D., <u>J. Am. Chem. Soc</u>. **62**, 3377 (1940).

(42) Fox, H. W., Taylor, P. W., Zisman, W. A., <u>Ind. Eng. Chem</u>. **39**, 1401 (1947).

(43) Fox, H. W., Hare, E. F., Zisman, W. A., <u>J. Colloid Sci</u>. **8**, 194 (1953).

(44) Fox, H. W., Hare, E. F., Zisman, W. A., <u>J. Phys. Chem</u>. **59**, 1097 (1955).

(45) Fox, H. W., Levine, O., private communication.

(46) Fox, H. W., Zisman, W. A., <u>J. Colloid Sci</u>. **5**, 514 (1950).

(47) <u>Ibid</u>., **7**, 109 (1952).

(48) <u>Ibid</u>., p., 428.

(49) Gardon, J. L., <u>J. Phys. Chem</u>. **67**, 1935 (1963).

(50) Gibbs, J. Willard, <u>Trans. Connecticut Acad</u>. **3** (1876-1878); "Collected Works," Vol. I, Longmans, Green, New York, 1928.

(51) Graham, D., J. Phys. Chem. **66**, 1815 (1962).
(52) Griffith, A. A., Phil. Trans. Roy. Soc. (London) **A221**, 163 (1920).
(53) Hare, E. F., Shafrin, E. G., Zisman, W. A., J. Phys. Chem. **58**, 236 (1954).
(54) Hare, E. F., Zisman, W. A., Ibid., **59**, 335 (1955).
(55) Harkins, W. D., Chem. Revs. **29**, 408 (1941).
(56) Harkins, W. D., Ewing, W. W., J. Am. Chem. Soc. **42**, 2539 (1920).
(57) Harkins, W. D., Feldman, A., Ibid., **44**, 2665 (1922).
(58) Harkins, W. D., Grafton, E. H., Ibid., **42**, 2534 (1920).
(59) Harkins, W. D., Loeser, E. H., J. Chem. Phys. **18**, 556 (1950).
(60) Hirshfelder, J. D., Curtiss, C. F., Bird, R. B., "Molecular Theory of Gases and Liquids," Wiley, New York, 1954.
(61) Jarvis, N. L., Fox, R. B., Zisman, W. A., Advan. Chem. Ser., No. 43, 317 (1963).
(62) Jarvis, N. L., Zisman, W. A., J. Phys. Chem. **63**, 727 (1959).
(63) Ibid., **64**, 150, 157 (1960).
(64) Johnson, R. E., Ibid., **63**, 1655 (1959).
(65) Jones, D. C., Saunders, L., J. Chem. Soc. 1951, 2944.
(66) Kaminski, A., Stanford Research Inst. Rept. 2, Contract N60ri-15402, Jan. 12, 1948.
(67) Karle, J., Brockway, L. O., J. Chem. Phys. **15**, 213 (1947).
(68) Klevens, H. B., Raison, M., J. Chim. Phys. **51**, 1 (1954).
(69) Kraus, G., Manson, J. E., J. Polymer Sci. 6 (5), 625 (1951).
(70) Langmuir, I., Chem. Revs. **6**, 451 (1929).
(71) Langmuir, I., "Collected Works," Pergamon Press, New York, 1960.
(72) Langmuir, I., J. Am. Chem. Soc. **38**, 2221 (1916).
(73) Langmuir, I., J. Franklin Inst. **218**, 143 (1934).
(74) Langmuir, I., Science **87**, 493 (1938).
(75) Langmuir, I., Trans. Faraday Soc. **15** (3), 62 (1920).
(76) Laplace, P., "Méchanique Céleste," Suppl. 10th Vol., 1806.
(77) Lester, G. R., J. Colloid Sci. **16**, 315 (1961).
(78) Levine, O., Zisman, W. A., J. Phys. Chem. **61**, 1068 (1957).
(79) Ibid., p. 1188.
(80) Loeser, E. H., Harkins, W. D., Twiss, S. B., Ibid., **57**, 251 (1953).
(81) McBain, M. E. L., Perry, L. H., J. Am. Chem. Soc. **62**, 989 (1940).
(82) Martinet, J. M., "Adsorption des Composés Organiques Volatiles par le Polytétrafluor Éthylene," Commissariat à l'Énergique Atomique, Rapport CEA 888, Centre d'Etudes Nucléaires de Socloy, 1958.
(83) Menter, J. W., Tabor, D., Proc. Roy. Soc. (London) **A204**, 151 (1951).
(84) Merker, R., Zisman, W. A., J. Phys. Chem. **56**, 399 (1952).
(85) Moilliet, J. L., J. Oil Colour Chemists' Assoc. **38**, 463 (1955).
(86) Mylonas, C., Proc. VII Intern. Congr. Appl. Mech., London, 1948.
(87) Quayle, O. R., Chem. Revs. **53**, 439 (1959).
(88) Ray, B. R., Anderson, J. R., Scholz, J. J., J. Phys. Chem. **62**, 1220 (1958).
(89) Scholberg, H. M., Guenthner, R. A., Coon, R. I., Ibid., **57**, 923 (1953).
(90) Scholz, J. J., Ray, B. R., Anderson, J. R., Ibid., **62**, 1227 (1958).
(91) Schulman, F., Zisman, W. A., J. Colloid Sci. **7**, 465 (1952).
(92) Shafrin, E. G., Zisman, W. A., "Hydrophobic Monolayers and Their Adsorption from Aqueous Solutions," p. 129, in "Monomolecular Layers," Am. Assoc. Advance. Sci., Washington, D. C., 1954.
(93) Shafrin, E. G., Zisman, W. A., J. Colloid Sci. **4**, 571 (1949).
(94) Ibid., **7**, 166 (1952).
(95) Shafrin, E. G., Zisman, W. A., J. Phys. Chem. **61**, 1046 (1957).
(96) Ibid., **64**, 519 (1960).
(97) Ibid., **66**, 740 (1962).
(98) Shuttleworth, R., Bailey, G. L., Discussions Faraday Soc. 1948, No. **3**, 16.
(99) Sumner, C. G., "Symposium on Detergency," p. 15, Chemical Publ. Co., New York, 1937.
(100) Tabor, D., Rept. Progr. Appl. Chem., Soc. Chem. Ind., London **36**, 621 (1951).
(101) Thompson, W., Phil. Mag. (4) **17**, 61 (1858).

(102) Thompson, W., Proc. Roy. Soc. (London) **9**, 255 (1858).
(103) Wenzel, R. N., Ind. Eng. Chem. **28**, 988 (1936).
(104) Young, Thomas, Phil. Trans. Roy. Soc. (London) **95**, 65 (1805).
(105) Zisman, W. A., "Relation of Chemical Constitution to the Wetting and Spreading of Liquids on Solids," p. 30 in "A Decade of Basic and Applied Science in the Navy," U. S. Government Printing Office, Washington, D. C., 1957.
(106) Zisman, W. A., "Constitutional Effects in Adhesion and Abhesion," in "Symposium on Adhesion and Cohesion," P. Weiss, ed., Elsevier, New York, 1962.
(107) Zisman, W. A., O'Rear, J. G., U. S. Patent 2,824,141 (Feb. 18, 1958).

Received April 16, 1963.

1964

KAROL J. MYSELS

UNIVERSITY OF SOUTHERN CALIFORNIA
LOS ANGELES, CA.

. . . for his outstanding contributions in original research and in guiding students in this branch of science. His work has led to intimate knowledge of the size, shape, surface structure, effective charge, and equilibrium relations of micelles formed by ionic association colloids, based on recognition and patient elimination of the effects of trace impurities and on extraordinarily precise measurements of their physical properties. Equally significant is his work on the formation and life history of soap films, leading to increased understanding of electrical double-layer effects and of van der Waals forces in colloidal systems.

The following Kendall Award Address is reproduced by permission from *The Journal of Physical Chemistry,* **68**, 3441 (1964). The Addendum has been written by Dr. Mysels for this volume.

Soap Films and Some Problems in Surface and Colloid Chemistry[1]

by Karol J. Mysels

Department of Chemistry, University of Southern California, Los Angeles, California 90007
(Received May 19, 1964)

Once their behavior is understood, soap films become a powerful tool for the study of surface and colloidal phenomena. This is because their geometry is well defined, their monolayer structure is relatively simple, and their behavior is easily observed. The formation and evolution of relatively thick films are controlled by ordinary hydrodynamics, without any indication of the existence of rigidified aqueous layers near the surface. Thinner films show the effect of both double-layer repulsion and van der Waals attractions. These forces balance at a thickness of the order of 100 Å. Some problems in the interpretation of recent measurements of these equilibrium thicknesses and some approaches leading to further information about these forces are discussed. Under certain conditions the films thin even further to about 45 Å. This is a thermodynamically stable state, which, though first observed by Newton, is still far from understood.

Soap films—the gossamer sheets which form a child's soap bubble, the pleasurable head on beer, or the distressing overflowing foam of the production vessel—have a respectable scientific history since the days when Robert Hooke[2] first called the attention of the Royal Society and of Newton to the optical phenomena which they exhibit. They have assisted in the development of the theory of optics,[3] of capillary forces,[4,5] and of minimal area problems[6]; they have served as delicate tools for detecting the magnetism of gases[7] and as analog computers in solving differential equations with complicated boundary conditions.[8] Today soap films serve science again in the elucidation of a number of problems in surface and colloid chemistry such as those of phase transitions in monolayers, of the structure of solvent in the neighborhood of a surface, of Gibbs film elasticity, of the magnitude of the double-layer repulsion, of the law of van der Waals attraction at

(1) Based on the Kendall Award Lecture presented at the 148th National Meeting of the American Chemical Society, Philadelphia, Pa., April, 1964. It reviews our work on films of ionic surfactants, which began under a 1-year grant from the American Chemical Society Petroleum Research Fund, was supported for several years by the Air Force Office of Scientific Research, and now continues with support from the National Science Foundation.

(2) R. Hooke, Communication to the Royal Society, March 28, 1672; T. Birch, "History of the Royal Society," Vol. III, A. Millard, London, 1757, p. 29.

(3) I. Newton, "Opticks," Book I, Part 2, expt. 4; Book II, Part 1, obs. 17–21, London, 1704.

medium range, and of factors governing specific ionic interactions.

Experimental Background. Just as the classical applications of soap films required the development of appropriate new techniques—since Newton first put a soap bubble under a bell jar to protect it from air currents and excessive evaporation[3]—so the recent developments have required new experimental approaches to produce the desired quantitative data. In addition, a new look was needed at the basic mechanisms by which a soap film is formed initially as a rather thick sheet of the order of 1 μ and then thins more or less gradually to an equilibrium structure whose thickness may be somewhat less than 50 Å.[9]

The basic condition for observing soap films over any length of time is that evaporation be absent so that closed vessels and good thermostats or minute circular films completely surrounded by a meniscus[10] are used. Plateau's[4,11] expedient of adding a hygroscopic agent, glycerine, to prevent bursting by evaporation has the basic disadvantage that the composition of the film changes all the time so that its behavior becomes much more complicated.

Soap films are almost always observed in reflected light so that the interference of the beams reflected by their front and rear surfaces produces colors or variations of intensity by which film thickness can be estimated and the motion of patches of different thicknesses observed. The direct and reflected beams may be either separate or combined by means of a semi-transparent mirror.

Visual observations are easiest when the films are large and flat. For this reason most of our work involved films several centimeters on a side or in diameter. Vertical films such as shown in Fig. 1 are most easily produced by submerging a rectangular frame in a solution and then withdrawing it partially. Horizontal films such as shown in Fig. 2 and 3 can be supported by the top of a funnel with a narrowed mouth. Very slight curvatures may then be produced by slight changes of air pressure inside the funnel.

Phase Transitions and Film Elasticity. There is a striking difference between two types of films. One, the mobile film, thins in minutes, shows turbulent motions along the edges, and has a horizontal layering of interference colors. The other, the rigid film, thins in hours, shows little motion, and has a generally irregular arrangement of colors. Most ionic surfactant solutions give mobile films. Rigid films are formed by a few combinations of surfactants capable of giving rigid monolayers[12] by tight packing of adsorbed molecules,[13] the two-dimensional equivalent of solid bodies. The transition between the two—the "melting" of

Figure 1. A rigid film (left) changes largely to a mobile one (right) as the total surface is rapidly expanded. Note that most fringes have become horizontal and that a remaining rigid area has risen to the top of the film. Film height is about 3 cm.

the rigid film—can be effected by a slight change of temperature[14] and also by an isothermal change in the surface area available per molecule as we found a few years ago with Dr. J. Skewis. The apparatus used was the same as devised for measuring Gibbs film elasticity.[15] A rectangular frame is rapidly raised from the solution or lowered into it. Both frames operate within a square bottle, 5 × 5 cm. As the total area of the solution is quite small, this motion of the second frame produces large percentage changes in the area available per adsorbed molecule before sur-

(4) J. Plateau, "Statique Experimentale et Theorique des Liquides Soumis aux Seulles Forces Moleculaires," Gauthier-Villars, Paris, 1873.

(5) J. W. Gibbs, *Trans. Connecticut Acad.*, **3**, 108, 343 (1876–8); "Collected Works," Vol. I, Longmans Green, New York, N. Y., 1928, 1931: (a) p. 300; (b) pp. 301–303; (c) p. 309.

(6) R. Courant and H. Robbins, "What is Mathematics?" Oxford University Press, New York, N. Y., 1941.

(7) M. Faraday, *Phil. Trans. Roy. Soc. London*, **141**, 7 (1851).

(8) L. Prandtl, *Physik. Z.*, **4**, 758 (1903); B. Johnston, *Civil Eng.* (N. Y.), **5**, 698 (1935).

(9) K. J. Mysels, K. Shinoda, and S. Frankel, "Soap Films—Studies of Their Thinning," Pergamon Press, New York, N. Y., 1959: (a) Chapter III; (b) p. 34–38, 85–88, pl. II; (c) p. 20–24; (d) p. 38–40; (e) p. 36; (f) p. 58–60; (g) p. 74–76; (h) Chapter V; (i) p. 69–71.

(10) A. Scheludko and D. Exerowa, *Kolloid-Z.*, **155**, 39 (1957).

(11) C. V. Boys, "Soap Bubbles and the Forces Which Mould Them," Society for Promoting Christian Knowledge, London, 1890, 1912, and recent reprints, Appendix.

(12) J. Ross and M. B. Epstein, *J. Phys. Chem.*, **62**, 533 (1958).

(13) A. Wilson, M. B. Epstein, and J. Ross, *J. Colloid Sci.*, **12**, 345 (1957).

(14) M. B. Epstein, J. Ross, and C. W. Jakob, *ibid.*, **9**, 50 (1954); M. B. Epstein, A. Wilson, C. W. Jakob, L. E. Conroy, and J. Ross, *J. Phys. Chem.*, **58**, 860 (1954).

(15) K. J. Mysels, M. C. Cox, and J. D. Skewis, *ibid.*, **65**, 1107 (1961).

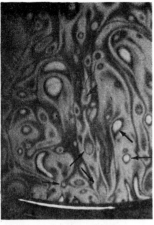

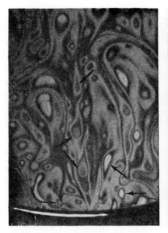

Figure 2. Marginal regeneration in an almost horizontal mobile film. Similarly oriented arrows indicate the same regions of the film in two successive stages which show the motion of thicker regions towards the border and that of the thinner ones away from it. The rim of the supporting funnel is at the bottom. The area shown is about 6 mm. across.

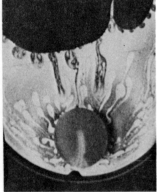

Figure 3. An almost horizontal mobile film. The round feature near the bottom edge is a droplet of solution some 4 mm. in diameter. The whole thick patch indicated by arrows on the left has been sucked in by the border of the drop as shown on the right. Note also the growth of the black film and the formation of thicker film areas by the displaced liquid.

face equilibrium is re-established. Hence the surface (or parts of it) loses its rigidity when the auxiliary frame is drawn out and becomes rigid again as it is lowered. Figure 1 shows selected frames from a color movie[16] recording the process.

The expansion and contraction of the area of the observed film as the auxiliary frame oscillates up and down can be evaluated from a photographic record. The corresponding variation of surface tension of the system can be recorded simultaneously. This gives[15] the two quantities needed to calculate the film elasticity E as defined by Gibbs.[5b]

$$E = 2d\sigma/d \ln s \qquad (1)$$

where σ is the surface tension, and s the area of the film.

Gibbs[5b] has also shown that in a two-component system this elasticity should be given by

$$E = 4\Gamma^2(d\mu/dG) \qquad (2)$$

where Γ is the surface density of the solute, μ its chemical potential, and G its total amount present per unit film surface. Verification of this relation requires experiments on rigorously purified systems since the presence of enough contaminant to form a fraction of a monolayer is enough to completely vitiate the results. We are now developing techniques[17] for working with pure surfaces, and as these are perfected it should become possible to test eq. 2 and perhaps use soap films to measure Γ and $d\mu$.

Thinning Mechanisms. Rigid films thin primarily by the gravitational outflow of the solution from between the two immobile surface layers, and this seems to obey the simple laws of hydrodynamics.[9a] As the constraining surfaces are so near each other, the ordinary viscosity of water reduces the rate of outflow to the point where it takes hours for the film to thin.

In mobile films the situation is more complex. Their rapid thinning involves the relative motion of whole patches of film, the two surfaces and the intralamellar solution moving together over easily visible areas and shearing against adjacent patches of the film. This motion is due to two different causes. The first is gravity[5c,9b] which causes the thinner (and therefore lighter) areas to move upwards replacing the ones that are thicker (and therefore heavier), thus leading to the horizontal stratification of colors. The other is the effect of the capillary suction at the border where there must always be a curved meniscus (the so-called Plateau border) through which the excess liquid flows down to the level of the bulk solution.[9d] This suction exerts a greater force upon a thick film than upon a thin film, thus causing the thick film to be pulled into the border while thin film is simultaneously pulled out of the border to replace the lost area of the film. This "marginal regeneration" mechanism[9c] is best observed in almost horizontal films where the effects of gravity are minimized.[9e] Figures 2 and 3 show the reality of marginal regeneration by stills from two color movies of such films. Whereas the details of the pulling-in process are difficult to analyze theoretically[9f] and have not yet been isolated experimentally, the pulling-out has been studied further as we shall see shortly.

(16) First shown at the 1960 Colloid Symposium at Lehigh University.

(17) P. Elworthy and K. J. Mysels, to be published.

As the film becomes thinner it reaches a point where it appears black because of the disappearance (or more exactly great reduction) of the reflected light as the two surfaces come close together. This low intensity of the reflected beam can still be measured photoelectrically and gives a sensitive measure of film thickness. To the naked eye, however, the film is invisible. This illusion is heightened by the sharp and abrupt transition between the uniformity of this black film and the colorful and smooth variations of the neighboring thicker film. Perrin likened it to the clean cut produced with a punch.[18] However, recent careful observations by Miss McEntee[19] have shown that a transition region can sometimes be seen in horizontal films. This boundary is frequently the site of a third thinning mechanism as the area of black film grows spontaneously at the expense of the thicker film. The film disproportionates, and the excess liquid, originally present in the area whose thickness decreases, is forced into a thicker welt which gradually grows in thickness and in area and then flows to a lower level by gravity (Fig. 3).[9g]

van der Waals Forces. The forces causing the growth of the black film must be van der Waals attractions between molecules, as has been emphasized by Overbeek.[20] Here these forces are exerted over distances of the order of the film thickness, *i.e.*, some hundreds of angstroms. This is a range in which these forces cannot be studied directly by most other techniques, yet it is also the range of significance in the interaction of colloidal particles.

Furthermore this intermediate range is of more general interest because it includes the transition region in which the distance dependence of van der Waals forces—or, more specifically, London dispersion forces—changes from the inverse seventh to the inverse eighth power. Hence, quantitative measurements in this domain would provide a severe test of the theories which have predicted correctly the short range as well as the retarded, long range behavior but do also specify[21] the still untested details of the transition region.

In this transition region, the distances are too long and the dispersion forces are too weak to be studied by their effect on the individual molecules. At the same time, the distances are too short and the forces are too strong to allow the application of the elegant and sensitive methods using macroscopic plates and lenses.[22] Yet, the simplest apparatus permits the direct observation of their effects in soap films.

At first sight it may be surprising that the same forces which cause the attraction of two glass plates across a thin film of vacuum should also cause the

thinning of a material film having air (which is practically vacuum) on both sides. Yet quantitative calculations show[20,23] that the pressures involved are exactly the same in both cases, and the following simple picture explains qualitatively why this is so.

Let us consider a transition region between thin and thick film in the range where van der Waals forces are significant as indicated in Fig. 4. A water molecule, such as A, situated in this region will be subject to attractions by all its neighbors, but within sphere I these effects all cancel each other since, for every molecule such as B, there is one B′ located diametrically opposite. Beyond this sphere, however, there are molecules such as C and D which do not have any opposites, precisely because the film is wedge-shaped. Their attractions therefore do not cancel and give a resultant force directed towards the thick film and tending to move our molecule away from the black film. Thus, the thin film becomes thinner, and the thick one thicker under the influence of van der Waals forces.

The rate of this growth of black film can be measured quite readily as has been done for several systems by Overbeek and McEntee[19]; it depends on such factors as the thickness of the thick film and the ionic strength, but the kinetics of the process still need clarification. When this is achieved we may have a very simple way of estimating van der Waals forces in this range.

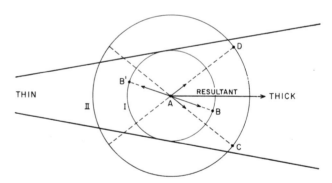

Figure 4. van der Waals forces in a transition region cause the intralamellar liquid to flow towards the thicker film.

(18) J. Perrin, *Ann. Phys.* (Paris), [9] **10**, 165 (1918).

(19) Unpublished results.

(20) J. Th. G. Overbeek, *J. Phys. Chem.*, **64**, 1178 (1960).

(21) H. B. G. Casimir and D. Polder, *Phys. Rev.*, **73**, 360 (1948); H. B. G. Casimir, *Proc. Koninkl. Ned. Akad. Wetenschap.*, **51**, 793 (1948); I. E. Dzyaloshinskii, E. M. Lifshits, and L. P. Pitaevskii, *Usp. Fiz. Nauk*, **73**, 381 (1961).

(22) *E.g.*, B. V. Derjaguin and I. I. Abrikosova, *Discussions Faraday Soc.*, **18**, 24 (1954); J. A. Kitchener and A. P. Prosser, *Proc. Roy. Soc.* (London), **A242**, 403 (1957); W. Black, J. G. V. de Jongh, J. Th. G. Overbeek, and M. J. Sparnay, *Trans. Faraday Soc.*, **56**, 1597 (1960).

(23) A. Scheludko and D. Exerowa, *Kolloid-Z.*, **168**, 24 (1960).

The Nonrigidity of Water near the Surface. Acting alone, the van der Waals forces would cause a self-accelerating thinning and bursting of the film. In fact, however, once the black-film stage is reached, many films remain unchanged for an indefinite time. The fact that the limiting thickness depends on the ionic strength has naturally led to the invocation of the double-layer repulsion theory. This is especially convincing when the surfactant is ionogenic, although water surfaces and those of solutions of nonionic surfactants probably always have a surface potential.[24] However, there has been also a massive body of arguments that a rigidification of water near the surface, perhaps an ice-like formation, is important in preventing the approach of the two surfaces.[25] A very direct refutation of this latter idea can be obtained by a careful study of the pull-out of soap films.[26]

Frankel has analyzed the hydrodynamics in the narrow transition region between the Plateau border and the film proper and has shown[9h, 27] that if one makes the assumptions that the surface of the solution is inextensible in this region and that the viscosity of the solution remains the same as in the bulk clear up to the surface, then it follows that the thickness δ is related to the velocity of pull-out, v, the viscosity, η, surface tension, γ, density, ρ, and gravity, g, by

$$\delta = 1.88 \frac{v^{2/3}\eta^{2/3}}{\gamma^{1/6}\rho^{1/2}g^{1/2}} \qquad (3)$$

Except for the numerical constant, this is the same formula which had been developed much earlier by Derjaguin[28] in connection with the very different problem of a liquid layer left on a solid inextensible substrate withdrawn from a liquid having a completely extensible surface and thus incapable of forming soap films.

The experimental test of Frankel's relation is somewhat difficult for thick (0.2–10 μ) films because of the need to extrapolate to zero film height where the experimentally accessible velocity of the frame approaches that of the film being withdrawn. Nevertheless, good agreement has been obtained for a variety of films having very different surface properties.[27] For thin films (<2000 Å.) the experimental problems are different, but, once these are overcome, it is possible to make very precise measurements.[26]

Equation 3 uses macroscopic, independently measured properties of the liquid to define a direct proportionality between the measured thickness and the two-thirds power of the velocity of pull-out as shown in Fig. 5. Any increased rigidity near the surface leads to an additional thickness, *i.e.*, to a positive intercept at the origin. Deviations from normal hydrodynamic

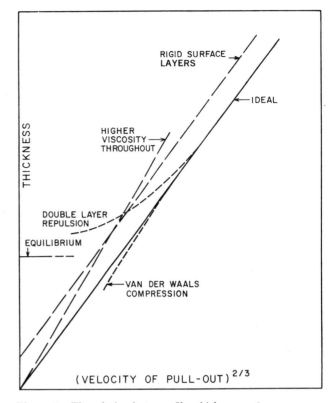

Figure 5. The relation between film thickness and rate of pull-out. Several types of deviations from the simple Frankel law are indicated schematically. At low speeds a thickness independent of rate indicates that a state of equilibrium has been reached.

behavior should alter the slope. Double-layer repulsions and van der Waals attraction should become noticeable for thin films and lead to larger and smaller thicknesses, respectively.[20, 29] In fact, measurements on rigid films below some 800–1000 Å. showed marked deviations from the straight line, but higher thicknesses gave linear results which could be extrapolated to give an intercept of the order of two surface monolayers.[26] This is as would be expected for these rigid films and shows that there are no solidified layers of

(24) *E.g.*, D. Exerowa and A. Scheludko, paper submitted at the 4th International Congress on Surface Active Substances, Brussels, September, 1964.

(25) J. C. Henniker, *Rev. Mod. Phys.*, **21**, 322 (1949); B. V. Derjaguin and A. S. Titijevskaya, *Discussions Faraday Soc.*, **18**, 27 (1954); G. J. Dasher and A. J. Mabis, *J. Phys. Chem.*, **64**, 77 (1960).

(26) J. Lyklema, P. C. Scholten, and K. J. Mysels, *ibid.*, in press.

(27) K. J. Mysels and M. C. Cox, *J. Colloid Sci.*, **17**, 136 (1962).

(28) B. V. Derjaguin and S. M. Levi, "Fiziko-khimiya Naneseniya Tonkikh Sloev na Dvizhushchuyusya Podlozhki," USSR Acad. Sci., Moscow, 1959; B. V. Derjaguin, *Zh. Eksperim. i Teor. Fiz.*, **15**, 9 (1945); see also B. V. Derjaguin and T. N. Voropayeva, *J. Colloid Sci.*, **19**, 113 (1964).

(29) J. Lyklema, *Rec. trav. chim.*, **81**, 890 (1962).

water within the experimental error of, say, 10 Å. for each surface, which is negligible in comparison with the equilibrium thicknesses.

Double-Layer Repulsion. These hydrodynamic experiments show, therefore, that in our systems the equilibrium thickness must depend only on repulsive forces (as opposed to structures). The obvious origin of such forces lies in the double-layer repulsion, *i.e.*, the total electric interaction of the two charged surfaces, as modified by the ion atmosphere of each. The theory of this interaction is well developed, based on the exact solution of the Boltzmann–Poisson equations with the assumption of point charges under the influence of the average potential.[30] Many proposed refinements[31] have only minor effects on the result. The repulsion increases very rapidly as the two charged planes approach each other beyond a certain point. The separation at which this repulsion becomes important increases rapidly as the ionic strength of the solution is lowered. The effect is shown schematically in Fig. 6 for two ionic strengths.

A necessary parameter in this calculation is the surface potential of the film. This is not easily accessible experimentally but fortunately is not of critical importance. A good estimate for this potential is 100 mv., based on electrokinetic measurements on micelles[32] which should have a surface similar to that of the soap films. A change of potential corresponds to a vertical shift of the repulsion lines of Fig. 6 by a factor of less than two, even if the potential is assumed to be infinite or only 60 mv., both highly unlikely possibilities. Such a vertical shift of these lines has only a slight percentage effect upon the thicknesses involved because of the almost vertical direction of the lines themselves.

Equilibrium Thickness. An evaluation of the van der Waals attractions can be also made on the basis of the estimated properties of water and gives values which are large for thin films but decrease so rapidly as the thickness increases that, for thick films, they become negligible compared to the hydrostatic suction due to the height of the film above bulk level. Figure 6 shows also the net resultant of these van der Waals and hydrostatic thinning forces. The intersections of this line with those representing the double-layer repulsion give, therefore, the theoretically anticipated thicknesses.

This theoretical development has essentially the same basis as has been used in the modern discussion of the diuturnity[33a] and flocculation of hydrophobic colloids.[33b] The main difference is that there it leads to the prediction of a kinetic process involving, in addition, the theory of the rate at which particles approach

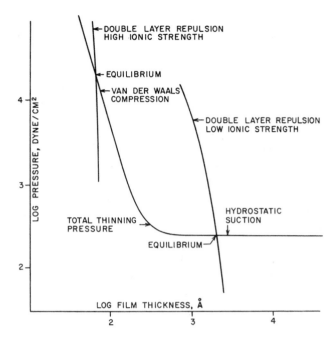

Figure 6. The intersection between double-layer repulsion and the resultant of van der Waals attraction and hydrostatic suction gives the equilibrium thickness.

each other, the more complicated equations for spheres, and the uncertain, and generally polydisperse, dimensions of the particles. In soap films, on the contrary, the geometry is that of simple planes, and both calculation and experiment can deal with an equilibrium state with all the attendant simplification.

The equilibrium thicknesses can be obtained as the limit of the thickness at very small velocities of pull-out as indicated in Fig. 5. This is a very reliable measurement because the thickness does not vary significantly over a broad range of experimentally accessible low rates.[34] The same result is obtained by allowing a thick film to drain spontaneously until it does not change further with time.

A very different technique has been developed by Scheludko[10] who forms a microscopic film completely surrounded by a meniscus by controlled withdrawal

(30) *E.g.*, J. Th. G. Overbeek in "Colloid Science," H. R. Kruyt, Ed., Elsevier Publishing Co., Amsterdam, 1952, Chapter VI.

(31) D. A. Haydon in "Recent Progress in Surface Science," J. F. Danielli, K. C. A. Pankhurst, and A. C. Riddiford, Ed., Academic Press, New York, N. Y., 1964, Chapter 3.

(32) D. Stigter and K. J. Mysels, *J. Phys. Chem.*, **59**, 45 (1955).

(33) (a) Typical hydrophobic colloids are unstable but very long lived, *i.e.*, diuturnal; (b) B. V. Derjaguin and L. Landau, *Acta Physicochim. URSS*, **14**, 633 (1941); *J. Exptl. Theoret. Phys.* (USSR), **11**, 802 (1941); E. J. W. Vervey and J. Th. G. Overbeek, "Theory of the Stability of Lyophobic Colloids," Elsevier Publishing Co., Amsterdam, 1948; H. R. Kruyt, "Colloid Science," Vol. I, Elsevier Publishing Co., Amsterdam, 1952.

(34) J. Lyklema and K. J. Mysels, to be published.

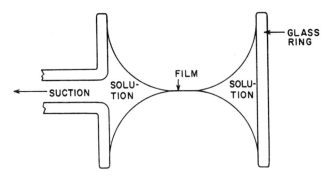

Figure 7. The principle of Scheludko's apparatus for producing films by gradual removal of liquid from a glass ring a few millimeters in diameter.

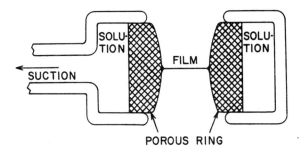

Figure 8. Apparatus for obtaining large hydrostatic suctions within a film produced by withdrawing liquid from a porous ring.

of liquid from a short piece of tubing, about 3 mm. in diameter, as indicated in Fig. 7. This technique gives excellent protection from evaporation and a rapid approach to equilibrium. The results obtained by these very different methods are quite concordant and agree qualitatively well with the theory. There are, however, quantitative discrepancies which are very disturbing. Thus, for example, the van der Waals attraction as calculated from these results seems to follow a third power dependence on distance for thicknesses[23,34] where it should be rather close to the fourth power.

Testing the Double-Layer Repulsion Theory. A test of the theory which involves both double-layer repulsion and van der Waals attraction simultaneously will always be complicated unless agreement is perfect since it is difficult to trace the source of any discrepancy. Hence, it would be better to test only one theory first, and that of the double-layer repulsion seems more easily isolated. Consideration of Fig. 6 shows that, at low ionic strengths, hydrostatic pressure is the dominant thinning force. A method for varying this pressure over a wide range should permit us to test the course of the double-layer repulsion curve independently of any van der Waals actions and at constant composition of the system.

Derjaguin and Titijevskaya[25] were able to cover a range of about 2 cm. of water (2×10^3 dynes/cm.) in an apparatus from which that of Fig. 7 was evolved and based on the same principle. The suction that can be applied in these instruments is limited by the capillary pressure of the orifice connecting the liquid to the outside; above this value air is drawn through the tube.

In order to overcome this limitation we are now developing a device for greatly extending this range of attainable pressures. It is shown schematically in Fig. 8. The film is formed in a tube of sintered glass

or porous porcelain which permits easy passage of the liquid from the interior of the film but can support large air pressures up to several atmospheres (10^6 to 10^7 dynes/cm.). Once experimental difficulties are ironed out, this should permit an extremely straightforward test of the double-layer repulsion theory.

Light Scattering. Another new approach[35] to the problem of measuring the forces determining the equilibrium thickness of these films is now being developed in Overbeek's laboratory. It is based on the measurement of the intensity of light, not reflected—as is done in thickness measurements—but scattered at angles other than that of the reflection. This scattering is due to minute fluctuations in thickness caused by thermal agitation and resisted by the same forces which yield the equilibrium value. When further developed, light scattering should prove just as powerful a tool for the determination of thermodynamic parameters in these systems as it is accepted to be in bulk solutions.

The Second Black Film. In some systems there is a further stage of thinness, the abrupt transition from the "first" black film to a "second" black film. Whereas, the former, as discussed above, has a variable thickness depending on ionic strength, the latter has a constant thickness of about 45 Å. At this thickness the film still contains several layers of water molecules between the two outer surfactant layers. At first we thought[9i] that this second black film—like Perrin's stratified films[18]—was caused always by evaporation, and others have suggested that it was due to impurities, but now further work by Drs. Scholten and Jones has shown[36] that its appearance is not an artifact but an equilibrium phenomenon depending on temperature and on ionic strength.

(35) A. Vrij, "Interdisciplinary Conference on Electromagnetic Scattering," M. Kerker, Ed., Macmillan and Co., New York, N. Y., 1963, p. 387; *J. Colloid Sci.*, **19**, 1 (1964).

(36) M. N. Jones, P. C. Scholten, and K. J. Mysels, to be published.

Our attention has been particularly attracted by the fact that this equilibrium between first and second black films is extremely sensitive to the nature of the counterions present. This is in marked contrast to the phenomena discussed heretofore, which are quite unspecific both with respect to the surfactant ion and with respect to the counterion. The second black film, on the other hand, exhibits the same kind of sensitivity to the specific nature of the counterion as so many colloidal and other phenomena, but this sensitivity seems to be greater by orders of magnitude. This makes it easy to discern the same kinds of influences that are apparent for example in the effect of counterions on the critical micelle concentration in bulk solutions of surfactants,[37] namely the role of the hydrated size of the ions and that of specific binding by hydrophobic bonds.

Thus here, as in many of the aspects which we have briefly reviewed, soap films offer an effective and unique tool to approach a problem in colloid and surface chemistry in addition to their intrinsic interest and aesthetic appeal.

---

(37) P. Mukerjee and K. J. Mysels, paper presented at the 131st National Meeting of the American Chemical Society, Miami, Fla., April, 1957, and unpublished results.

ADDENDUM 1972

Karol J. Mysels

I. Micelles

It was somewhat surprising and embarrassing to realize in Philadelphia, when my Kendall Award address was ready for delivery with slides and movies all organized, that the citation of the Award which was handed to me emphasized micelles rather than the soap films about which I was going to talk the next day. It was too late to include any of that work in the talk and I did not feel that it should then be inserted into the printed version, which is the article reproduced above. This addendum presents the opportunity to outline this work and to list the main references, including some recent ones, for those whom it may interest.

Initially, these studies were concerned with a new approach of tagging the micelle with a water insoluble but readily solubilized dye, so that micellar behavior alone, rather than the average of micelles and monomers, could be observed[1]. Specially designed instruments permitted using this approach to determine micellar electrophoretic mobilities[2] and diffusion coefficients[3]. This, together with light scattering studies[4] and application of the electrophoretic method to radioactively-tagged counterions[5], permitted an evaluation of the properties of micelles of sodium dodecyl sulfate. The results were somewhat dependent on the state of the monomeric solution bathing the micelles and this led to studies of the critical micellization concentration (cmc) and of the methods for determining it[6,7], and to a considerable effort to pinpoint the non-ideality of these solutions by investigation of dimerization[8] and of mixed micelles[9].

Whereas cmc's are now well understood and have been included among properties collected by the National Standard Reference Data System[10], the non-ideality still seems quite elusive. For example, some unpublished, very precise vapor-osmometry measurements of Dr. Huisman indicated that at 37°C the dimerization of sodium dodecyl sulfate was not detectable. Later experiments were mainly designed to show that the popular phase-separation theory of micelle formation was only a zeroth approximation. Surface tension[11] and dialysis[12] measurements agreed that the activity of the solute continues to increase with total concentration above the cmc at about the rate expected from an ordinary equilibrium. In connection with dialysis, the original dye-tagging technique made it possible to show that micelles did not cross the cellophane membrane and to explain how they nevertheless could help greatly the transport of a slightly soluble substance across it[12,13].

II. Soap Films

The following references, incomplete in the original article, can now be given 17[11], 24 is more completely discussed later[14], 26[15], 34[16], 36[17], and 37[7].

The article made the point that methods were needed to measure separately the effect of electrocratic repulsion and of van der Waals attraction because of the difficulties in interpreting the balance of the two. It also gave in Fig. 8 the suggestion of a method for observing the electric component alone. This experiment was quite successful[18] and gave observed slopes of the logarithm of outside pressure (and therefore of double layer repulsion) versus film thickness in excellent agreement with the theoretical prediction of $-\kappa$, the inverse of the "thickness" of the double layer. This result indicates that the important, low potential part of the theory of double layer repulsion is indeed directly applicable to these systems. One may also say that this is the most direct measurement available of $1/\kappa$, the so-called Debye length of electrolyte solutions.

The complementary measurement of the van der Waals contribution became possible through the measurement of the film contact angle, the macroscopic angle at the point where the meniscus or Plateau border joins the film[19]. Its existence is the result of the difference of surface free energies between the bulk liquid and the film or, what is equivalent, the difference of their surface tensions. These tensions, being unequal, can only balance if the two surfaces join at an angle. Here it is the free energy that is significant, i.e. the integral of the forces discussed in connection with Fig. 6 of the article. The steepness, noted there, of the electric force-versus-distance line makes its integral small for an equilibrium film as compared to that of the much less steep line of van der Waals forces. Hence, it is mainly the latter integral that is measured by the contact angle. These angles are generally small and do not exceed 1° except for films less than about 80 Å thick, which are formed by solutions of ionic surfactants only if their ionic strength is quite high. Thus the most easily measurable angles are in systems whose interpretation is also most complicated. Lack of good agreement with theory was therefore not very disturbing at first. In Newton black films (as the "second" black films are now to be called in contrast to the common or formerly "first" ones), the film contact angles can become quite large and they show very nicely the specific ionic effects that were mentioned in the article as affecting the equilibrium between the Newton and the common black films.

Recently, Princen and Frankel described a very sensitive method of measuring film contact angles based on a detailed interpretation of the diffraction pattern of a parallel beam passing through the boundary of the film[20]. It permitted significant measurements on films as thick as 600 Å[21] but the results are still not easily understandable. Whether the difficulty lies in properly taking into account the microscopic details of the contact angle region[22], or whether there are some significant deficiencies in the applicable theory of van der Waals forces, or whether the results are simply distorted by the presence of contaminants, remains to be seen. If the latter were the case it would account also for contradictory results obtained on equilibrium thicknesses in different laboratories[23].

The problem of purification of surfaces of these surfactant solutions is a difficult one, though it is obvious that significant data can only be obtained on sufficiently pure surfaces. The method of foaming followed by foam removal, which was alluded to in the article, is very useful[11] but an objective criterion is required to establish when sufficient purification has been attained[24]. The rate of change of the surface tension of a freshly expanded surface may permit an estimate of the surface purity of a system and of the surface activity of the contaminants. This is because of the fact that when diffusion controls the rate of adsorption, the surfactant will reach adsorption equilibrium much faster than an impurity present at much higher dilution and also because such a much more dilute impurity is harmless unless it is much more surface active than the main component[24,25]. The first experiments along these lines only served to show that it is not easy to obtain and maintain such surface purity for solutions of surfactants[25].

The Gibbs elasticity of films, in connection with which the purity problem was mentioned in the article, can now be measured in a much simpler way by a method developed by Prins, Arcuri and van den Tempel[26].

The bursting of soap films has always been an erratic and often irritating impediment to their study and, of course, underlies the important practical problem of foam stability. However, a few high speed flash pictures of the bursting process, taken to see if film and foam stability could be investigated, revealed that the process itself was most interesting: The rapidly growing hole in the bursting film is preceded by an even more rapidly growing aureole, a thickening in which the film is accelerated by a gradient of surface tension. The latter decreases as the two monolayers on the surface of the film are compressed. This aureole can be studied by several methods[27,30], and the mechanics of its behavior analyzed in considerable detail[31,29]. The results show that the surface tension may drop considerably below 20 dynes/cm so that these monolayers are compressed far beyond the normally attainable surface concentrations. In addition, the corresponding surface pressure–area per molecule curve suffers a curious reversal of curvature under these conditions which indicates that unexpectedly strong cohesive forces become effective as the surfactant ions become tightly crowded in the monolayer[29]. It is also likely that the rate of desorption from these supercompressed monolayers can be measured by studying the film bursting process[29,30]. Soap films are thus becoming a tool in the study of monolayers under conditions inaccessible to the Langmuir trough or ordinary surface tension measurements.

Thus in the years since 1964, soap films have continued to fulfill their promise as an effective and unique tool in colloid and surface chemistry and remain a challenging subject of research.

Bibliography

1. Hoyer, H. W., and K. J. Mysels, *J. Phys. Colloid Chem.*, **54**, 966 (1950); Mysels, K. J., *J. Colloid Interface Sci.*, **23**, 474 (1967)

2. Hoyer, H. W., K. J. Mysels and D. Stigter, *J. Phys. Chem.*, **58**, 385 (1954); Stigter, D., and K. J. Mysels, *J. Phys. Chem.*, **59**, 45 (1955)

3. Mysels, K. J., and D. Stigter, *J. Phys. Chem.*, **57**, 104 (1953); Stigter, D., R. J. Williams and K. J. Mysels, *J. Phys. Chem.*, **59**, 330 (1955)

4. Princen, L. H., and K. J. Mysels, *J. Colloid Sci.*, **12**, 594 (1957); Mysels, K. J., and L. H. Princen, *J. Phys. Chem.*, **63**, 1696 (1959); Princen, L. H., and K. J. Mysels, *J. Phys. Chem.*, **63**, 1781 (1959)

5. Mysels, K. J., and C. I. Dulin, *J. Colloid Sci.,* **10**, 461 (1955)

6. Mukerjee, P., and K. J. Mysels, *J. Am. Chem. Soc.,* **77**, 2937 (1955); Williams, R. J., J. N. Phillips and K. J. Mysels, *Trans. Farad. Soc.,* **51**, 728 (1955); Mysels, K. J., and P. Kapauan, *J. Colloid Sci.,* **16**, 481 (1961); Mysels, E. K., and K. J. Mysels, *J. Colloid Sci.,* **20**, 315 (1965); Mysels, E. K., and K. J. Mysels, *J. Colloid Interface Sci.,* **38**, 388 (1972)

7. Mukerjee, P., K. J. Mysels and P. Kapauan, *J. Phys. Chem.,* **71**, 4166 (1967)

8. Mukerjee, P., K. J. Mysels and C. I. Dulin, *J. Phys. Chem.,* **62**, 1390 (1958); Mykerjee, P., and K. J. Mysels, *J. Phys. Chem.,* **62**, 1400 (1958)

9. Mysels, K. J., and R. J. Otter, *J. Colloid Sci.,* **16**, 462 (1961); **16**, 474 (1961)

10. Mukerjee, P., and K. J. Mysels, *"Critical Micelle Concentrations of Aqueous Surfactant Systems"*, NSRDS-NBS 36, Superintendent of Documents, Washington, D. C., 1971

11. Elworthy, P. H., and K. J. Mysels, *J. Colloid Interface Sci.,* **21**, 331 (1966)

12. Mysels, K. J., P. Mukerjee and M. Abu-Hamdiyyah, *J. Phys. Chem.,* **67**, 1943 (1963); Abu-Hamdiyyah, M., and K. J. Mysels, *J. Phys. Chem.,* **71**, 418 (1967)

13. Mysels, K. J., in *"Pesticidal Formulations Research"*, Advances in Chemistry Series No. 86, American Chemical Society, Washington, D. C., 1969, pp. 24-38

14. Exerowa, D., *Kolloid Z. Z. Polym.,* **232**, 703 (1969)

15. Lyklema, J., P. C. Scholten and K. J. Mysels, *J. Phys. Chem.,* **69**, 116 (1965)

16. Lyklema, J., and K. J. Mysels, *J. Am. Chem. Soc.,* **87**, 2539 (1965)

17. Jones, M. N., K. J. Mysels and P. C. Scholten, *Trans. Farad. Soc.,* **62**, 1336 (1966)

18. Mysels, K. J., and M. N. Jones, *Discuss. Farad. Soc.,* **42**, 42 (1967)

19. Mysels, K. J., H. F. Huisman and R. Razouk, *J. Phys. Chem.,* **70**, 1339 (1966); Huisman, F., and K. J. Mysels, *J. Phys. Chem.,* **73**, 489 (1969)

20. Princen, H. M., and S. Frankel, *J. Colloid Interface Sci.,* **35**, 386 (1971)

21. Mysels, K. J., and J. W. Buchanan, *J. Electroanal. Chem. & Interfac. Electrochem.,* **37**, 23 (1972)

22. De Feijter, J. A., and A. Vrij, *J. Electroanal. Chem. & Interfac. Electrochem.,* **37**, 9 (1972)

23. Bruil, H. G., *Meded. Landbouwhogesch. Wageningen,* **70**, 9 (1970)

24. Mysels, K. J., and A. T. Florence, in *Clean Surfaces: Their Preparation and Characterization for Interfacial Studies,* George Goldfinger (ed.), Marcel Dekker, New York, 1970, pp. 227-268

25. Mysels, K. J., and A. T. Florence, *J. Colloid Interface Sci.,* in press

26. Prins, A., C. Arcuri and M. van den Tempel, *J. Colloid Interface Sci.,* **24**, 84 (1967)

27. McEntee, W. R., and K. J. Mysels, *J. Phys. Chem.,* **73**, 3018 (1969)

28. Mysels, K. J., and J. A. Stikeleather, *J. Colloid Interface Sci.,* **35**, 159 (1971)

29. Frens, G., K. J. Mysels and B. R. Vijayendran, *Spec. Discuss. Farad. Soc.,* **1**, 12 (1970)

30. Florence, A. T., and G. Frens, *J. Phys. Chem.,* **76**, 3024 (1972)

31. Frankel, S., and K. J. Mysels, *J. Phys. Chem.,* **73**, 3028 (1969)

1965

GEORGE D. HALSEY, JR.

UNIVERSITY OF WASHINGTON
SEATTLE, WASH.

. . . for his series of original and incisive papers dealing with the creep of textile fibers; for his major advances in our knowledge and understanding of physical adsorption of gases on solids, including a prediction of the form of isotherms and the variation of isoteric heats of adsorption with coverage on a uniform surface; for his pioneering work in the measurement of the interaction of a single inert gas atom with surfaces; and especially for his originality, ingenious experimentation, and impressive theoretical insight.

The following review has been written "in the spirit of the award time" for this volume by Prof. Halsey. The intercalated article is reproduced by permission from *The Journal of Physical Chemistry*, **71**, 4012 (1967).

Details of Monolayer Formation

G. D. Halsey, Jr.
University of Washington
Seattle, Washington

General Concepts of Physical Adsorption. The ideas of Polanyi *(a)* were instrumental in explaining the phenomenon of local condensation of an unsaturated vapor in the neighborhood of a solid surface. The potential energy of interaction of atoms in the solid near the surface with the gas molecules creates an external field. This field affects the density to produce a surface excess, or in other words, adsorption. Any consideration of adsorbate structure that accompanies this theory views the vapor near the surface as a dense gas with a density gradient but with the same amorphous structure as the vapor itself. Experience (as well as theory) has shown that this structural model is only valid when the potential is quite small and varies slowly with distance. To account for the facts, the surface potential must be retained, but it must be considered as acting on specific structures of adsorbed molecules quite foreign to the gas phase or perhaps any bulk phase of the adsorbed gas.

These structures may be observed in a variety of ways; directly by x-ray observations, by measurements of heat of adsorption or of heat capacity, or finally, by analysis of the adsorption isotherm. We will discuss this last and most conventional approach in greater detail.

Overall Picture of Physical Adsorption. In the introductory chapter of geography textbooks, there is often a map of a mythical country that illustrates in the one picture all the features of a real map; mountains, peninsula, island and swamp. Such a hypothetical map of physical adsorption in the form of an isotherm is illustrated in Figure 1. At lowest pressures, there may be a clear-cut Henry's law region, but, more often than not, the isotherm *appears* to be tangent to the vertical axis. This behavior is the result of extremely tenacious adsorption on a small portion of the free surface or in the interior of the adsorbent. On a more uniform and perfect surface or at a higher temperature, this region will be replaced by a linear portion of the isotherm whose slope corresponds to the Henry's law constant of the strongest sites.

At somewhat higher coverage or amount adsorbed, the isotherm bends and becomes convex to the pressure axis.

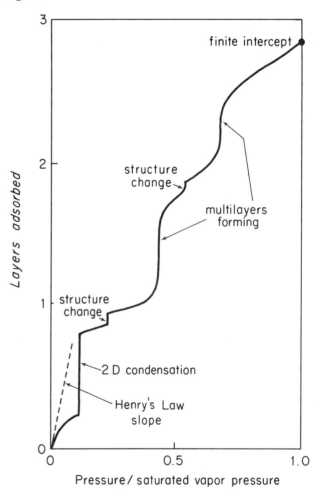

Figure 1. An exemplary adsorption isotherm.

The strongest sites or most active portion of the surface is covered, and the pressure increases more rapidly. Suddenly, the pressure remaining fixed or nearly so, a great increase in adsorption may take place, which corresponds to a two-dimensional condensation. This behavior is dependent on the lateral energy of interaction between adsorbed molecules, and is a cooperative phenomenon, which at least simulates a first-order phase transition. The transition is between a dilute, highly dispersed "phase" into a condensed or concentrated "phase". Therefore the amount of adsorbate required must be a substantial fraction of that required to form a monolayer. Thus we may tentatively assume that the height of the step is a rough measure of the monolayer volume. There are two exceptions possible however; first if the condensation is taking place only on one part of the surface, and second, if we are quite close to the critical temperature for two-dimensional condensation, in which case, the height of the step will be very sensitive to temperature changes.

Following any large-scale condensation, a further slow increase in density of the film with pressure will ensue. There may then be a second abrupt step, of much smaller magnitude than the first one. This may be viewed either as an analog of the liquid-solid transition, or a change in film structure of some other kind. Then the density of the film increases slowly in the region of one completed monolayer. At this point it is almost impossible to ascribe the increase in adsorption to increase in the density of the monolayer because material may now be accumulating in the second layer. A second large-scale step gives clear indication that adsorption is proceeding into the second layer. However, there is no reason to rule out a phase change in the first layer caused by the adhesion of the second layer. This effect can cause a slight bump, which can also be buried in the second layer step and thus add to it.

Adsorption then increases rapidly as saturation is approached. Further steps may or may not be obscured by the condensation of bulk material. The intercept of the isotherm with the pressure axis at saturation is in many cases finite, however. One explanation of this finite adsorption at saturation is a different crystal structure in the layers vs the bulk solid, if adsorption takes place below the triple point of the bulk adsorbate. Another explanation lies in the finite contact angle of the liquid adsorbate and the solid adsorbent *(b)*.

Model for Adsorption on a Free Surface. Studies of the interaction of simple isolated molecules with typical surfaces *(c)* show that the interaction energy between the molecule and surface (the so-called vertical interaction energy) is large compared with kT at the temperatures where adsorption measurements of the type discussed here are carried out. A typical value of E_{vert}/kT will be of the order of 20. Thus the adsorbed molecule can be considered to be trapped in a potential well. If the potential does not vary over the surface, the molecule is capable of free translation in the two dimensions parallel to the surface. We say then that its lateral energy is independent of its position on the surface. Such a model can be rigorously developed in terms of translational and vibrational degrees of freedom, but it is customary to introduce another way of viewing the problem.

Two Dimensional Abstraction. Many problems in physics have two dimensional solutions, in some cases markedly simpler than the real three-dimensional problem. These solutions are unreal, in a strict sense, but can be applied on a more or less trial basis to surface problems. In particular, a two-dimensional imperfect gas can be used as a model for a three-dimensional gas strongly attracted to a surface. It should be emphasized that rigorous arguments about the two-dimensional abstraction loose their rigor when applied to the surface states. Agreement between the experimental observations and the theoretical predictions are usually plausible arguments for the validity of the two-dimensional model.

A third physical case has much less to do with surface phenomenon; that is the two-dimensional growth of a crystal or other structure in three-dimensions. This case may apply to structures like liquid crystals, but not to an adsorbed film supported by a rigid solid.

Models that Show the Simplest Phase Changes. Once the two-dimensional abstraction is employed as an approximation, considerations are limited to formations of a single layer of adsorbed atoms. With this limitation, what can cause a phase change? In principle, the two-dimensional analog of hard spheres, "hard discs" might be expected to show a change from a disordered to ordered structure as the surface density increases. However, at the temperatures where monolayers are conveniently studied $E_{vert}/kT \gg 1$. A sort of rough corresponding states argument about the lateral interaction energy E_{lat} would then indicate that $E_{lat}/kT > 1$, even if in general $E_{vert} > E_{lat}$. Only in the case of helium would one expect that $E_{vert} \gg E_{lat}$. To give some idea of what to expect with simple adatoms such as argon on oxide surfaces, if all pair-wise interaction energies were equal to the same value E_o, then the total lateral interaction and the vertical energy would both be about the same $E_{lat} = E_{vert} = 3E_o$. Therefore, in order to make a realistic model, some account must be taken of lateral attraction, and this can be done by replacing the hard discs with an attractive potential extending beyond the repulsive disc radius. In the paper which follows, a model of this sort is treated.

Simple Attractive-Disk Monolayer Isotherms with Phase Transitions

by F. Tsien and G. D. Halsey, Jr.

Department of Chemistry, University of Washington, Seattle, Washington 98105   (*Received May 29, 1967*)

Monolayer adsorption isotherms for the model of disks showing repulsion and inverse power attraction are discussed. The van der Waals (mobile) and Fowler (immobile) isotherms are put in compatible form. One adjustable parameter reflects the assignment of the reduced value of the van der Waals parameter a. Solid, liquid, and gaslike phases are identified and critical and triple points are calculated. The theory is applied to the data for krypton on exfoliated graphite, with reasonable success.

1. Introduction

Theoretical treatment for adsorption isotherms has been given in detail both for mobile and localized cases. Stebbins and Halsey[1] treated the phase transitions between the localized and mobile isotherms for hard disks on a structureless surface. We shall consider molecules here as disks with attractive force varying according to an inverse power law. The surface is assumed to be structureless to simplify the treatment.

2. Mobile Monolayers

The two-dimensional analog of the van der Waals equation had been used to represent a mobile monolayer.[2,3] It can be expressed as

$$\left(\phi + \frac{a'N^2}{A^2}\right)(A - Nb) = NkT \qquad (2.1)$$

where ϕ is the spreading pressure, A is the area of the film, N is the number of molecules adsorbed, and a' and b are parameters. b is sometimes referred to as the "co-area" per molecule, in analogy to the co-volume in the van der Waals equation. We shall choose b so that the area occupied by an adatom in the completed monolayer is

$$b = \sqrt{3}/2(r^*)^2 \qquad (2.2)$$

where r^* is the distance between the centers of atoms in closest array.

If the adatom has no permanent dipole, the correction term a' is then the attraction which arises from the London dispersion forces. In calculating a', we assume a uniform radial distribution function outside of $r = r^*$, and zero inside. The molecular interaction will be de-

scribed on a pairwise basis and in terms of the Sutherland potential

$$u(r) = -\epsilon^*(r^*/r)^6 \qquad (r \geqslant r^*) \qquad (2.3)$$

$$u(r) = \infty \qquad (r < r^*)$$

We can then calculate a,[4] where $a = a'/kT$

$$a = \frac{1}{2}\int_{r^*}^{\infty} \frac{u(r)}{kT} 2\pi r dr \qquad (2.4)$$

$$= \frac{\pi}{4} \frac{(r^*)^2}{kT} \epsilon^* \qquad (2.5)$$

and

$$\frac{a}{b} = \frac{\pi}{2\sqrt{3}} \frac{\epsilon^*}{kT} \qquad (2.6)$$

We shall, in the following calculations, assume the same temperature dependence, but will express a as

$$\frac{a}{b} = \frac{C}{2T^*} \qquad (2.7)$$

where C is an adjustable parameter, and the reduced temperature

$$T^* = kT/\epsilon^* \qquad (2.8)$$

With this relation for a/b, the two-dimensional equation of state eq 2.1 can be related to the adsorption isotherm

(1) J. P. Stebbins and G. D. Halsey, Jr., *J. Phys. Chem.*, **68**, 3863 (1964).

(2) T. L. Hill, *Advan. Catalysis*, **4**, 211 (1952).

(3) S. Ross and J. P. Olivier, "On Physical Adsorption," Interscience Publishers, Inc., New York, N. Y., 1964.

(4) T. L. Hill, *J. Chem. Phys.*, **14**, 441 (1946).

via the Gibbs adsorption equation which takes the form

$$d\phi = kT\Gamma \, d \ln p \qquad (2.9)$$

where p is the gas pressure and Γ is the surface concentration per unit area. Equations 2.1 and 2.9 give us the mobile isotherm

$$\ln p = -\ln k_v + \ln \frac{\theta}{1-\theta} + \frac{\theta}{1-\theta} - \frac{2a}{b}\theta \qquad (2.10)$$

or

$$\ln p = -\ln k_v + \ln\left(\frac{\theta}{1-\theta}\right) + \left(\frac{\theta}{1-\theta}\right) - \frac{C}{T^*}\theta \qquad (2.11)$$

where θ is the fractional coverage. k_v can be written as

$$k_v = \frac{b}{\Lambda^2} f_\perp [\exp(-\chi/kT)] e^{\mu^0/kt} \qquad (2.12)$$

where μ^0 is the standard chemical potential, χ is the minimum energy required to evaporate an adsorbed atom from its lowest energy state in the monolayer, $f_\perp$ is the partition function for motion normal to the surface, and

$$\Lambda = \frac{h}{(2\pi m k T)^{1/2}} \qquad (2.13)$$

3. Localized Monolayers

We assume that the sites are fixed and that each adsorbed molecule interacts with its nearest neighbors only. Each lattice site has z nearest neighbor sites. The approximate adsorption isotherm for a random distribution as derived by Fowler and Guggenheim[5] (referred to as the Fowler isotherm from now on), after rearranging and taking $z = 6$ for a triangular close packed two-dimensional lattice, has the form

$$\ln p = -\ln k_F + \ln\left(\frac{\theta}{1-\theta}\right) - \frac{6}{T^*}\theta \qquad (3.1)$$

where

$$k_F = f_x f_y f_\perp [\exp(-\chi/kT)] e^{\mu^0/kT} \qquad (3.2)$$

and f_x and f_y are the partition function for motion parallel to the surface. Equation 3.1, combined with the Gibbs adsorption equation, yields

$$\frac{\phi A}{NkT} = \frac{1}{\theta} \ln\left(\frac{1}{1-\theta}\right) - \frac{3}{T^*}\theta \qquad (3.3)$$

4. Corresponding States

For convenience, we define K as[6]

$$K = k_v/k_F \qquad (4.1)$$

which, after cancellation, becomes

$$K = \left(\frac{b}{\Lambda^2}\right) \bigg/ f_x f_y \qquad (4.2)$$

$$= \frac{b}{\Lambda^2} \bigg/ \left[\frac{\exp(-\theta_E/2T)}{1 - \exp(-\theta_E/T)}\right]^2 \qquad (4.3)$$

where θ_E is the Einstein "characteristic temperature." θ_E will be approximated in the following fashion. The Lennard-Jones (6–12) potential is summed over the nearest neighbors in the hexagonal lattice and then differentiated twice with respect to r. This gives

$$\theta_E = \frac{1}{2\pi}\frac{h}{k}\sqrt{\frac{216\epsilon^*}{m(r^*)^2}} \qquad (4.4)$$

$$= \frac{3\sqrt{6}}{\pi}\frac{\epsilon^*}{k}\Lambda^* \qquad (4.5)$$

where

$$\Lambda^* = \frac{h}{r^*\sqrt{m\epsilon^*}} \qquad (4.6)$$

θ_E values for argon and krypton, calculated by this method, are 46 and 36°K, respectively.[7] These values are close to the ones used by McAlpin and Pierotti,[8] which are 45 and 34°K, respectively. K can now be expressed as

$$K = \frac{\left[\dfrac{\sqrt{3}\pi T^*}{\Lambda^{*2}}\right]}{\left[\dfrac{\exp\left(-\dfrac{3\sqrt{6}}{2\pi}\dfrac{\Lambda^*}{T^*}\right)}{1 - \exp\left(-\dfrac{3\sqrt{6}}{\pi}\dfrac{\Lambda^*}{T^*}\right)}\right]^2} \qquad (4.7)$$

The adsorption isotherms can be expressed in a much simpler form if we define

$$p^* = pk_v \qquad (4.8)$$

The mobile isotherm is then

$$\ln p^* = \ln\left(\frac{\theta}{1-\theta}\right) + \frac{\theta}{1-\theta} - \frac{C}{T^*}\theta \qquad (4.9)$$

(5) R. H. Fowler and E. A. Guggenheim, "Statistical Thermodynamics," Cambridge University Press, London, 1949.

(6) We have used the same form of definition for K as by Stebbins and Halsey. However, K in their eq 31 and 32 as well as in Figure 4 should be replaced by $1/K$.

(7) Values of ϵ^* and r^* are taken from J. O. Hirschfelder, C. F. Curtiss, and R. B. Bird, "Molecular Theory of Gases and Liquids," John Wiley and Sons, Inc., New York, N. Y., 1954.

(8) J. J. McAlpin and R. A. Pierotti, *J. Chem. Phys.*, **41**, 68 (1964).

and the Fowler isotherm becomes

$$\ln p^* = \ln K + \ln\left(\frac{\theta}{1-\theta}\right) - \frac{6}{T^*}\theta \quad (4.10)$$

5. Classical Case

The classical law of corresponding states is correct when the mass of the molecule is sufficiently large, or when Λ^* is negligibly small. In such a case, we can expand the exponentiation term $e^x \approx 1 + x$ so that $1 - e^{-x} \approx x$. The ratio K can then be approximated as

$$K = \frac{\dfrac{\sqrt{3}\pi T^*}{\Lambda^{*2}}}{\left[\dfrac{1}{1-\exp\left(-\dfrac{3\sqrt{6}}{\pi}\dfrac{\Lambda^*}{T^*}\right)}\right]^2} \quad (5.1)$$

after expansion

$$K = \frac{\dfrac{\sqrt{3}\pi T^*}{\Lambda^{*2}}}{\left(\dfrac{1}{\dfrac{3\sqrt{6}}{\pi}\dfrac{\Lambda^*}{T^*}}\right)^2} \quad (5.2)$$

$$\approx 30/T^* \quad (5.3)$$

so that our classical Fowler isotherm then becomes

$$\ln p^* = \ln 30 - \ln T^* + \ln\left(\frac{\theta}{1-\theta}\right) - \frac{6}{T^*}\theta \quad (5.4)$$

6. Critical Conditions

Much work has been done on the two-dimensional condensation for the mobile isotherms[3] and the Fowler isotherm.[5] Designating the reduced critical temperature by T_c^*

$$T_c^*(\text{mobile}) = \frac{4}{27}C \quad (6.1)$$

$$T_c^*(\text{Fowler}) = 1.5 \quad (6.2)$$

so that for T^* lower than $^4/_{27}C$ and 1.5 first-order phase changes occur for both the mobile and Fowler isotherms. Under these conditions, we will show that first-order phase changes occur between the mobile and Fowler isotherms.

7. Phase Transitions

We will refer this section to Figure 1, and the subscripts A, B, C, etc., to the points A, B, C, etc. For the mobile isotherm ABCDEFGH, the criteria for phase transition are satisfied when the two phases have equal pressure, and also when they have equal spreading pressure. Expressed in mathematical form, they are

$$\ln p_B^* = \ln p_G^* \quad (7.1)$$

and

$$\phi_B = \phi_G \quad (7.2)$$

The expression $\phi_B = \phi_G$ is equivalent to saying that the area BDEB is equal to the area EFGE. The spreading pressure can be calculated via the Gibbs adsorption equation

$$d\phi = kT\Gamma \, d\ln p \quad (7.3)$$

$$= kT\Gamma \frac{\partial \ln p}{\partial \theta}\, d\theta \quad (7.4)$$

Integration of eq 7.4 gives

$$\phi = \int_0^\theta kT\Gamma \frac{\partial \ln p}{\partial \theta}\, d\theta \quad (7.5)$$

when applied to the mobile isotherm

$$\phi_G - \phi_B = \Gamma_\infty kT\left[\frac{1}{1-\theta_G} - \frac{C}{2T^*}\theta_G^2 - \left(\frac{1}{1-\theta_B} - \frac{C}{2T^*}\theta_B^2\right)\right] = 0 \quad (7.6)$$

where Γ_∞ is the maximum surface concentration as the pressure approaches infinity.

The criteria for phase transitions are also satisfied between the mobile and Fowler isotherms. Taking the two hypothetical isotherms ABGH (mobile) and IJKO (Fowler), we will find that at points H and N

$$\ln p_H^* = \ln p_N^* \quad (7.7)$$

and

$$\phi_H = \phi_N \quad (7.8)$$

where, after integration

$$\phi_H = \Gamma_\infty kT\left[\frac{\theta_H}{1-\theta_H} - \frac{C}{2T^*}\theta_H^2\right] \quad (7.9)$$

and

$$\phi_N = \Gamma_\infty kT\left[-\ln(1-\theta_N) - \frac{3}{T^*}\theta_N^2\right] \quad (7.10)$$

At sufficiently low coverage, the stable isotherm is the mobile isotherm, and at extremely high coverage, the stable isotherm is the Fowler isotherm. Therefore, the stable isotherm path must then be ABEGHNO. It can also be shown that, at any given p^*, the stable isotherm path has a lower chemical potential than the other isotherm paths. We shall designate the expanded phase AB, the condensed phase GH, and the ordered phase NO as the two-dimensional gas phase, liquid phase, and solid phase, respectively.

Under certain conditions, when the mobile transition pressure is close to the Fowler transition pressure, there is also a phase transition between the unstable mobile isotherm and the stable Fowler isotherm. Thus the unstable two-dimensional gas–solid transition is represented by the path CM.

When T^* gets sufficiently low, another type of phase transition takes place. The equilibrium between the two isotherms takes place before the mobile transition. The stable isotherm path now follows PQLO which has the lower chemical potential. OL is then the two-dimensional gas–solid transition.

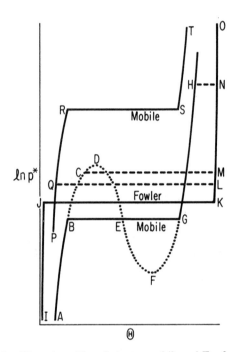

Figure 1. Phase transitions between mobile and Fowler isotherms, indicated by long dotted lines, as explained in text.

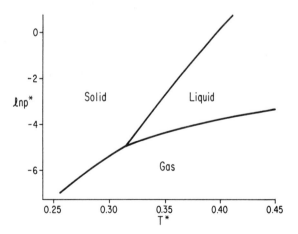

Figure 2. Two-dimensional reduced phase diagram for $C = 4.2$.

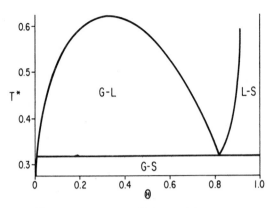

Figure 3. Plot of coverage at which transition occurs as a function of reduced temperature for $C = 4.2$.

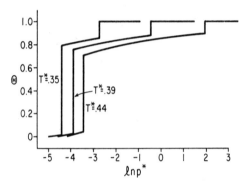

Figure 4. Adsorption isotherm for reduced temperature of 0.35, 0.39, and 0.44. $C = 4.2$.

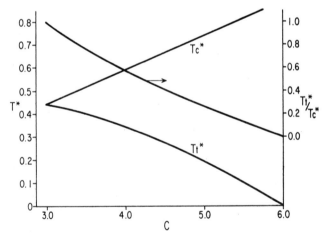

Figure 5. Plot of reduced critical temperature and reduced triple point as a function of C. The ratio of triple point and critical temperature is also shown.

A typical phase diagram is shown for $C = 4.2$ (Figure 2). The corresponding coverage for which the phase transitions occur as a function of reduced temperature is shown in Figure 3. Some typical isotherms are shown in Figure 4. For analysis of data, however, C is available as an adjustable parameter. It determines the values of the reduced triple point and critical temperatures, and thus the ratio of the unreduced temperatures. To show the dependence on C, T^* of the critical point and of the triple point are plotted as a function of C, in Figure 5. The ratio of the critical temperature to the triple point is also shown.

8. Application to Data for Krypton on Graphite

The recent data of Thomy and Duval[9] resemble the isotherms of Figure 4 and appear to be suitable for analysis. Although these data do not quite reach a two-dimensional triple point at their lowest temperature of 77.4°K, the short vertical riser that appears to correspond to a "solid–liquid" equilibrium is almost on top of the longer "gas–liquid" riser at that temperature. A short extrapolation suggests a triple-point temperature of about 77°K. The observed critical temperature is about 86°K. The ratio of 0.9 (Figure 5) yields a value of $C = 3.2$ These results may be contrasted with the values for bulk krypton where the ratio of triple-point to critical temperature is 0.55 for a value of $C = 4.1$.

From the reduced value of the critical temperature at $C = 3.2$, ϵ^*/k is calculated to be 180°. This result is close to the free gas pair interaction energy of 171°,[7] but rather far from the pair interaction calculated for the case of krypton on a graphite surface (144°).[8]

9. Discussion

Both the mobile and immobile isotherms that are used in this analysis are of the lowest degree of approximation to the exact isotherms, and at least are consistent. The application of a Sutherland potential to the mobile case and a harmonic oscillator model to the solid is not consistent. At high temperature ($T^* = 30$), eq 5.3 predicts a reversal in the stability of the low density mobile film for this reason. This temperature would be of the order of 5000°K for krypton and is thus not physically important.

Reference to Figure 5 indicates that if $C < 3$ no liquid ever forms, and the transition is always from mobile gas to solid, at any value of T^*. There are thus no critical or triple-point temperatures below this value. For $C > 6$, the liquid phase is stable down to $T^* = 0$, and only the critical temperature remains. It should be noted that if we use eq 2.3 to calculate C, a value of 1.8 is obtained. This value is below the range of interest, and this is the reason we have left C as an empirical parameter. A similar difficulty is encountered if the equation is used in three-dimensional form to estimate van der Waals constants.

The existence of an unstable transition from an over-compressed gas to the solid (Figure 1) may have a bearing on the difficulty of observing the two-step gas–liquid–solid condensation. If the equilibrium solid-liquid region is typically as short as it is for krypton on graphite, the triple point could easily be lost if nucleation for the liquid is difficult. Although a true critical point would then be missing, a crude sampling of data would indicate that one existed. All that actually would happen is that a wide step would change into a narrow one just below $\theta = 1$. This phenomenon would occur near the lattice critical temperature, and the narrow step might be missed in the onset of second-layer formation, and so one could be misled into seeing "T_c."

Note that aside from this possibility, a direct manifestation of a lattice critical temperature is not present in our results. In order to find such a temperature, one must explore the region where K in eq 5.3 is less than unity, and thus the underlying surface favors a lattice gas.

(9) A. Thomy and X. Duval, Colloques Internationaux Du Center National de La Recherche Scientifique, No. 152, 1965.

Influence of Underlying Lattice. The approximate theory of phase transitions that we have given assumes a lattice structure for the localized monolayer, but this assumption does not imply that there is an underlying lattice structure. On the contrary, the theory is appropriate only if the underlying structure can be neglected.

When the underlying structure plays a role, the changes that can take place include other more complex phenomena, even if the adsorbed gas is a simple monatomic gas. Of course, if the adsorbed molecules are complex, a whole new source of complications presents itself.

Newer Work Showing Transitions. In recent years, Dash and his coworkers have investigated the adsorption of helium by means of very sensitive heat capacity measurements. In particular, Bretz and Dash *(d)* have studied both isotopes of helium on graphite at about one-half of a monolayer coverage in the range 1 to 4°K. They discovered very sharp peaks in the heat capacity at 3° that were quite sensitive to coverage. Both isotopes showed the effect but He[4] showed a much sharper peak.

They conclude that above the transition, the helium is essentially a two-dimensional gas, and that below the transition specifically ordered structures are formed. These are presumably in registry with the underlying graphite lattice.

These results, the first of their kind, are only illustrative of the many types of transitions that can take place on a real surface. Aside from experimental difficulties, the only reason these "anomalies" are rare, is the extreme rarity of perfect enough surfaces. Otherwise, surface heterogeneities will act to smear out and obscure many phenomena *(e)*.

Because of the requirement of a clean, homogeneous surface, most of the significant work in recent years has been limited to the graphitic surface. Larher has extended the range of surfaces that can be successfully studied by cleaving layer-like halides of seven metals, and measuring rare gas isotherms on the (111) planes of these solids. He shows that there is a definite effect caused by the incompatibility of the underlying lattice with the lattice of the solid rare gas, which can be readily varied by choice of metal halide and of rare gas.

Additional References

a. Polanyi, M., *Verh. Deutsch, Phys. Ges.,* **16**, 1912 (1914) quoted in "Physical Adsorption of Gases", D. M. Young and A. D. Crowell, Butterworths, Washington, 1962

b. Adamson, A. W., *Journal of Chem. Educ.,* **44**, 710 (1968)

c. Ono, S. and Kondo, S., *Handbuch der Physik,* **X**, 231 (in English) ed. S. Flügge Springer, Berlin (1960)

d. Bretz, M. and Dash, J. G., *Phys. Rev. Letters,* **27**, 647 (1971)

e. Roy, N. N. and Halsey, G. D., *Journal of Low Temp. Phys.,* **4**, 231 (1971)

f. Larher, Y., Thesis, Orsay (1970)

1966

ROBERT S. HANSEN

IOWA STATE UNIVERSITY
AMES, IO.

. . . for his imaginative application of the principles of thermodynamics and kinetics to adsorption processes; for his development of powerful new methods for studying adsorption and rates of reaction at interfaces; for his pioneering theoretical and experimental work on capillary ripples; and for the stimulation he imparted to all around him in his vigorous and enthusiastic pursuit of colloid science.

Surface Films and the Propagation of Capillary Ripples

Robert S. Hansen
Iowa State University
Ames, Iowa

Introduction

Pliny the Elder[1] commented that divers spread oil on the water surface to improve their vision, presumably by damping surface waves. This and such maxims as "pour oil on troubled waters" indicate that the wave damping effect of surface films has been recognized for centuries. The quantitative theory of the propagation characteristics of surface waves, and experimental techniques for testing this theory with suitable precision have had much shorter histories and in fact at present are under active development. They are the subjects of this presentation.

The theory of surface waves received considerable attention in the development of mathematical hydrodynamics during the 19th century. Stokes[2] and Rayleigh[3] both wrote on gravity waves, and Kelvin[4] on capillary waves; these treatments, however, all dealt with ideal liquids and therefore did not attack the wave damping problem. Stokes[2] did indicate methods for estimating how the viscosity of a real liquid could modify wave properties characteristic of ideal liquids. Early editions of Lamb's[5] treatise on hydrodynamics give good approximate treatments for the effect of inextensible surface films in damping capillary waves on real liquids.

The surface elastic modulus, $\epsilon = \dfrac{d\gamma}{d \ln A}$, where γ is the surface tension and A the surface area, characterizes the resistance of a surface to stretching. Levich[6] was the first to show how wave propagation characteristics could be calculated for arbitrary ϵ. Subsequent treatments by Dorrestein[7], Goodrich[8], Hansen and Mann[9] and van den Tempel and van de Riet[10] have been roughly similar in concept; van den Tempel and van de Riet give a particularly clear presentation of the theoretical framework.

Formal solutions to the wave propagation problem have an abstract appearance, and their physical implication has not always been clear. One of the most interesting characteristics of capillary waves is the fact that the damping passes through a maximum as the surface elastic modulus increases. Levich did not notice this, his text contains in the damping formula an error in sign so that no maximum is implied by it, and Levich states explicitly that none will occur. Dorrestein[7] appears to have been the first to recognize that the theory implied a maximum and Hansen and Mann[9] gave numerical solutions to the theory showing the maximum graphically.

There are two major experimental techniques at present capable of measuring wavelength and damping coefficient under experimental conditions justifying application of the linearized theories so far developed and with precision suitable for testing them. These are the optical method of Brown[12] and the electromechanical method of Mann and Hansen[13].

The theoretical and experimental problems are illustrated in Fig. 1. Let the plane $z = 0$ represent the liquid-air interface at rest, with liquid below and air above. Imagine a wave propagated in the $+x$ direction by a probe whose mean position is along the y axis but which oscillates up and down with a frequency ν and amplitude a, so that its z coordinate at any time t is a $\cos \omega t$. The simplest kind of wave we might expect will lead to a liquid-air interface whose z coordinate at time t and position x is given by

$$\zeta = ae^{-\alpha x} \cos(\kappa x - \omega t) \qquad (1)$$

where α is the damping coefficient, κ is the wave number ($\kappa = 2\pi/\lambda$, where λ is the wavelength), and $\omega = 2\pi\nu$ is the angular frequency of the wave. We have three major questions to answer, namely:

1) Under what circumstances may we expect the wave to be of the simple form indicated by Eq (1) (rather than, for example, an infinite Fourier series of similar terms)?

2) How do α and κ depend on ω, on bulk properties of the liquid, and on properties of the interface?

3) How can we measure α, κ, and ω with precision suf-

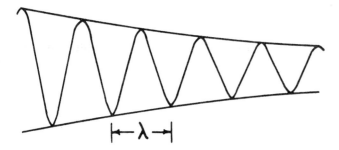

RIPPLE CHARACTERISTICS

1. WAVE LENGTH $\lambda = \dfrac{2\pi}{k}$

2. DAMPING COEFFICIENT a
$\zeta = a\cos(kx - \omega t)e^{-ax}$
FOR 200 c.p.s. WAVES
$\lambda \sim 0.2\,cm,\ a \sim 0.5\,cm^{-1}$
$k \sim 30\,cm^{-1}$

Figure 1. Capillary wave propagating to the right, with characteristics to be measured and explained indicated. Reproduced by permission from "Progress in Surface and Membrane Science, Vol. 4", J. F. Danielli, M. D. Rosenberg, D. A. Cadenhead, Editors; Academic Press, Inc., N.Y., 1971, pp. 1-56.

ficient to be informative under conditions where the wave will be represented by Eq (1)?

Theoretical

An isotropic Newtonian fluid moves according to the Navier-Stokes Equation.

$$\rho\left(\frac{\partial \vec{v}}{\partial t} + \vec{v}\cdot\nabla\vec{v}\right) = -\nabla p + \mu\nabla^2\vec{v} + \rho\vec{g} \tag{2}$$

where ρ, $\vec{v}$, p, and μ are the density, velocity, pressure, and viscosity of the liquid at (x, y, z, t) and $\vec{g}$ is the acceleration of gravity. This can be easily recognized as a form of Newton's equation of motion $m\vec{a} = \vec{F}$ (m = mass, $\vec{a}$ = acceleration, $\vec{F}$ = force) by multiplying Eq (2) through by a volume element $\Delta V = \Delta x\Delta y\Delta z$. The left side is now of the form $m\vec{a}$. Clearly $\rho\Delta V$ is the mass of the volume element ΔV. Considering the x component of $\frac{\partial \vec{v}}{\partial t} + \vec{v}\cdot\nabla\vec{v}$, i.e. $\frac{\partial v_x}{\partial t} + v_x\frac{\partial v_x}{\partial x} + v_y\frac{\partial v_x}{\partial y} + v_z\frac{\partial v_x}{\partial z}$, we can recognize it, for the element ΔV, as of the form $\frac{\partial v_x}{\partial t} + \frac{\partial v_x}{\partial x}\frac{dx}{dt} + \frac{\partial v_x}{\partial y}\frac{dy}{dt} + \frac{\partial v_x}{\partial z}\frac{dz}{dt}$ if we think momentarily of x, y, and z in the derivatives $\frac{dx}{dt}$, $\frac{dy}{dt}$ and $\frac{dz}{dt}$ as coordinates of the (moving) center of mass of ΔV. This component is hence the x component of the acceleration of ΔV, and in general the quantity $\frac{\partial \vec{v}}{\partial t} + \vec{v}\cdot\nabla\vec{v}$ is the acceleration of ΔV. The first term

on the right side, when Eq (2) is multiplied by ΔV, is the force on ΔV due to the pressure change across it, the second is the force due to viscous drag on the faces of the element ΔV, and last is the force of gravity on ΔV.

Continuity requires

$$\frac{\partial \rho}{\partial t} + \vec{\nabla}\cdot(\rho\vec{v}) = 0 \tag{3}$$

This can be seen again by multiplying the equation by ΔV; the first term in the resulting equation is the rate of increase of mass of the volume element ΔV, the second is the net flow of mass out of ΔV. For incompressible fluids the density is independent of time, and hence

$$\vec{\nabla}\cdot\vec{v} = 0 \tag{4}$$

We shall treat liquids as incompressible, and therefore require their motions to satisfy both Eqs (2) and (4).

Equation (2) is nonlinear because of the $\vec{v}\cdot\vec{\nabla}\vec{v}$ term which appears in it, and has been so far analytically intractable because of it. It is therefore attractive to design experiments which make the nonlinear term negligible. If the displacement is of the form given by Eq (1), $\frac{\partial v}{\partial t}$ will be of the order $a\omega^2$, $\vec{v}\cdot\nabla\vec{v}$ of order $a^2\kappa\omega^2$, and $\frac{\mu}{\rho}\nabla^2\vec{v}$ of order $\frac{\mu}{\rho}a\omega\kappa^2$. Hence the term $\vec{v}\cdot\vec{\nabla}\vec{v}$ will be small compared to $\frac{\partial v}{\partial t}$ if $\kappa a \ll 1$, but the more demanding requirement that $\vec{v}\cdot\vec{\nabla}\vec{v}$ be small compared to $\frac{\mu}{\rho}a\omega\kappa^2$ appears to require $\kappa a \ll \frac{\mu\kappa^2}{\rho\omega}$, i.e. that κa be small compared to a quantity which, in the capillary wave case, proves to be about 10^{-2}. This is obviously an order of magnitude treatment, and a detailed analysis indicates that the $(\vec{v}\cdot\vec{\nabla})\vec{v}$ term can be neglected with satisfactorily small error if experiments are so designed that $\kappa a < 10^{-2}$, $(a/\lambda) \sim 10^{-3}$ or less. When this term is neglected we obtain the *linearized* Navier-Stokes equation

$$\rho\frac{\partial \vec{v}}{\partial t} = -\vec{\nabla}p + \mu\nabla^2\vec{v} + \rho\vec{g} \tag{5}$$

We now resolve $\vec{v}$ into two components
$$\vec{v} = \vec{v_1} + \vec{v_2} \tag{6}$$
such that
$$\rho\frac{\partial \vec{v_1}}{\partial t} = -\vec{\nabla}p + \rho\vec{g} \tag{7}$$
$$\rho\frac{\partial \vec{v_2}}{\partial t} = \mu\nabla^2\vec{v_2} \tag{8}$$
with
$$\vec{\nabla}\cdot\vec{v_1} = \vec{\nabla}\cdot\vec{v_2} = 0 \tag{9}$$
to satisfy Eq (4). We define a complex wave number k by
$$k = \kappa + i\alpha \tag{10}$$
so that the real part of the complex surface wave
$$\zeta = \zeta_o e^{i(kx - \omega t)} \tag{11}$$
is of the form of Eq (1). We now consider the $\vec{v_1}$ and $\vec{v_2}$ motions separately.

From Eq (7) we find immediately that $\vec{\nabla} \times \vec{v}_1 = 0$, hence $\vec{v}_1 = -\vec{\nabla}\varphi$ where φ is a scalar potential function. To satisfy Eq (9) also, we must have

$$\nabla^2 \varphi = 0 \qquad (12)$$

The solution of Eq (12) periodic in x and t satisfying $\varphi \to 0$ as $z \to -\infty$ is

$$\varphi = Ae^{kz} \, e^{i(kx - \omega t)} \qquad (13)$$

Using this result in Eq (7), we find

$$\vec{\nabla} p = \rho \frac{\partial}{\partial t} \vec{\nabla}\varphi + \rho \vec{g} \qquad (14)$$

or

$$p = \rho \frac{\partial \varphi}{\partial t} - \rho g z \qquad (15)$$

$$= -i\rho\omega\varphi - \rho g z \qquad (16)$$

(the vector $\vec{g}$, pointing in the $-z$ direction, has been replaced by the scalar g).

From Eq (8) we deduce that $\vec{v}_2$ is derivable from a vector potential, i.e. $\vec{v}_2 = -\vec{\nabla} \times \vec{\psi}$. If $\vec{\psi}$ has *only* a y component $\vec{v}_2$ will have no y component as is physically required. This form for $\vec{v}_2$ automatically satisfies $\nabla \cdot \vec{v}_2 = 0$, and Eq (8) becomes

$$\rho \frac{\partial \psi}{\partial t} = \mu \nabla^2 \psi \qquad (17)$$

This is satisfied by

$$\psi = Be^{mz} \, e^{i(kx - \omega t)} \qquad (18)$$

with

$$m^2 = k^2 - \frac{i\rho\omega}{\mu}$$

and (to satisfy $\psi \to 0$ as $z \to -\infty$) the root of Eq (18) with positive real part is taken.

Now both the $\vec{v}_1$ and $\vec{v}_2$ motions satisfy the linearized Navier-Stokes equation, are periodic in x and t, and vanish as $z \to -\infty$, so this will also be true of any linear combination of them. It is interesting to note that they are very different kinds of motion. Let ξ and ζ be the x and z displacements. We have, for example

$$\xi_1 = \int v_{1x} dt = -\int \frac{\partial \varphi}{\partial x} dt = -ik\int \varphi \, dt = \frac{k}{\omega}\varphi \qquad (19a)$$

Using Eq (13) and taking the real part

$$\xi_1 = \frac{Ak}{\omega} e^{kz} \cos(kx - \omega t) \qquad (19b)$$

Similarly

$$\zeta_1 = \frac{Ak}{\omega} e^{kz} \sin(kx - \omega t) \qquad (20)$$

$$\xi_2 = \left(\frac{\rho}{\mu\omega}\right)^{1/2} Be^{\left(\frac{\rho\omega}{2\mu}\right)^{1/2} z} \cos\left(\theta + \frac{\pi}{4}\right) \qquad (21)$$

$$\zeta_2 = \frac{k}{\omega} Be^{\left(\frac{\rho\omega}{2\mu}\right)^{1/2} z} \cos\theta \qquad (22)$$

where

$$\theta = kx - \left(\frac{\rho\omega}{2\mu}\right)^{1/2} z - \omega t \qquad (23)$$

and the approximation $m \approx \left(\frac{\rho\omega}{2\mu}\right)^{1/2} (1 - i)$ has been used. Eqs (19b) and (20) represent the particle orbit in the $\vec{v}_1$ motion parametrically; it is the circle

$$\xi_1^2 + \zeta_1^2 = \frac{A^2 k^2}{\omega^2} e^{2kz} \qquad (24)$$

Similarly Eqs (22) and (23) represent an ellipse parametrically. In capillary ripple experiments involving water-air interfaces typical magnitudes are $k \sim 30 \text{ cm}^{-1}$, $\rho \sim 1 \text{ gm/cm}^3$, $\mu \sim 10^{-2}$ poise and so $m^2 \sim -\frac{i\omega\rho}{\mu} \sim -10^5 \, i \gg k^2$. Hence the amplitude component parallel to the surface in the v_2 motion is about ten times as great as that perpendicular to the surface. Further, the amplitude of the orbit in the v_1 motion decays by a factor e each 300 microns from the surface, but the same factor e decay in the v_2 motion requires only about 40 microns. These characteristics are all illustrated in Fig. 2. It should be apparent from this figure first, that the v_2 motion will be extremely sensitive to surface properties, and second, that the depth of the liquid will be relatively unimportant for experiments in the $\omega \sim 1000 \text{ sec}^{-1}$ range so long as it is greater than a few mm.

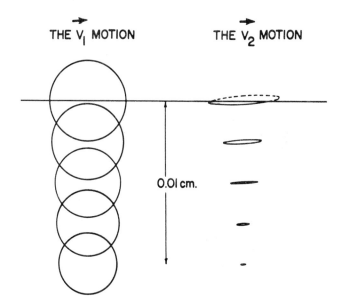

THE V_1 MOTION   THE V_2 MOTION

0.01 cm.

Figure 2. Particle orbits for two characteristic motions of a fluid in small amplitude oscillation. Reproduced by permission from "Progress in Surface and Membrane Science, Vol. 4", J. F. Danielli, M. D. Rosenberg, D. A. Cadenhead, Editors; Academic Press, Inc., N.Y., 1971, pp. 1-56.

The actual combination of v_1 and v_2 motions depends on the boundary conditions we have not yet considered, i.e. those concerned with the liquid-air interface, and all of the surface chemistry enters the problem mathematically through these boundary conditions. The dividing (mathematical) surface has no mass, and hence (to avoid an infinite acceleration) must have no net stress on it. Hence the stress due to the motion of the liquid must be exactly balanced by stresses due to the surface forces. The stress boundary conditions (also linearized) are for the normal stress

$$\gamma \frac{\partial^2 \zeta}{\partial x^2} = \left[-p + 2\mu \left(\frac{\partial v_z}{\partial z} \right) \right]_{z = \zeta} \quad (25)$$

and for the tangential stress

$$\epsilon \frac{\partial^2 \xi}{\partial x^2} = \mu \left[\frac{\partial v_z}{\partial x} + \frac{\partial v_x}{\partial z} \right]_{z = \zeta} \quad (26)$$

The left side of Eq (25), in linearized form, is γ/R where R is one principal radius of curvature of the surface (the other is infinite); if $\frac{\partial^2 \zeta}{\partial x^2} > 0$ the center of curvature is on the air side of the interface. The pressure reference is taken as zero on the air side of the interface, and $(p)_{z = \zeta}$ is then the pressure on the liquid side of the interface. Hence for all velocities equal to zero, Eq (25) reduces to the Kelvin Equation $\Delta p = \gamma(1/R_1 + 1/R_2)$ for the pressure change across a curved interface. The quantity $2\mu \frac{\partial v_z}{\partial z}$ is an additional normal stress developed by virtue of the viscosity in a moving viscous fluid. The left side of Eq (26) arises, in the simplest case, from the stress due to a surface tension gradient resulting from differential stretching of the surface. For where $ds = (d\xi^2 + d\zeta^2)^{1/2}$ and γ_o is the surface tension of the unstretched surface, we can imagine the surface tension at any point developed in power series of the form

$$\gamma = \gamma_o + \epsilon \frac{ds}{dx} + \epsilon_2 \left(\frac{ds}{dx} \right)^2 + \dots \;;$$ to linear terms it follows that

$$\frac{d\gamma}{dx} = \epsilon \frac{d}{dx} \frac{ds}{dx} \approx \epsilon \frac{d}{dx} \frac{d\xi}{dx} = \epsilon \frac{d^2 \xi}{dx^2}.$$ If the surface is characterized by a surface viscosity it may be shown that this contributes a stress which is also proportional to $\frac{d^2 \xi}{dx^2}$, and hence can be simply handled by proper interpretation of the parameter ϵ (in particular, by treating it as a complex modulus of the form $\epsilon = \epsilon' - i\epsilon''$). The right side of Eq (26) is the shear stress exerted on the surface (tangentially directed) by virtue of the liquid motion.

The boundary conditions in Eqs (25) and (26) are to be satisfied at the surface, i.e. at $z = \zeta$. Except for the term p in Eq (25) (in which the quantity $\rho g z$ in Eq (16) must be explicitly included as $\rho g \zeta$ at the surface) all other quantities, to the level of approximation appropriate to the linearized theory, can be evaluated at $z = 0$. Doing this, we obtain from Eq (25)

$$\rho \omega^2 A - (\rho g k + \gamma k^3)(A + iB) + 2\mu k \omega (ikA - mB) = 0 \quad (27)$$

and from Eq (26)

$$\mu \omega [2k^2 A + i(m^2 + k^2)B] + k^2 \epsilon(ikA - mB) = 0 \quad (28)$$

These two homogeneous linear equations in the coefficients A and B can have a non-trivial solution only if the determinant of their coefficients is zero. The resulting determinantal equation can be solved to establish, for example, the

complex number $k = \kappa + i\alpha$ (both real and imaginary parts) as a function of ρ, ω, g, γ, μ, and ϵ. The ideal liquid result of Kelvin can be obtained from Eq (27) by setting $B = 0$, $\mu = 0$ to obtain

$$\rho \omega^2 = \rho g k + \gamma k^3 \text{ (ideal liquid)} \quad (29)$$

The waves described (approximately) by this equation are called gravity waves if $\rho g k \gg \gamma k^3$, and capillary waves if $\gamma k^3 \gg \rho g k$. We shall neglect $\rho g k$ compared to γk^3 in Eqs (27) and (29) for simplification in the following discussion, since experiments to be discussed will be capillary waves; for precise calculations it should be included and its inclusion adds only minor computational difficulty.

Hansen and Mann[9] have studied solutions to the equation resulting from setting the determinant of the coefficients of A and B in Eqs (27) and (28) equal to zero (and neglecting $\rho g k$ in Eq (27)) over wide ranges of relevant parameters. They used dimensionless groups $y_1 = \frac{\rho \omega^2}{\gamma k^3}$, $y_2 = \frac{\alpha}{\kappa}$, $u_1 = \frac{\mu \omega}{\gamma \kappa}$, and $u_2 e^{i\theta_2} = \frac{\epsilon}{\gamma}$, and solved the equation by successive approximations using $y_1 - 1$, y_2, and u_1 as parameters of smallness. In this way they obtained for u_2 small

$$y_1 = 1 - (2u_1)^{3/2} + u_2 \cos\theta_2 + 4u_2 u_1^{1/2} \sin(\theta - \tfrac{1}{4}\pi) \quad (30a)$$

$$y_2 = \frac{4}{3} u_1 - \frac{1}{3}(2u_1)^{3/2} - \frac{1}{3} u_2 \cos\theta_2 + \frac{4}{3} u_2 u_1^{1/2} \cos(\theta_2 - \tfrac{1}{4}\pi) \quad (30b)$$

and for u_2 large

$$y_1 = 1 - \frac{1}{2}(2u_1)^{1/2} + \frac{1}{4} u_1 - \frac{u_1}{u_2} \sin\theta_2 + 4 \frac{u_1^{3/2}}{u_2} \sin(\theta_2 - \tfrac{1}{4}\pi) \quad (31a)$$

$$y_2 = \frac{1}{6}(2u_1)^{1/2} + \frac{11}{36} u_1 + \frac{1}{3} \frac{u_1}{u_2} \cos\theta_2 - \frac{4u_1^{3/2}}{3u_2} \cos(\theta_2 - \tfrac{1}{4}\pi) \quad (31b)$$

Hansen and Mann also solved the equation numerically for fixed u_1, and various values of θ_2 over the entire range of u_2 with results shown in Figs. 3 and 4. The curves for $\theta_2 = 0$ correspond to a real elastic modulus ($\epsilon'' = 0$ in $\epsilon = \epsilon' - i\epsilon''$), indicating maxima in both y_1 and y_2 for u_2 near 0.1 The curves for $\theta_2 = 45°$ correspond to negative ϵ'', and I know of no model according to which this situation is likely to arise physically. Negative values of θ_2 can result from diffusional interchange, surface viscosity, or other monolayer relaxation processes. In particular, for diffusion-limited surface-to-bulk interchange the linearized theory gives in good approximation

$$u_2 e^{i\theta_2} = \frac{\epsilon'_o}{\gamma} \left[1 + \left(\frac{D}{\omega} \right)^{1/2} e^{\pi i/4} \left(\frac{dC}{d\Gamma} \right)_{C_o} \right]^{-1} \quad (32)$$

in which D is the diffusion coefficient, Γ (moles/cm^2) the surface density of adsorbed species, C_o (moles/cm^3) the bulk adsorbate concentration, and ϵ'_o the elastic modulus that would obtain in the absence of diffusional interchange.

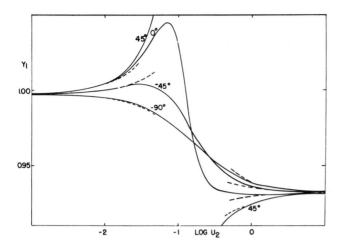

Figure 3. Dependence of $y_1 = \dfrac{\rho\omega^2}{\gamma\kappa^3}$ on $u_2 e^{i\theta_2} = \dfrac{\epsilon' - i\epsilon''}{\gamma}$ for $u_1 = \dfrac{\mu\omega}{\gamma\kappa} = 10^{-2}$. Dashed curves are from Eqs (30-1). Reproduced by permission from J. Appl. Phys., **35**, 152 (1964).

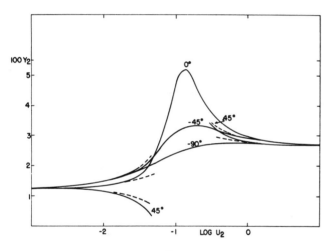

Figure 4. Dependence of $y_2 = \dfrac{\alpha}{\kappa}$ on u_2 and θ_2 for $u_1 = 10^{-2}$. Dashed curves are from Eqs (30-1). Reproduced by permission from J. Appl. Phys., **35**, 152 (1964).

Plainly values of θ_2 in the range of $-45° < \theta_2 < 0$ are consistent with diffusional interchange, and Figs. 3 and 4 show how the maxima are sharply suppressed, both for y_1 and y_2, as θ_2 becomes negative.

Experimental

To document the difference between y_1 and 1 expected from Eq (30a) wavelengths must be measured to at least 0.1%, and to justify anything beyond the leading terms for y_2 in Eqs (30b) and (31b), damping coefficients must be measured to within a few percent. To justify the linearized theory, the measurements must be made on waves whose amplitude is about 10^{-3} times the wavelength. The electro-

mechanical instrument of Mann and Hansen[13] is capable of meeting all of these requirements; its principle is illustrated in Fig. 5. A needle probe (the generator probe) lying in the surface is connected to a loudspeaker bellows; an alternating current drives the bellows and hence the probe. A second probe (the receiver probe) lying in the surface parallel to the first probe is connected to a phonograph cartridge, so when the second probe oscillates as the result of wave motion an electrical signal is generated by the cartridge and subsequently amplified. This amplified signal is measured for two properties. First, an oscilloscope is used to compare its phase with that of the signal to the loudspeaker. This is done by placing one signal on the x input and one on the y input of the oscilloscope and observing the resulting Lissajous figure. For example, a straight line Lissajous figure passing through first and third quadrants indicates that the two signals are in phase. The receiver probe can then be translated until the Lissajous figure is repeated, and the translation distance is the wavelength. Second, the voltage of the amplified signal is measured with a suitable voltmeter, and the variation in voltage as the receiver probe is translated can be measured to obtain the damping coefficient.

Fig. 6 shows, using a linear variable differential transformer (LVDT) in place of the generating probe, how it was proved that the probe displacement was proportional to the speaker voltage (and what the proportionality constant was, so that probe amplitude could be precisely controlled), and how it was proved that the amplified signal from the receiver probe was also proportional to the speaker voltage (and hence to the generator probe amplitude).

Introduction of the receiver probe introduces a standing wave character to the propagating ripples, i.e. not only the ripples generated at the generator probe, but also those reflected from the receiver probe must be considered in analyzing the dependence of receiver probe output on its distance from the generator probe. Figures 7 and 8 illustrate this point; at large separations the output signal decays exponentially with distance, but at short separations a strong periodic variation is superimposed on the exponential decay.

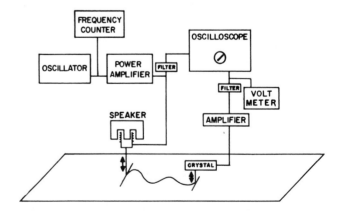

Figure 5. Block diagram of apparatus for measuring ripple wavelength and amplitude (hence damping).

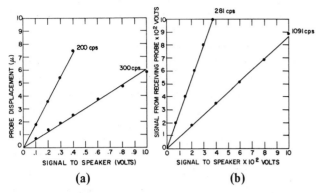

(a)   **(b)**

Figure 6. Demonstration of proportionality between receiving probe vibration amplitude and receiving transducer output showing (a) sending probe amplitude proportional to voltage to speaker (b) receiver output voltage proportional to voltage to speaker. Reproduced by permission from J. Colloid and Interface Sci., 18, 757 (1963).

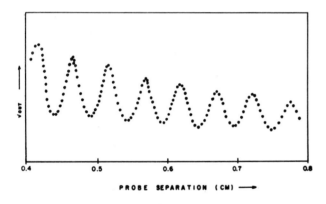

Figure 7. Dependence of receiver output on probe separation at small separations. Reproduced by permission from J. Colloid and Interface Sci., 18, 757 (1963).

The effect of the reflected wave can be treated quantitatively[13], so that data illustrated in Fig. 8 can be used to obtain both wavelength and damping coefficient.

Figure 9 shows the dependence of surface tension on concentration for aqueous solutions of octanoic acid and of lauryl amine hydrochloride as obtained by the capillary wave method[14]. Comparison data of Matijevic and Pethica[15] are given for the former case, and the agreement is plainly excellent. Similarly excellent agreement between surface pressures measured by capillary ripple and equilibrium (Langmuir balance) techniques is shown for spread monolayers in Fig. 10.

The variation in damping coefficient with surface concentration (implicitly, with surface elastic modulus) for spread monolayers is illustrated in Fig. 11 for two monolayer types, and the damping maximum is clearly revealed in each. Conversely, measured damping coefficients can be used to calculate the surface elastic modulus, and provide sensitive measures of the modulus particularly in the neighborhood of the damping maximum. Figure 12 compares surface modulus measurements obtained by capillary ripple and equilibrium (Langmuir balance) methods.

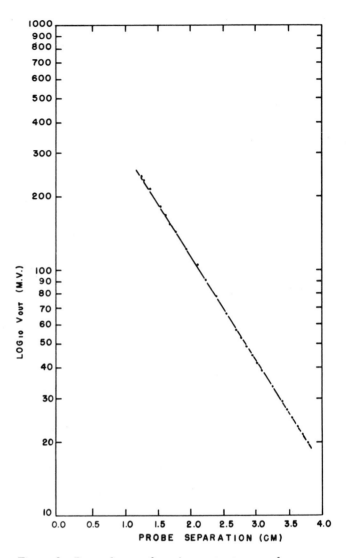

Figure 8. Dependence of receiver output on probe separation at large separations. Reproduced by permission from J. Colloid and Interface Sci., 18, 757 (1963).

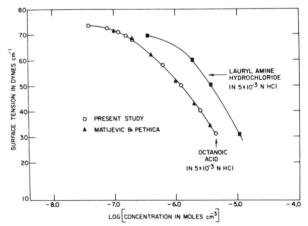

Figure 9. Dependence of surface tension on concentration for octanoic acid and lauryl amine hydrochloride according to capillary ripple measurements. Equilibrium data for the former system are included for comparison[14,15]. Reproduced by permission from J. Colloid and Interface Sci., 22, 32 (1966).

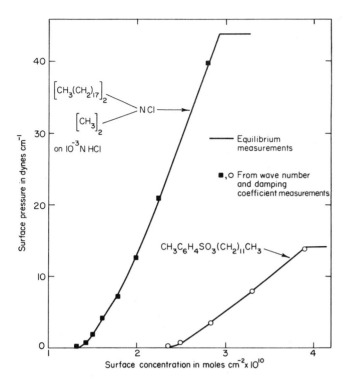

Figure 10. Comparison of surface pressure results from capillary ripple and equilibrium measurements for spread monolayers[14]. Reproduced by permission from J. Colloid and Interface Sci., 22, 32 (1966).

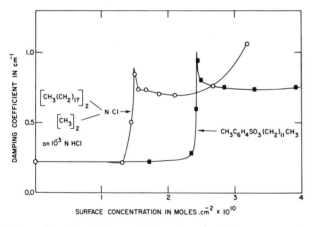

Figure 11. Dependence of damping coefficients on surface concentration for spread monolayers of distearyl dimethyl ammonium chloride and N-dodecyl-p-toluyl sulfonate[14]. Reproduced by permission from J. Colloid and Interface Sci., 22, 32 (1966).

Figure 13 provides a particularly striking illustration of the effect of surface properties or damping coefficient behavior for adsorbed monolayers. The damping coefficient maximum occurs near the concentration leading to half coverage ($\Gamma \sim \frac{1}{2} \Gamma$ max), and this concentration occurs at a lower value the longer the hydrocarbon chain (it decreases about a factor 3 for each methylene group). The concentrations at which the maxima occur are too low for appreciable diffusional interchange to occur over the wave period

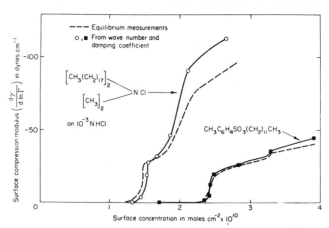

Figure 12. Comparison of surface compression modulus results from capillary ripple and equilibrium measurements. The surface elastic modulus $\epsilon = - \dfrac{d\gamma}{d\ln\Gamma}$ for these systems[14]. Reproduced by permission from J. Colloid and Interface Sci., 22, 32 (1966).

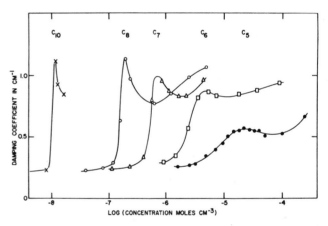

Figure 13. Dependence of damping coefficient on fatty acid concentration in N/200 HCl for decanoic (C_{10}), octanoic (C_8), heptanoic (C_7), hexanoic (C_6) and pentanoic (C_5) acids[16]. Reproduced by permission from J. Colloid and Interface Sci., 23, 319-28 (1967).

in the cases of octanoic and decanoic acid; in this circumstance the maxima are expected theoretically to have the same height, as indeed they do. The width of the damping peak provides a measure of interaction between tails (as measured, for example, by the interaction parameter in the Frumkin adsorption isotherm). This interaction should be greater for decanoic acid than for octanoic acid, and theory then indicates that the former peak should be narrower than the latter; Fig. 13 shows that this expectation is satisfied. At higher concentrations those portions of the surface stretched at a given instant (and thereby having a higher surface tension due to the reduction of adsorbed surfactant density) tend to have their surface tension lowered by diffusion from surfactant from bulk to surface and subsequent adsorption. Conversely, those portions compressed tend to lose adsorbed molecules to bulk by diffusion. Hence diffusional interchange acts to reduce surface tension gradients, and so also to reduce the effective surface elastic modulus and the damping coefficient. The lowering is quite noticeable with heptanoic acid, more so with hexanoic acid, and with pentanoic acid the maximum has almost disappeared. The progressive broadening of the damping peak as chain length (and hence interaction between tails) decreases is evident throughout the series. This behavior can be interpreted quantitatively; indeed the lowering of the damping coefficient can be used to calculate the solute diffusion coefficient, and values of the diffusion coefficients of pentanol, pentanoic acid and hexanoic acid obtained in this way are in reasonable agreement with literature values. Diffusion coefficients obtained in this way for heptanoic and octanoic acid are about an order of magnitude high; this may be because the maximum depression is too slight for accurate measurement, but it may conceivably reflect the operation of a relaxation mechanism other than diffusion for reducing surface tension gradients developed by ripple propagation.

Conclusion

The linearized hydrodynamic theory of surface waves, when applied to data obtained in experiments of low amplitude-to-wavelength ratio so as to validate linearization assumptions, provides an adequate interpretation of the data. Surface tension and surface elastic modulus values obtained from capillary wave measurements are in good quantitative agreement with conventionally obtained values. Damping coefficient behavior, including damping coefficient maximum and its suppression by diffusional interchange, are all in general accord with theoretical expectation.

As to future developments in this field, there are several possibilities which at least interest me considerably.

First, there seem to me no inherent difficulties, either theoretically or experimentally, in generalizing the treatment of surface waves to waves at liquid-liquid interfaces; we are working on this problem at present.

Second, the theory predicts a maximum in the dimensionless group $y_1 = \gamma k^3/\rho\omega$ as elastic modulus increases. This has not yet been documented and will require wavelength measurements of accuracy well under 0.5% to document it, but this seems well within instrumental capabilities.

The third prospect is, I think, the most interesting. Capillary ripple measurements for a given system at a given frequency provide two numbers—the wavelength and the damping coefficient, from which two surface properties may be inferred. If the surface tension is measured independently, then the two surface properties can be the real and imaginary parts of the surface elastic modulus, i.e. ϵ'

and ϵ'' in $\epsilon = \epsilon' - i\epsilon''$. Much as the dependence of real and imaginary parts of the dielectric constant on frequency can be used to reveal relaxation mechanisms in solid dielectrics, so similar data may reveal relaxation processes in surface films. We have had heretofore almost no method to get at questions of this sort.

Acknowledgements

A professor's research depends critically on his graduate students and postdoctoral associates who are never-ending sources of enthusiasm and fresh insights. Mine have been particular sources of pleasure and inspiration to me, and I want first to express my appreciation to them collectively. The work I have discussed today involved two of them particularly. Our laboratory's work on capillary waves began with the dissertation research of Dr. J. Adin Mann, Jr.; our own version of the hydrodynamic theory of capillary waves and the development of the electromechanical instrument for measuring their properties developed in the course of this research. My second major collaborator was Dr. J. Lucassen, who did much to sharpen my understanding of the theory of these waves and whose elegant experiments are illustrated in Figs. 9-13.

I am deeply grateful to the U. S. Atomic Energy Commission which, through its Ames Laboratory at Iowa State University, has supported my research throughout my professional career.

References

1. *Pliny the Elder Historia Naturalis,* Lib. II, Cap. 106
2. G. G. Stokes, *Trans. Cambridge Phil. Soc.,* **8,** 441 (1847)
3. Lord Rayleigh, *Phil. Mag. (6),* **21,** 177 (1911)
4. Lord Kelvin, *Phil. Mag. (4),* **42,** 368 (1871)
5. H. Lamb, *"Hydrodynamics",* 6th Ed., Dover Publications, New York, 1945
6. V. G. Levich, *Acta Physicochim U.R.S.S.,* **14,** 307, 321 (1941)
7. R. Dorrestein, *Proc. K. Ned. Akad. Wet,* **B54,** 260 (1951)
8. F. C. Goodrich, *Proc. Roy. Soc.* (London), **A260,** 490, 503 (1961)
9. R. S. Hansen and J. A. Mann, *J. Applied Phys.,* **35,** 158 (1964)
10. V. van den Tempel and R. P. van de Riet, *J. Chem. Phys.,* **42,** 2769 (1965)
11. V. G. Levich, *"Physicochemical Hydrodynamics"* (English translation by Scripta Technica, Inc.), Prentice Hall, Inc., Englewood Cliffs, N. J. (1962)
12. R. C. Brown, *Proc. Phys. Soc.* (London), **48,** 312, 323 (1936)
13. J. A. Mann and R. S. Hansen, *J. Colloid Sci.,* **18,** 757, 805 (1963)
14. J. Lucassen and R. S. Hansen, subsequently published, *J. Colloid and Interface Sci.,* **22,** 32 (1966)
15. E. Matijević and B. A. Pethica, *Croat. Chem. Acta,* **29,** 431 (1957)
16. J. Lucassen and R. S. Hansen, subsequently published, *J. Colloid and Interface Sci.,* **23,** 319 (1967)

ADDENDUM 1972:

CAPILLARY WAVES SINCE 1966

Robert S. Hansen
Iowa State University
Ames, Iowa

Extensive reviews devoted to capillary waves have been published by Lucassen-Reynders and Lucassen[1] and by Hansen and Ahmad[2] (1969 and 1971 respectively). A review of surface-driven phenomena by Levich and Krylov[3] gives excellent brief discussions of many hydrodynamic phenomena in which surface properties are involved, including waves. The field has been a busy one since 1966.

Of the three "problems of the future" listed in the conclusion of my 1966 presentation, two have been fairly satisfactorily resolved.

Hansen, Lucassen, Bendure and Bierwagen[4] have carried out the hydrodynamic analysis of waves at interfaces between two liquids, have shown that the electromechanical technique, with suitable probe modification, can be used to measure wavelengths and damping coefficients at such interfaces, and have shown results to be reasonably in accord with theory at water-heptane and polarized mercury-aqueous electrolyte interfaces. Measurement precision was less than at water-air interface. It also proves to be the case that, for two liquids with the same densities and viscosities, propagation characteristics of capillary waves will be independent of elastic modulus, and it is therefore scarcely surprising that increasing surfactant concentration at the water-heptane interface is not associated with a pronounced damping maximum.

The theoretically-expected maximum in $y_1 = \dfrac{\rho \omega^2}{\gamma k^3}$ was documented by Hansen and Bendure[5] for tetradecanol, hexadecanol, and distearyl dimethyl ammonium chloride monolayers on water, using a variable frequency modification of the electromechanical technique permitting differences in y_1 to be measured to one part in 10^3 quite readily.

Progress with resolution of real and imaginary parts of the elastic modulus has been much less encouraging; the imaginary part of the modulus, ϵ'', is extremely sensitive to errors in γ, κ, and α; except in the neighborhood of the damping maximum with solutes known to be undergoing diffusional interchange we have not been able, in our laboratory, to find a case where capillary ripple measurements could unambiguously prove that ϵ'' was other than zero.

Thiessen and Scheludko[6] and Thiessen and Schwartz[7] have developed an optical method for measuring properties of standing waves in a cylindrical vessel.

Laser beat frequency spectroscopy has been applied to the scattering of light from surfaces[8,9,10,11]. The application to the study of surface waves[11] involves the fact that light of frequency ω_0 can be inelastically scattered by a surface wave of frequency ω_r, so that the scattered wave has frequencies $\omega_0 \pm \omega_r$, and the frequency shift can be detected by "beating" scattered and non-scattered beams of light. The idea involves measuring the beat frequency spectrum for a wave of known wavelength; the maximum in the spectrum peak yields the ripple frequency and the damping coefficient can be obtained from the width of the peak at half height. So far the precisions in wavelength, frequency, and damping coefficient are all at least an order of magnitude poorer than for the electromechanical method, but they are improving and the method holds promise for investigations up to 10^6 hertz.

The most interesting development over the past five years has been Lucassen's discovery of longitudinal waves. Lucassen was the first to recognize that the equation resulting from setting the determinant of the coefficients of A and B in Eqs (27) and (28) equal to zero had a second physically significant root (i.e. other than the transverse wave root for $k = \kappa + i\alpha$ which was the subject of my presentation). He investigated the motion associated with this root (it is chiefly a compression-rarefaction wave in the plane of the surface, hence the "longitudinal" designation), devised equipment to generate this motion and proved that the resulting wave had the expected properties. In retrospect, Lucassen's idea can be explained approximately in a very simple way. The normal and tangential stress boundary conditions, Eqs (25) and (26), involve $\gamma \dfrac{\partial^2 \zeta}{\partial x^2}$ and $\epsilon \dfrac{\partial^2 \xi}{\partial x^2}$ respectively in striking symmetry. The Kelvin Equation (which is the first approximation to the transverse wave problem) if only the *normal* stress boundary condition is considered, and the $\vec{v}_2$ motion is ignored ($B = 0$). Correspondingly, the first approximation solution for the longitudinal wave results if only the *tangential* stress boundary condition is considered, and the $\vec{v}_1$ motion is ignored ($A = 0$). The first approximation solution for wave number κ and damping coefficient α for the longitudinal wave (and this approximation will suffice for most applications) is

$$\kappa + i\alpha = \rho^{\frac14} \, \omega^{\frac34} \, \mu^{\frac14} \, \epsilon^{-\frac12} \, e^{\pi i/8}$$

Both wave number and damping coefficient are sensitive to the elastic modulus ϵ for all values of ϵ, not just those near a damping maximum as is the case for transverse waves.

References

1. E. H. Lucassen-Reynders and J. Lucassen, *Advances in Colloid and Interface Science*, **2**, 347 (1969).

2. R. S. Hansen and J. Ahmad in J. F. Danielli, M. D. Rosenberg and D. A. Cadenhead, Editors *"Progress in Surface and Membrane Science"*, **4**, 1. Academic Press, N.Y. 1971.

3. V. G. Levich and V. S. Krylov, *Ann. Review of Fluid Mechanics,* **1**, 293 (1969).

4. R. S. Hansen, J. Lucassen, R. L. Bendure and G. P. Bierwagen, *J. Colloid and Interface Sci.,* **26**, 198 (1968).

5. R. S. Hansen and R. L. Bendure, *J. Phys. Chem.,* **71**, 2889 (1967).

6. D. Thiessen and A. Scheludko, *Kolloid-Z.,* **218**, 139 (1967).

7. D. Thiessen and P. Schwartz, *Z. Physik Chem.,* **236**, 363 (1967).

8. R. H. Katyl and K. U. Ingard, *Phys. Rev. Lett.,* **23**, 752 (1969).

9. M. A. Bouchiat, J. Meunier, and J. Brossel, *Compt Rend. Acad. Sci.,* Paris, Ser. B **266**, 255 (1968).

10. M. A. Bouchiat and J. Meunier, *Phys. Rev. Lett.,* **23**, 752 (1969).

11. J. A. Mann, J. F. Baret, F. J. Dechow and R. S. Hansen, *J. Colloid and Interface Sci.,* **37**, 14 (1971).

12. J. Lucassen, *Trans. Faraday Soc.,* **64**, 2221, 2230 (1968).

1967

STANLEY G. MASON

PULP AND PAPER RESEARCH INSTITUTE OF CANADA
MONTREAL, P. Q.

. . . for his contributions to the physical and colloid chemistry of lignin and cellulose; for his advancement of the study and theory of the hydrodynamics of dispersions of small particles and for the imaginative and far-reaching applications of this work; and for his exemplary pursuit of colloid science which is an inspiration to his students and colleagues alike.

Kinetic Theory of Flowing Dispersions

S. G. Mason

Department of Chemistry, McGill University
and
Pulp and Paper Research Institute of Canada
Montreal, Canada

Abstract

When a dilute suspension is sheared the particles rotate, are acted upon by surface forces, and translate relative to one another. As a result particles can assume preferred orientations, can be deformed, and can undergo collisions with one another. This lecture summarizes an extensive series of theoretical and experimental studies of the rotations, deformations, orientations and n-body collisions ($n \geqslant 2$) of a variety of particle shapes, types and sizes in dilute and concentrated suspensions undergoing laminar shear flow. Equations for 2-body collision frequencies, mean free paths, mean particle displacement etc. have been derived (most of which have been confirmed by direct experimental observation) which are very similar to those in the classical kinetic theory of gases except that many of the shear-kinetic phenomena show microscopic reversibility. The significance of some of the phenomena to a number of areas of colloid chemistry is indicated.

1. Introduction

I thank my good friend and colleague, David Goring for his generous introduction, a host of past and present graduate students associated with me in the McGill Chemistry Department and the Pulp and Paper Research Institute of Canada who did most of the work for which this award has been granted to me, so many of my friends who have come from so many places to participate in this Symposium, the Kendall Company for making the award possible and finally the American Chemical Society who selected me for this honor.

Nearly twenty years ago, one of the technical committees of the Canadian Pulp and Paper Association suggested as a research topic for one or more of my graduate students at McGill, a basic study of the flocculation of fibers in the papermaking process and its influence upon the uniformity of paper. At that time, the view was held that under the conditions of papermaking, flocculation was governed by molecular and electrostatic interactions, according to the well-known Derjaguin-Landau-Verwey-Overbeek (or DLVO) theory which was developed for colloidal sols. However, since fibers, and particularly wood pulp fibers, are large in size, thus preventing them from having the appreciable Brownian movement which is an essential requirement of the DLVO theory, we could not subscribe to the idea that these mechanisms were applicable to flocculation of fibers in pulp suspensions. We confirmed this in a series of researches (which I will not discuss today) and demonstrated instead that fiber flocculation is essentially a fluid mechanical, and not a molecular, phenomenon. So it happened that I as a physical chemist with interests in colloid and surface chemistry found myself projected into an area combining fluid mechanics with physical chemistry (in Russia it is called physico-chemical hydrodynamics) which has led us to a very interesting series of investigations which have proven to be of interest not only to the pulp and paper industry, but to a number of other fields, such as polymer science and technology, geology, and physiology. Some idea of the widespread interest and application of the principles we have been studying is given by the sources of financial support of this work, including the pulp and paper industry, the Defence Research Board of Canada, the National Heart Institute of the National Institutes of Health in this country, the Petroleum Research Fund and others.

What I outline will, I hope, demonstrate, especially to those of you who are from the universities, that there are technological problems which, when treated with imagination and a reasonable degree of freedom, are every bit as challenging and just as suitable for post-graduate work as the so-called "pure" research topics which sometimes seem to be deliberately chosen by my academic colleagues just because they have no conceivable application.

I would like to discuss the behavior of various kinds of particles suspended in liquids subjected to two very simple kinds of shear flow, namely Couette flow, to be defined in a moment, and Poiseuille flow which is laminar flow through a circular tube. I want to show that in such systems there are many parallels between the movements and interactions of the particles in the suspensions and those of molecules as treated in the kinetic theory of gases, except that those which I shall describe can be observed directly, and are much more orderly; indeed some of them show "microscopic reversibility" in contrast to molecular systems which show "microscopic irreversibility" of one of the keystones, incidentally, of statistical mechanics.

We have studied various particle shapes but mainly spheres, rods and discs, both rigid and flexible. Our interest in rods stems mainly from our interest in fibers; and in discs from our interest in the behavior of red blood cells in flowing blood. The behavior of these particles has proven to be of interest not only to papermaking, which I will by-pass this morning, but also to the rheology and stability of emulsions, latexes, aerosols, and other suspensions; to theories of viscosity of colloidal systems and solutions of macromolecules; to polymer physical chemistry and plastics technology; and to haemodynamics which deals with the flow behavior of blood and particularly of the red blood cells in living systems.

I would like to outline some of the phenomena and to mention several applications. I will present the material in a purely descriptive way, with the aid of cine film, but I wish to emphasize from the outset that nearly everything I mention is based on reasonably rigorous quantitative theory, some of which we have developed ourselves over the years.

2. Particles in Couette Flow

I shall start by considering what happens to particles when they are subjected to simple shear (or Couette) flow which is familiar to all of you in connection with the concept of viscosity (Fig. 1). Here we have two parallel plane surfaces which move relative to one another with the distance between them fixed. It is convenient to consider that one of the planes moves in one direction, the top one let us say to the right, and the bottom one to the left, with the liquid containing the suspended particles between them. A shearing action is thereby established in which the velocity of the liquid is represented by the vectors on the left of the diagram. Below the dotted line, which is a stationary layer, the liquid moves to the left at rates indicated by the lengths of the arrows, and above it to the right. If we put a particle in the liquid, centered in the stationary layer, the center will remain fixed in space but it will, however, rotate; this rotation is of considerable importance.

If the particle is spherical it will spin at a steady rate because of the inherently rotational nature of Couette flow. Here (Fig. 1, top) is a streamline of the liquid which is initially horizontal and on approaching the particle from the top is deflected and then returns to its original position. Similarly a streamline below the particle center is deflected

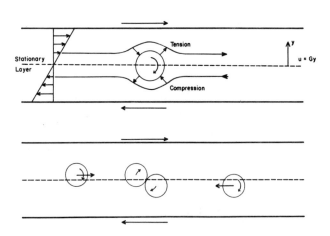

Figure 1. *Couette flow (schematic) showing disturbance to streamlines around a rotating sphere (top) and approach of two spheres to form a collision doublet (bottom). G is the velocity gradient and u is the velocity of the fluid in the horizontal direction; y is the displacement in the vertical direction of the figure.*

symmetrically. The result is that the liquid, which is viscous, drags the upper part of the particle to the right and the lower part to the left, thus causing it to spin. The spin rate can be predicted for simple particle shapes. In addition, the particle is acted upon by forces generated by the liquid as the result of the disturbances to its flow which can squeeze deformable particles out of shape, as we shall see.

If we have more than one particle in the system, for example two spheres (Fig. 1, bottom), one of which is above the stationary layer so that it spins clockwise and translates to the right, and the other below the stationary layer so that it moves to the left, the two will approach one another, and will appear to collide if the geometry is favorable. I shall have something to say about such two body collisions.

We have a number of experimental devices for studying particles in Couette flow, and have, in fact, constructed five different designs of apparatus; these devices enable us to view particles along various coordinate axes. In one of these (Fig. 2) the two bounding surfaces are metallic and are electrically insulated from one another, so that we can simultaneously apply electrical fields at right angles to the direction of flow.

I wish to describe the behavior of a particle in this kind of flow when it is observed along the axis normal to the plane of Fig. 1 and I shall illustrate with a series of movie sequences taken through a microscope arranged as in Fig. 2

Film Sequence 1: Rigid Particles. Starting with a tiny polystyrene sphere containing air bubbles, you can clearly see the steady clockwise rotation as the particle is held in the stationary layer and thus remains in the microscope field. Next we see a rigid disc describing what at first sigh appears to be a very complicated rotational orbit, with three variable angular velocities, one about each of the three prin-

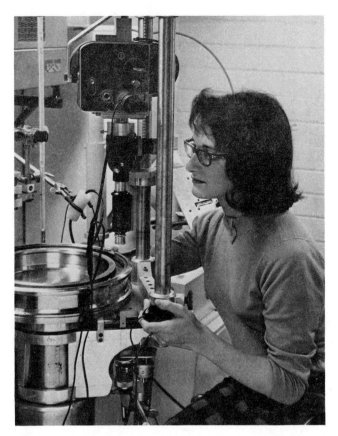

Figure 2. Couette apparatus (Mark II) consisting of two concentric metal cylinders which rotate in opposite directions. The shear field is established in the liquid suspended in the annular gap.

ciple axes of the particle. Now it is important to realize that we have a quantitative theory which enables us to predict all three angular velocities at all parts of the rotational orbit. They all vary periodically, and as a result we get preferred orientations when we have a large number of discs in the suspension. A rigid rod, shown next, executes a very similar orbit, although it may look somewhat different to you. Here by using polarized illumination we can see spin about the axis of the rod in this particular position, and then, as it precesses like a disturbed gyroscope, one end of it moves closer to you than the other. The important thing to observe is that angular velocity is a minimum when the rod is aligned with the flow and is a maximum when at right angles. It is clear from the variation in the angular velocity which you can see, that, when we have a large number of rods, there will be preferred orientation in the direction of flow. This preferred orientation is of importance in a number of situations, including some prevailing in papermaking. It helps, for example, to give paper its characteristic anisotropy, resulting from the preferred orientation of fibers in the machine direction. These three scenes demonstrate that in this kind of flow single particles rotate regardless of their shape.

I would now like to say a few words about what happens to particles which can deform under the action of the stresses generated by the liquid. The simplest case to consider is that of an immiscible liquid drop (or an air bubble); in the

absence of flow it is kept spherical by surface tension. When we have Couette flow, indicated here (Fig. 3), a theoretical analysis shows that over in the first quadrant we have compression, in the second quadrant we have tension, in the third compression, and in the fourth tension, which tend to squeeze and pull and drop into an ellipsoidal shape oriented at 45° to direction of flow. This deformation is resisted by the interfacial tension.

The theory for small values of the deformation D_s was developed a number of years ago by Sir Geoffrey Taylor of Cambridge University and his equations are at the bottom of the slide (Fig. 3) and involve the velocity gradient, or rate of shear, the viscosity of the suspending fluid and the interfacial tension. One can use the equation for D_s to calculate the interfacial tension. In fact we now use shear deformation as a routine method of measuring interfacial tension, and find it particularly convenient for extremely viscous fluids; it may even be useful for molten polymers, although we have not tested this application.

The next slide (Fig. 4) shows a rod which rotates entirely in the plane of the screen. Here we have a similar situation, except that we need only consider the forces acting along the particle axis. In this first quadrant the axial force is compressive and in the second it is tensile. We developed a

SHEAR DEFORMATION

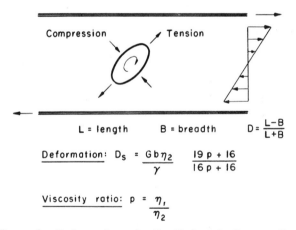

$$L = \text{length} \qquad B = \text{breadth} \qquad D = \frac{L-B}{L+B}$$

$$\text{Deformation:} \quad D_s = \frac{G b \eta_2}{\gamma} \quad \frac{19 p + 16}{16 p + 16}$$

$$\text{Viscosity ratio:} \quad p = \frac{\eta_1}{\eta_2}$$

Figure 3. Deformation of a liquid drop in Couette flow. Taylor's deformation equations are given at the bottom. G is the rate of shear, b the initial drop radius, η_2 the viscosity of the medium, and γ the interfacial tension.

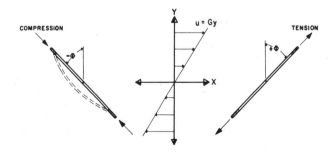

Figure 4. Buckling of a rod in Couette flow (Reference 4).

theory which enables us to calculate the magnitude of the force, and have combined it with the classical Euler theory of critical buckling, to calculate the velocity gradient at which the axial compressive force is high enough to cause the particle to buckle. We have observed critical buckling of both natural and synthetic fibers, and with the aid of our theory can calculate the bending modulus provided that we know the fiber dimensions. This can be done right down to very small particles and conceivably to molecules. The axial force, either tensile or compressive, reaches a maximum when the rod is at 45 degrees to the direction of flow and is zero when at right angles. Compression tends to buckle fiber and tension to pull it straight. It is conceivable that the tension can be made great enough to break the particle. Recently the theory has been used by Zimm and others to calculate the tensile strength of macromolecules such as DNA.

Film Sequence 2: Deformable Particles. Now let us look at the next sequence of the film which illustrates several of these points. We start out with a liquid sphere at rest. On starting the gradient we develop compression here and tension there, and as you see, there is circulation inside the drop. When the gradient is stopped the particle reverts to the spherical shape. The gradient is started again and is progressively increased, and as we see with this particular drop the deformation becomes very large as the drop is pulled out like a piece of taffy until it eventually breaks up into a myriad of droplets. Next we see a wood pulp fiber which is deforming away beyond the critical buckling condition, and is undergoing a characteristic rotational orbit which we call a snake orbit. It is one of four basic classes of orbit that wood pulp fibers and other flexible threads undergo. We have used these principles in developing a rather simple technique of measuring the flexibility spectrum of wood pulp fibers.

Recently we became interested in forming organized aggregates of particles, one of which is this linear chain of spheres suspended in a dielectric liquid (Fig. 5). These

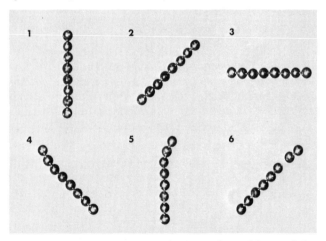

Figure 5. Linear aggregate of spheres formed by applying an electric field aligned vertically in the field (1). After removing the electric field, Couette flow occurs with resultant rotation (2, 3) and buckling (4, 5) and breaking of the chain (Reference 10).

spheres have been brought together in the Couette apparatus at rest by applying an electrostatic field; this induces dipoles in the particles in such a way that they attract one another and they line up, or "fibrate". When the electric field is removed we have a string of pearls without a string connecting them. They are merely close together. When we apply a velocity gradient it is astonishing that they rotate but stay in a linear array just like a rigid rod (Fig. 4). Detailed analysis of the rotation shows that the angular velocity of the aggregate is the same as if they were rigidly cemented together. As the velocity gradient is increased the axial compression produces buckling in one quadrant, and the tension in the next which can break the chain.

Another kind of linear aggregate is one in which spheres are held together by suspending them in a liquid such as silicone oil and then adding a drop of water which rapidly spreads and forms menisci between the particles so that they are held together by capillary forces like fibers in a wet web (Fig. 6). By doing this we think we may have created a classical "string of pearls" model of a linear polymer. We have also created thereby a model fiber of zero stiffness, and a structure similar to a rouleau of red blood cells.

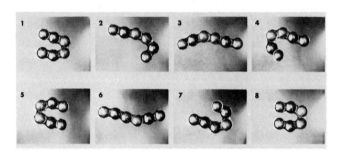

Figure 6. Chain of spheres held together by capillary forces and undergoing snake rotation in Couette flow. The water menisci are clearly visible (Reference 10).

Film Sequence 3: Fibrous Aggregates. Here we see the rotation of both kinds of aggregate with the electric field off and the shear field on. In the first scene we see the rotation, reminiscent of a rigid rod, of an aggregate (like that in Fig. 5) at such low gradients that it remains linear at all orientations during its rotation. Next we see a linear aggregate held together by a drop of water (as in Fig. 6) whose meniscus you can clearly see. The stresses here which are higher than they were in the previous scene cause the aggregate, which we can now consider to be a coherent particle, to execute a snake rotation just like the pulp fiber. Now we must do this slowly because the particle is so weak that if we did it at the velocity gradients to which the fiber was subjected, we would pull it apart. This particular aggregate has six spheres in it. We have gone as high as twenty spheres, but find that as we increase the number of spheres it becomes more difficult to keep them aligned; we tend to get three-dimensional aggregates which, incidentally, are also interesting, but for other reasons which I don't have time to go into.

Film Sequence 4: Two-body Collisions: Now I would like to turn to two-body interactions and here we see a collision between two rigid spheres, which has the characteristic of being completely reversible, so that when we reverse the flow we go through all the same configurations again and again.

We have done a lot of work on two-body collisions with many kinds of particle, but mainly with spheres, rods, and discs.

I would like to say a few words about collisions of rigid spheres (Fig. 7). What you see in the film are two spheres which translate towards one another along nearly straight lines until they appear to hit. Then they lock together and rotate like a rigid dumbell until they reach the mirror image of the initial angle of collision, when they separate. The

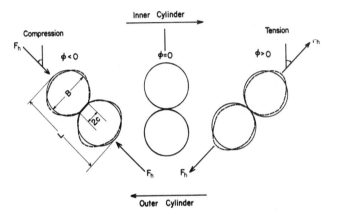

Figure 7. Collision doublet of spheres (schematic).

fact that we can consider that the particles move in a straight line until they hit makes it possible to calculate many of the details of such collisions. Thus we can calculate the collision frequency in much the same way as was done many years ago for bimolecular collisions in a gas, and more recently reactions between sub-atomic particles and atomic nuclei, where the concepts of rectilinear approach and collision cross section are used. In our systems, however, the collision frequencies can be measured because we can see the particles. We can look at a selected particle in the microscope and actually count the number of times it collides with other particles and have obtained very good agreement with theory. Let us consider a sphere which is not in the stationary layer so that it moves horizontally on the screen until it encounters another sphere and becomes displaced vertically while undergoing collision and is restored to its original vertical position; the mean free path between collisions is readily calculated and shows reasonable agreement with experimental values. The collisions are sticky, that is, the particles stay together for a while before they separate, after a calculable doublet lifetime. Since there is a range of possible geometries of approach, we get a distribution of doublet lives, given by a rather simple equation which has shown very good agreement with values measured with a stopwatch. Later I will discuss another characteristic of the path, the amount of vertical (or lateral) displacement which

the particle undergoes, and here again we get good agreement between theory and experiment.

One kind of collision that has intrigued us for some years is that which would occur if two particles could hit head-on, which if we limit consideration to two body interactions, is impossible since both move at the same velocity so that one can't catch up the other. As I will demonstrate in a moment, it is possible to bring them into this configuration by forming a triplet or higher order multiplet. When this happens we get planetary movement of one sphere about the other.

Film Sequence 5: Head-on Doublet. In this particular sequence one sphere is closer to you than the other so that the sphere centers are not in the same vertical plane, but this is not the point. You will notice that they are rolling relative to one another as contrasted with the previous doublet where there was rigid rotation. The line joining their centers continued to have clockwise angular velocity, so they pull each other back together and form a closed orbit which for each particle center is nearly elliptical.

Film Sequence 6: Triple Collisions. We frequently find these non-separating or permanent doublets in dilute suspensions and think that they are formed by three body collisions. Here we see a triplet in which one sphere has pushed the other two into this non-separating configuration where they rotate about one another without ever separating, unless we reverse the flow. Just like the doublet we saw earlier, it is reversible so that by reversing the flow we can bring back the first sphere and destroy the permanent doublet.

This leads me to the subject of flow reversibility, which I must admit hasn't much to do with papermaking, but has nevertheless intrigued us greatly. We have performed a variety of experiments, but I will mention only one or two. We have, for example, formed a structure which might be a close array of discs, or rods, and then have subjected them to Couette flow. They start rotating as an aggregate and eventually they separate and move in what appear to be uncorrelated orbits. But because these particles are rigid all of these processes are reversible, so that on reversing the flow we can put the structure back together again, an impressive thing to see.

Film Sequence 7: Reversibility Phenomena. The next part of the film shows a rather trivial but striking demonstration of this and was, as a matter of fact, the very first experiment along these lines that we did and is easy to grasp. We have done a lot of more sophisticated things since this time. We have taken a rather crude Couette apparatus and have written "PPRIC" in dye in the liquid in the annulus. The first time we did this was at Christmas, and we wrote in "Merry Xmas" in red. Here we have an organized structure which shear now destroys. When the inner cylinder is rotated a few times—it can be rotated several hundred times with the same effect—and then is reversed to restore the letters (Fig. 8), except for some imperfections which arise from faulty machining of the apparatus and from the fact that the dye undergoes some molecular diffusion, which of course is irreversible. There are many suspensions which

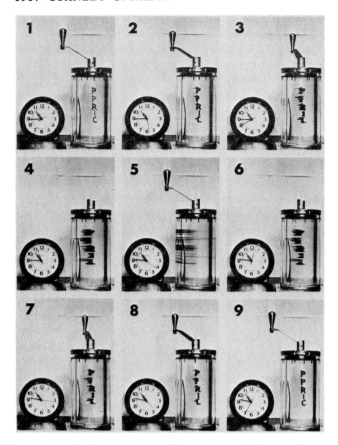

Figure 8. Reversibility experiment in Couette flow. After frame 5 the direction of rotation of the inner cylinder was reversed. Note that the clock continued to advance.

do not show this reversible behavior and we are now beginning to understand why they don't but I haven't time to go into this.

3. Particles Flowing Through Tubes

I would like instead to turn to the second kind of flow, steady flow through a tube or Poiseuille flow, and say a few words about it. Here (Fig. 9) we have flow of a Newtonian liquid through a circular tube with the liquid sticking to the wall, thus giving rise to telescopic flow. If we plot the velocity of the liquid across the tube we find that it is parabolic, zero at the wall and reaching a maximum at the axis. Because the translational velocity varies radially this is a shear flow, but it turns out that unlike Couette flow where it is constant, the velocity gradient varies linearly with radial distance from the tube axis where it is zero and reaching a maximum at the wall as shown on the slide (Fig. 9).

What happens to a particle in this kind of flow? If the particle is small compared with the diameter of the tube we can say a great deal purely on the basis of what we already know about Couette flow. If, for example, we take a sphere which is somewhat smaller relative to the tube than is indicated in the slide (Fig. 9), say less than one-twentieth of the tube diameter, then the particle center will move in accordance with the parabolic profile. In the immediate neighborhood of the particle we will have a Couette field which will

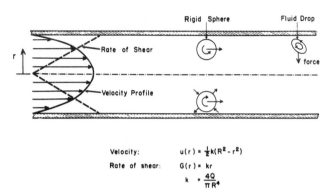

Poiseuille Flow – Variable Rate of Shear

$$\text{Velocity:} \quad u(r) = \tfrac{1}{2}k(R^2 - r^2)$$
$$\text{Rate of shear:} \quad G(r) = kr$$
$$k = \frac{4Q}{\pi R^4}$$

Figure 9. Poiseuille flow (schematic). R is the tube radius and Q is the volumetric flow rate.

make the particle spin clockwise if it is below the axis, counterclockwise if above (Fig. 9). It will be acted on by all the forces that we had in Couette flow, so that if it is deformable, it will be squeezed out of shape. For example, if it is a liquid drop it will be deformed into an ellipsoid which looks like this (Fig. 9, upper right) and is rotating as it goes down the tube. We can apply Taylor's theory and calculate the deformation, or alternatively the interfacial tension. We can observe burst and break-up exactly the same as in Couette flow.

If we introduce a rod or a disc its rotational orbit is exactly as predicted from Couette flow, thus enabling us to say a great deal about preferred orientations in dilute suspensions flowing through tubes. In view of some work that we had done previously on pulp suspensions, we were very interested to see if the center of particle tended to move relative to the wall. We have known for some time that pulp fibers move away from the wall, and that this has a profound effect on the flow behavior of pulp suspensions. The same appears to be true in the flow of blood; it has been known for many years that red blood cells can move away from the wall, but no one knew why.

It was to know more about radial migration that we went into Poiseuille flow experiments, and what we found is very easily summarized as follows. When the velocity of the fluid is very low, so that we are in what is called the Stokes flow regime, a rigid particle regardless of shape maintains a fixed radial position, that is it does not tend to move either away from the wall or towards it. When it is deformable, it always moves away from the wall no matter what the shape and eventually reaches the tube axis. We have developed a theoretical explanation for this behavior which is based on the hydrodynamic interaction between the deformed particle, actually a liquid drop, and the wall.

We have developed experimental techniques for looking at particles in this kind of flow which involve essentially a microscope which moves along at exactly the same speed as the particle which we wish to examine, and resorting to a few optical tricks which enable us to look right into the suspension without optical distortion (Fig. 10). In a moment I

Figure 10. Travelling microscope for following particles in Poiseuille flow.

am going to show you a movie sequence of a very concentrated suspension flowing through a tube, in which we have gone up to concentrations of spheres as high as 60 percent by volume. We are able to look right through these suspensions by using tiny plastic spheres, mainly polymethyl methacrylate, whose refractive index we can match to within better than five digits of the refractive index of the suspending liquid, so that in effect we have a transparent suspension. Then we throw in a few polystyrene spheres which have a different refractive index and these we see.

Film Sequence 8: Steady Poiseuille Flow. Here is a single rigid disc first with the microscope fixed and then moving along with the particle, and we see the disc undergoing exactly the same rotational orbit as we saw earlier in the Couette flow. One interesting thing about this orbit is that we only see one face of the disc and never the other, just like the moon. All three of the angular velocities can be predicted accurately from existing theory, just as in Couette flow. Next we see a liquid drop which is deformed into an ellipsoid. It has another tiny drop inside so that you can see the rotation. As it moves down the tube we see that it is being pushed away from the wall, and since it moves towards smaller gradients, the deformation decreases. Next we look at a 42 percent suspension of rigid spheres but all we see are the polystyrene spheres, which constitute about 0.1 percent. Close analysis of this film will show that some of the spheres are rotating and are in frequent collision with others which makes them move in erratic paths. The next slide (Fig. 11) shows the results obtained by this technique in a 17 percent suspension of spheres. The tracer spheres provide a measure of the translational velocities, and as we see the profile is parabolic so that it follows Poiseuille's law. Remember that at this concentration there is a tremendous amount of particle interaction. Not only do we get frequent two-body collisions but we also get three-body, four-body collisions, five-body collisions, and so on, all of which we can see. This means that a sphere moves along what appears to be a random path. Here (Fig. 11) are three measured

translational paths, along each of which the particle also rotates, although erratically. A rather extraordinary thing is that because we have rigid spheres, and because we have no net migration of particles from the wall, we have complete reversibility. Thus if we reverse the flow by pumping backwards, each sphere will retrace its path. In other words, unlike a staggering drunkard, it remembers everywhere it has been and every spin it has made. This is a rather difficult experiment to do, but having done it, it is rather impressive, although in retrospect not surprising. The next slide (Fig. 12) shows what happens at a higher concentration (33 percent).

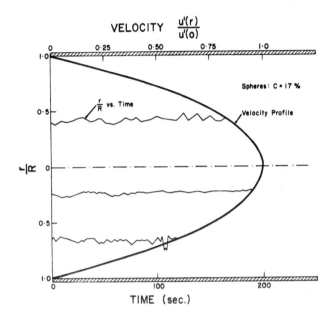

Figure 11. Velocity profile and particle paths in 17 percent suspension of rigid spheres (Reference 11).

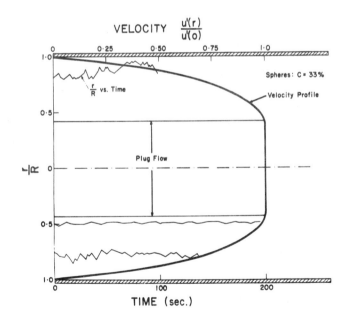

Figure 12. Velocity profile and particle paths in 33 percent suspension of rigid spheres (Reference 11).

The velocity profile is no longer parabolic, but becomes flat near the tube center. The relative velocity profile is independent of the rate of flow, so that if we calculate the viscosity of this suspension we find that it is independent of the flow rate, and hence of the rate of shear, so that we should say that the suspension is Newtonian. But if it is Newtonian, it should give a parabolic profile, so that we have a paradox in that the system is at the same time Newtonian and non-Newtonian. I don't have time to go further into this but let me say that the system is actually Newtonian. The blunting of the profile arises from hydrodynamic interaction between the particles and the wall, which produces a pseudo-slip at the wall. In the central part we have plug flow as also exhibited by dilute pulp suspensions at low flow rates and by other particulate systems. In plug flow the velocity gradient is zero so that all of the particles move at exactly the same velocity and without any rotation. In some situations this is an advantage, and in others a disadvantage. In the peripheral layers, where we have a velocity gradient, and with it particle interactions, we have erratic paths, and rotation, and all reversible if the particles are rigid.

We have gone a good deal further into the question of the movement of particles relative to the wall. As we have seen the particles move away from the walls if they are flexible. But there are ways to make *rigid* particles move away from the wall, for example, by using a visco-elastic suspending medium. Another way, applicable to a Newtonian suspending liquid, is to go to a high Reynolds number so that the particle and liquid inertial effects become appreciable; here we get an effect roughly equivalent to that of a curving ball which causes migration. One way of doing this is to subject the system to oscillatory flow, and the next part of the film shows what I mean.

Film Sequence 9: Oscillatory Flow. Here we have oscillatory flow back and forth and the inertial effects are evident from the phase differences in velocities across the tube. The periodic profiles are, of course, no longer parabolic but they can be calculated and agree with what we now see. We predicted that because of the high velocity reached in this sinusoidal cycle that a suspended rigid sphere would be pushed away from the wall, as we now see. In this way we can cause a profound change in the way the suspension flows. Here is the same sort of thing at higher concentrations, when we see the development of what is called a particle-free or "plasma" layer next to the wall which can act as a lubricating layer. The last part of this film simply shows the equilibrium plasma layer reached after some time when further inward migration of particles is prevented by particle crowding.

This migration has a number of interesting consequences, of which I would like to mention one. By subjecting a suspension to oscillatory flow, we can radically change the way it flows. Here (Fig. 13) is a suspension of rigid spheres 22 percent by volume which initially gave a parabolic profile in steady flow. Then the steady flow was stopped, and

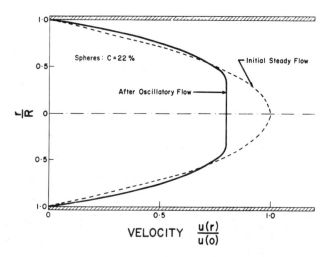

Figure 13. Change in velocity profile caused by development of plasma layer by oscillatory flow (Reference 12).

the suspension subjected to oscillatory flow which caused the development of a plasma layer at the wall; then we restored the steady flow and measured the velocity profile again. You can see that it is markedly changed with pronounced blunting in the center (Fig. 13) with a region of plug flow. We predicted that the energy input required to move the suspension at a given rate is reduced. In other words, the apparent viscosity of the system would be decreased. This was indeed observed to be the case, the reduction in this experiment being about one-third.

I will end up on the note that these oscillatory flows either by themselves or superimposed on a steady flow, to give what is called pulsatile flow, may have a number of important consequences in the flow of various suspensions, such as blood, latexes, polymer suspensions, and possibly even in the pipeline transport of wood chips, a problem under active study in the Pulp and Paper Institute at the present time.

4. Postscript (June 1972)

I In the period following this lecture, many aspects of the work described above have been expanded. Some are described in References *12–24*. Particular attention is drawn to the extensive summary in Reference *21*.

Pertinent Literature References

1. *The Behavior of Suspended Particles in Laminar Shear*
   S. G. Mason and W. Bartok
   Rheology of Disperse Systems, Pergamon Press (1959)
2. *The Hydrodynamic Behaviour of Papermaking Fibres*
   O. L. Forgacs, A. A. Robertson and S. G. Mason
   Pulp and Paper Mag. Can., **59,** 117-128 (1958)
3. *Particle Motions in Sheared Suspensions X. Orbits of Flexible Thread-Like Particles*
   O. L. Forgacs and S. G. Mason
   J. Colloid Science, **14,** 473-491 (1959)
4. *The Flexibility of Wood-Pulp Fibers*
   O. L. Forgacs and S. G. Mason
   TAPPI, **41,** 695-704 (1958)

5. *Particle Motions in Sheared Suspensions XIV. Coalescence of Liquid Drops in Electric and Shear Fields*
   R. S. Allan and S. G. Mason
   J. Colloid Sci., **17**, 383-408 (1962)

6. *The Flow of Suspensions Through Tubes I. Single Spheres, Rods and Discs*
   H. L. Goldsmith and S. G. Mason
   J. Colloid Sci., **17**, 448-476 (1962)

7. *Axial Migration of Particles in Poiseuille Flow*
   A. Karnis, H. L. Goldsmith and S. G. Mason
   Nature, **200**, 159-160 (1963)

8. *Particle Motions in Sheared Suspensions XVIII. Wall Migration (Theoretical)*
   C. E. Chaffey, H. Brenner, and S. G. Mason
   Rheologica Acta, Band 4, Heft 1, 64-72 (1965)
   Correction: *Rheologica Acta,* **6**, 100 (1967)

9. *The Flow of Suspensions Through Tubes IV. Oscillatory Flow of Rigid Spheres*
   B. Shizgal, H. L. Goldsmith and S. G. Mason
   Can. J. Chemical Engineering, **43**, 97-101 (1965)

10. *Ordered Aggregates of Particles in Shear Flow*
    I. Y. Z. Zia, R. G. Cox and S. G. Mason
    Proc. Roy. Soc., **A300**, 421-441 (1967)

11. *The Kinetics of Flowing Dispersions I. Concentrated Suspensions of Rigid Particles*
    A. Karnis, H. L. Goldsmith and S. G. Mason
    J. Colloid and Interface Sci., **22**, 531-553 (1966)

12. *The Flow of Suspensions Through Tubes V. Inertial Effects*
    A. Karnis, H. L. Goldsmith and S. G. Mason
    Can J. Chem. Eng., **44**, 181-193 (1966)

13. *The Flow of Suspensions Through Tubes VII. Rotation and Deformation of Particles in Pulsatile Flow*
    M. Takano, H. L. Goldsmith and S. G. Mason
    J. Colloid and Interface Sci., **23**, 248-265 (1967)

14. *Particle Motions in Sheared Suspensions XIX. Viscoelastic Media*
    A Karnis and S. G. Mason
    Trans. Soc. Rheol., **10**, 571-593 (1966)

15. *The Kinetics of Flowing Dispersions II. Equilibrium Orientations of Rods and Discs (Theoretical)*
    E. Anczurowski and S. G. Mason
    J. Colloid and Interface Sci., **23**, 522-532 (1967)

16. *Particle Behaviour in Shear and Electric Fields V. Effect on Suspension Viscosity*
    C. E. Chaffey and S. G. Mason
    J. Colloid and Interface Sci., **27**, 115-126 (1968)

17. *The Flow of Suspensions Through Tubes IX. Particle Interactions in Pulsatile Flow*
    M. Takano, H. L. Goldsmith and S. G. Mason
    J. Colloid and Interface Sci., **27**, 268-281 (1968)

18. *Some Model Experiments in Hemodynamics III.*
    H. L. Goldsmith and S. G. Mason
    in *Hemorheology: Proc. 1st Int. Cong. Hemorheology,* (A. L. Copley, ed.), Pergamon Press (London) (1968)

19. *The Flow Behaviour of Particulate Suspensions*
    S. G. Mason and H. L. Goldsmith
    Ciba Foundation Symposium on Circulatory and Respiratory Mass Transport, 1969, pp. 105-125, (edited by G. E. W. Wolstenholme and Julie Knight)

20. *Model Particles and Red Cells in Flowing Concentrated Suspensions*
    H. L. Goldsmith and S. G. Mason
    5th European Conference on Microcirculation, Gothenburg 1968, Bibli, Anat. No. 10, 1-8 (Karger 1969)

21. *The Microrheology of Dispersions*
    H. L. Goldsmith and S. G. Mason
    in *Rheology: Theory and Applications* (F. R. Eirich, ed.) Vol. 4, Chapter 2, 85-250, Academic Press, New York (1967)

22. *Particle Motions in Non-Newtonian Media I. Couette Flow*
    F. Gauthier, H. L. Goldsmith and S. G. Mason
    Rheol. Acta, **10**, 344-364 (1971)

23. *Particle Motions in Non-Newtonian Media: II. Poiseuille Flow*
    F. Gauthier, H. L. Goldsmith and S. G. Mason
    Trans. Soc. Rheology, **15:2**, 297-330 (1971)

24. *The Kinetics of Flowing Dispersions: V. Orientation Distributions of Cylinders in Newtonian and non-Newtonian Systems*
    F. Gauthier, H. L. Goldsmith and S. G. Mason
    Kolloid-Z. u. Polymere, **248**, 1000-1015 (1971)

1968

ALBERT C. ZETTLEMOYER

LEHIGH UNIVERSITY
BETHLEHEM, PA.

*. . . for his important contributions to the basic under-
standing of interfaces from heats of wetting, physical and
chemical adsorption, adsorption from solution and heterogeneous nucleation; for his imaginative and far-sighted
development of the science and technology of printing ink;
and for his outstanding leadership in the advancement of
colloid and surface chemistry as educator, organizer, editor,
writer and speaker.*

The following Kendall Award Address is reproduced by permission from *The Journal of Colloid and Interface Chemistry*, **28**, 343 (1968). Copyright © 1968 by Academic Press Inc.

Hydrophobic Surfaces[1]

A. C. ZETTLEMOYER

Center for Surface and Coatings Research, Lehigh University, Bethlehem, Pennsylvania 18015

Hydrophobic surfaces are characterized by high contact angles with water, often in the range of 40 to 110 degrees, and low heats of immersion, -6 to -90 ergs/cm². Current theories of the interaction across interfaces indicate that the water and the hydrophobic substance interact only through dispersion forces. Thus, the water doesn't spread because only the $\gamma_{H_2O}^d$, which is 22 ergs/cm² not the total 72 ergs/cm², and only the γ_S^d are responsible for the interfacial free energy.

The Frenkel-Halsey-Hill (FHH) plots indicate that the interaction of nitrogen or water with hydrophobic surfaces is less in the first layer and greater than for polar surfaces in the second layer and beyond. Furthermore, the break in the FHH plots for nitrogen on graphitic surfaces at 0.4 relative pressure strongly suggests that the packing is loose, one nitrogen to each ring, at the low apparent monolayer coverage; as the relative pressure is increased this layer then fills. Hence the many reported surface areas and other surface properties of graphitic solids, such as Graphon, must be altered to conform with the 20 A²/N_2 molecule to be used in the area estimation by the BET method.

Hydrophilic sites invariably present on "real" hydrophobic surfaces are interesting both theoretically and practically. Adsorption of water and other polar molecules on these sites is energetically weak and increases with increasing temperature. The molecules form clusters around the first-down molecules at high relative pressures; the adsorbed molecules are highly entropic and the heat of immersion rises with increasing precoverage. Such sites are responsible for the heterogeneous nucleating ability of essentially hydrophobic "cloud seeders" such as silver iodide and hydrophobic silicas.

INTRODUCTION

Some Definitions. That a surface is hydrophobic means, in simplest terms, that water does not spread on it. Instead, the water stands up in the form of drops and a contact angle can be measured from the plane of the surface, tangent to the water surface at the three phase boundary line. The resolution of forces at the junction as:

$$\gamma_{SV} = \gamma_{SL} + \gamma_{LV} \cos \theta \qquad [1]$$

(see Fig. 1) was first proposed by Thomas Young (1) in 1805 (Dupré, 1869).[2] Young was interested, however, in drops of mercury resting on "metals."

The physical significance of the contact angle is quite clear. For example, to develop water repellency we desire the contact angle to be as large as possible. Most people think of fabric in connection with water repellency and here it is a matter of capillary action. The fabric remains porous on treatment and it is really not waterproof. Water will run through if sufficient hydrostatic head is imposed. Indeed, this approach is taken to test the water repellency of fabrics (3). It is curious, too, that the pores of our skin are hydrophobic, but the pores

---

[1] Kendall Award Address, American Chemical Society, San Francisco April 1, 1968.

[2] The author is aware that Eq. [1] enjoys a more sophisticated derivation based on surface free energies although this approach may be less pedagogically palatable to the uninformed. A detailed general thermodynamic proof has been given by Johnson (2).)

"breathe" because water vapor passes through.

The equation of Laplace for nonwetting gives the pressure difference across a curved meniscus as might form when the water does not wet the fibers; see Fig. 2. The equation is:

$$\Delta P = \frac{2\gamma_{LV}\cos\theta}{r}, \qquad [2]$$

or using Eq. [1]:

$$\Delta P = \frac{2(\gamma_{SV} - \gamma_{SL})}{r}. \qquad [3]$$

For water repellency ΔP must be negative; i.e., $\cos\theta$ negative or $\gamma_{SL} > \gamma_{SV}$. Equation [3] shows that the surface tension or free

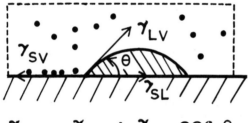

$$\gamma_{SV} = \gamma_{SL} + \gamma_{LV}\cos\theta$$

FIG. 1. Schematic representation of the contact angle formed by a liquid drop on a solid surface which it fails to wet, and of the resolution of the forces involved at the triple boundary line; Young's equation.

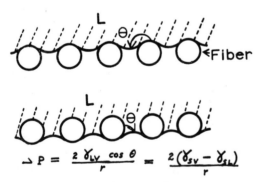

$$\rightarrow P = \frac{2\gamma_{LV}\cos\theta}{r} = \frac{2(\gamma_{SV} - \gamma_{SL})}{r}$$

Fig. 2. Schematic representation of a liquid in contact with a porous fiber surface and the mathematical description of the pressure needed to force the liquid through the fiber structure; Laplace equation.

energy of the liquid is not directly involved. Low-energy or waxy coatings are employed to make γ_{SV} as low as possible, although very little improvement is achieved by the use of perfluorinated derivatives instead of cheaper coatings. Also, the smaller the pore size r, the greater is the hydrostatic head required to push water through the mesh of the fabric.

Two other effects may tend to increase the apparent contact angle to a greater value than the true one. The first is surface roughness which may be enhanced by the repellent coating; if the true contact angle is greater than 90°, then the angle can be increased by the roughness. According to the Wenzel relation (4):

$$\cos\theta_{\text{app}} = R\cos\theta_{\text{true}}, \qquad [4]$$

where R is the ratio of actual to apparent or projected area. Of course, if θ is less than 90°, roughness appears to enhance wetting. The second effect becomes evident from an expansion of the Young Eq. [1] to encompass the contribution of two different types of surfaces or surface heterogeneities occupying fractions f_1 and f_2 of the surface:

$$\gamma_{LV}\cos\theta_{\text{app}} = f_1(\gamma_{S_1V} - \gamma_{S_1L}) \\ + f_2(\gamma_{S_2V} - \gamma_{S_2L}). \qquad [5]$$

If f_2 is the fraction of open area, then γ_{S_2V} is zero and γ_{S_2L} is simply γ_{LV}. Dividing through by γ_{LV} leads to

$$\cos\theta_{\text{app}} = f_1\cos\theta_1 - f_2; \qquad [6]$$

so, as $\cos\theta_{\text{app}}$ is diminished by a greater fraction of open space, the apparent contact angle is larger. These effects have been explored by Wenzel (5), Baxter and Cassie (6), and Dettre and Johnson (7). Also, with regard to surface heterogeneities, per cent replacements of CH_2 groups in polyethylene by chlorine and fluorine (8) showed a linear change of $\cos\theta$ vs. atom per cent in some systems. The model implied by Eq. [6] is that $\cos\theta$ is averaged, but perhaps the situation is more complex so that θ or some involved function of θ should be averaged. In fact, certain systems explored by Ellison and Zisman suggested such complications.

Here is where the trouble lies. At first glance, the contact angle phenomenon appears straightforward. But for almost all real surfaces, both physical and chemical heterogeneities complicate matters. We do not yet know, for example, what size patches of chemical heterogeneities must be to influence the contact angle. Only a few modest attempts at determining the effect of chemmical heterogeneities on contact angles have been made (9). Some understanding of the effects of physical heterogeneities has been gained by the recent studies of Dettre and Johnson (10). Of course, Zisman's group (11) had earlier established, with carefully prepared uniform films, the effects of different terminal groups such as CH_3-, CH_2F-, CHF_2-, and CF_3- in close-packed array. The mass of results obtained, especially with various series of monofunctional compounds, is a valuable addition to our arsenal of knowledge of wetting. Nevertheless, the effects on wetting of the heterogeneities on real surfaces remain challenging problems of surface chemistry.

Perhaps the best technique to assay composite surfaces depends on the measurement of heats of immersion. It is interesting that the evolution of a heat of wetting was reported by Leslie (12) in 1802, even earlier than Thomas Young resolved the forces involved in contact angles. Although there were earlier attempts to put such effects on a unit area basis (13), sound values were achieved only when the BET and "B point" methods (14) gave us reasonably valid surface areas (1936–40). The heat of immersion technique requires the use of powdered samples since the magnitude of the heat evolved on breaking solids into liquids varies usually only from 10's to 100's of ergs/cm². A simple calorimeter for measurements on powders possessing areas over 5 m²/gm is depicted in Fig. 3. Writing h_i for the enthalpy

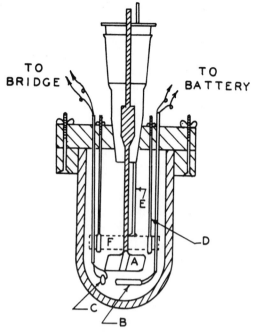

FIG. 3. Thermistor calorimeter for measurement of heat of immersion. A, stirrer; B, heater; C, thermistor; D, sample holder; E, breaking rod; F, sample tube. A sensitive resistance bridge such as the Mueller bridge and a galvanometer of sensitivity of the order of 10^{-10} amp allows the determination of heats of the order of 0.01 cal. The calorimeter is placed inside an air thermostat. Evolution of heat up to about 10 min can be followed by means of a simple arrangement of this type. Submarine-type calorimeters enclosed in large water thermostats are needed, for greater sensitivity, or to follow heat effects over longer periods.

change per unit area $\Delta H_1/\Sigma$, we have:

$$h_i = \frac{\Delta H_I}{\Sigma} = h_{SL} - h_S \simeq e_i, \qquad [7]$$

and the relations to relate h_i with γ's are obviously:

$$\frac{\Delta G_I}{\Sigma} = \gamma_{SL} - \gamma_S = \gamma_i \, g_i \qquad [8]$$

and

$$h_i = \gamma_i - T \frac{d\gamma_i}{dT}. \qquad [9]$$

We shall come back to Eq. [9] later. It will

suffice to point out here that the energetics of the heterogeneities can be mapped by successive measurements as a function of precoverage from the vapor of the same substance as the wetting liquid. The difference between the heats of immersion of the base and partially coated sample is:

$$h_{i(SL)} - h_{i(SfL)} = \int_0^{d\Gamma} q_{st} + \Gamma h_L \quad [10]$$
$$\simeq \Gamma(\epsilon_A - \epsilon_L).$$

The last expression gives the energy change of the adsorbate itself in moving from the bulk liquid to the solid surface if the solid is negligibly perturbed and lateral interactions are similar in the adsorbed film to those in the bulk liquid. This expression has been used in this Laboratory (15) to estimate the force field emanating from solids. Equation [10] also relates the integral heats of immersion to the differential heats which can be obtained from multitemperature isotherms through a Clapeyron-Clausius approach.

Several characteristic shapes are found for the heats of immersion $h_{i(SfL)}$ versus the precoverage θ; these heats of immersion isotherms have been classified in this Lab-

oratory (16). We shall return to the subject of heterogeneities on hydrophobic surfaces in the final section.

Values of contact angles and heats of immersion for hydrophobic surfaces are presented in Table I.

Two other terms require introduction. The one is the spreading or film *pressure* π. Conceptually, it is easily introduced for insoluble films on water where the film bombards any barrier in its way as it attempts to reduce the free energy of the uncovered surface beyond. Thus:

$$\pi = \gamma_{LV} - \gamma_{LF}, \quad [11]$$

where γ_{LF} refers to the film-covered water. For the case of a vapor adsorbed on a solid, Bangham and Razouk (17) used the Gibbs adsorption equation to obtain at saturation (or in equilibrium with the liquid):

$$\pi_e = \gamma_S - \gamma_{SV} = \frac{RT}{V\Sigma} \int_0^{P_0} v \, d \ln P, \quad [12]$$

where γ_S is the surface free energy of the bare solid and V is the molar volume. So one way to estimate π_e is to measure the adsorption isotherm of the vapor. Furthermore, we can write for multitemperature

TABLE I

Properties of Some Hydrophobic Surfaces

Solid	Water		π_e (vapor)	γ_c	$\gamma_s{}^d$	$-h_i{}^d$(Liq.) ergs/cm²
	θ	$-h_i$				
Teflon (PTFE)	108	6	0 (C_{16}-C_{10})	18.5	19–24	33–34 (C_{16}-C_{10})
			3.2 (C_8)			39 (C_8)
			5.7 (C_6)			47 (C_6)
			4.9 (C_4) 0°C			
			5.9 (C_2) −90°C			
Polypropylene	95		12 (N_2) 70°K	24 (est.)	26–29	45 (calc. AM)
			13 (A) 90°K			
Polyethylene	94		14 (N_2)	31	35	53 (calc. AM)
Polymethylmethacry-late (Plexiglas)	80			39	49	67 (calc. AM)
Graphon	82	26	6 (H_2O)	60–65 (est.)	70	51 (N_2)
			30 (PrOH)			91 (BuOH)
			24 (C_6H_{12})			82 (C_6H_{12})
			28 (C_7)			91 (C_7)

data:

$$\pi_e - T\frac{d\pi_e}{dT} = h_{i(SfL)} - h_{i(SL)}. \quad [13]$$

Such data are indeed rare for water on hydrophobic surfaces.

The second term to be introduced is the spreading *coefficient* first used by Harkins (18). In Fig. 4, the spreading coefficient of water over a solid $S_{L/S}$ is simply the reduction in surface free energy on losing the bare solid surface and forming the new solid/water and water/vapor interface. Therefore, per unit area:

$$S_{L/S} = \gamma_S - \gamma_{SL} - \gamma_{LV}, \quad [14]$$

where an initial or clean surface is considered to be operating. Clearly, this free energy change will be positive as written for spontaneous spreading. For water on hydrophobic surfaces, $S_{L/S}$ will be negative. Its relation to the contact angle is easily shown with γ_S replaced by $(\gamma_{SV} + \pi_e)$ and then $(\gamma_{SV} - \gamma_{SL})$ by $\gamma_{LV} \cos\theta$ through Eq. [1]:

$$S_{L/S} = \pi_e + \gamma_{LV}(\cos\theta - 1). \quad [15]$$

The spreading pressure π_e is expected to be small or negligible for water on a hydrophobic surface. As a matter of fact, we would expect it to be lower, the more hydrophobic the surface and the greater the contact angle. Intuitively, both should go hand in hand. Equation [15] tells us what

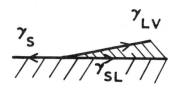

$$S_{L/S} = \gamma_S - \gamma_{SL} - \gamma_{LV}$$

$$= \pi_e + \gamma_{LV}(\cos\theta - 1)$$

FIG. 4. Schematic representation and mathematical description of the tendency of a liquid to spread on the bare surface of a solid.

we expect then: the spreading coefficient will be negative for water on hydrophobic solids, and the more so the higher the contact angle. We shall examine the nature and the magnitudes of π_e later.

All this above does not really help us understand hydrophobicity explicitly. It took almost 160 years after Leslie and Young to understand high contact angles and low heats of immersion of hydrophobic solids. Take the graphite/water system: the contact angle is 82°–86° and the heat of immersion is 26 ergs/cm². How is it that this solid possessing a free surface energy of 70 ergs/cm² interacts so weakly with a polar liquid? We shall examine this question next. Then, adsorption over the hydrophobic matrix by a variety of substances will be summarized. We shall learn that the application of the Frenkel-Halsey-Hill equation fits well adsorption isotherms on these surfaces in the 0.2 to 0.9 range of relative pressure, but with slopes differing from those for hydrophilic and high-energy surfaces. Finally, attention will be given to the hydrophilic sites which are invariably present on hydrophobic solids. We shall find that water adsorbed on these sites displays curious properties and, at the same time, we shall find that this phenomenon is of immense practical significance.

THE NATURE OF INTERFACIAL TENSIONS OR FREE ENERGIES

GEOMETRIC MEAN AVERAGING

One unifying scheme for treating physical interactions at interfaces has recently been developed by Fowkes (19). It is a geometric mean (GM) technique based on dispersion forces. Some intermolecular forces contributing to surface tension, such as the metallic bond or the hydrogen bond, depend on the specific chemical nature of the material being considered. London dispersion forces, on the other hand, exist in all types of matter and are always attractive forces. They arise from the interaction of fluctuating electric dipoles which induce dipoles in neighboring atoms or molecules. The effects of the fluctuating dipoles cancel out, but not that of the induced dipoles. It should be mentioned at the outset that Fowkes' scheme would appear to be an oversimplification which proves useful probably because of the cancellation of effects not taken into account. Girifalco and Good (20) had previously used the geometric mean in a somewhat different way with some success.

In a liquid such as a hydrocarbon, the intermolecular forces are only dispersion forces and we can write $\gamma_{HC} = \gamma_{HC}^d$. In a liquid such as mercury, both dispersion forces and metallic bonding are important to interatomic forces and $\gamma_{Hg} = \gamma_{Hg}^d + \gamma_{Hg}^m$. Similar equations for a polar liquid involve both a polar and a dispersion contribution to the total surface tension: $\gamma = \gamma^d + \gamma^p$. These very simple additivity rules are surprisingly useful.

Let us consider an interface with two phases in contact. For the moment we shall deal with two immiscible liquids. The interfacial free energy or the interfacial tension can be expressed in terms of the surface free energies or tensions of the two individual phases by:

$$\gamma_{12} = \gamma_1 + \gamma_2 - 2\sqrt{\gamma_1{}^d\gamma_2{}^d}. \quad [16]$$

For liquids, the interfacial or surface free energy (in ergs/cm² in the cgs system) is the same as the interfacial or surface tension (in dynes/cm) because ordinarily the surface atoms or molecules of a liquid are mobile. The atoms of a solid are not. When a fresh surface is formed, then, only the liquid can rearrange itself rapidly enough for the measured surface tension and surface free energy to be the same. By definition, if a surface is cut by a plane normal to it, then it will be necessary to apply some external force (defined as surface stress if per unit length) so the atoms on either side of the cut remain in equilibrium. Half the sum of the two surface stresses along mutually perpendicular cuts is the *surface tension. The surface free energy*, on the other hand, can be regarded as the work of bringing a molecule (or atom) or a

unit area of them from the interior of a material to the surface. This work arises from the unbalancing of the net forces on the molecule as it goes from the interior to the surface.

Then for liquids, but not usually for solids, Eq. [16] can be discussed in terms of either surface tensions or free energies. The geometric mean contained in it is not unusual in physical interactions; but here it is based on the dispersion part of the surface free energies only, the γ^d's, and arises as indicated in Fig. 5. Fowkes' major contribution was that the free surface energies due to the several intermolecular forces are additive. As mentioned previously, Girifalco and Good (20) already had used another sort of geometric mean averaging based on an even earlier use for other physical properties. Take liquid 1, first without a second condensed phase in contact. Surface concentration of the molecules is lower than in the bulk as a manifestation of the pull inward due to the surface tension γ_1. At equilibrium, and except near the critical temperature, the vapor phase is obviously much less dense than the liquid. When a second condensed phase 2 is brought in contact with the first, the γ_1 tension is reduced

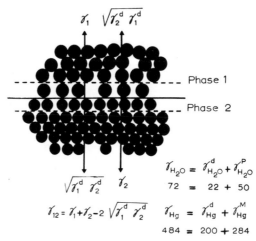

FIG. 5. At the interface between phases *1* and *2* the resultant force field is made up of components arising from bulk attractive forces in each phase and from the London dispersion forces operating across the interface itself.

by the pull across the interface in the opposite direction. This tension is found, in the simplest cases, to depend almost entirely on the dispersion force interaction. It is expressed as the geometric mean $\sqrt{\gamma_1^d \gamma_2^d}$ and so the γ_1 is reduced to $\gamma_1 - \sqrt{\gamma_1^d \gamma_2^d}$. A similar argument for the second phase leads to the new tension $\gamma_2 - \sqrt{\gamma_1^d \gamma_2^d}$. Upon addition, Eq. [16] is obtained.

A saturated hydrocarbon in contact with water is a simple case to consider. The interfacial tensions of many such systems are well known, as of course are the γ_1 and γ_2's. Furthermore, as noted above, for the hydrocarbon γ_1 and γ_1^d will be the same. Not so for water. The γ_2 for water (72 dynes/cm at 25°C) will be much larger than γ_2^d because the large dipole of water contributes heavily to γ_2. By fitting known values of surface and interfacial tensions into Eq. [1], Fowkes (19) finds $\gamma_{H_2O}^d$ is 21.8 ± 0.7 dynes/cm. Then γ^p, the dipole or hydrogen bond contribution, is $72 - 22 = 50$ dynes/cm. Very often only the γ^d contributes appreciably across an interface with another condensed phase, so that for water, interactions with another phase often depend mainly on a small number ($\gamma^d = 22$) not on the larger (total) number ($\gamma = 72$).

Consider the mercury/water interface. The surface tension of mercury is 484 dynes/cm at 25°C. It is the metallic bond which is responsible for the high surface tensions or surface free energies of metals, which are commonly hundreds or even thousands of dynes/cm or ergs/cm². Using the same technique as for hydrocarbon/water interfaces with Eq. [16] and the measured values of interfacial tensions for hydrocarbon/mercury interfaces, we find γ_{Hg}^d to be 200 ± 7. Thus, the appropriate value for γ_{12} for the water/mercury interface is readily calculated by application of Eq. [16] to be 425 ± 4 dynes/cm. The experimental 426 ± 1 dynes/cm agrees well with the calculated value confirming the usefulness of this GM approach and also confirming the observation that often only dispersion forces can be considered to interact appreciably across an interface.

ARITHMETIC MEAN AVERAGING

A surprising result of using the geometric mean relationship to predict intermolecular forces acting at a metal-organic liquid interface is that the interaction of dipoles with the metal often appears to be negligibly small. Furthermore, there is no indication of an interaction of the metal (mercury) with the pi-electrons of aromatic compounds even though contact potential measurements and heats of adsorption suggest a considerable effect (19).

J. Lavelle of our Laboratory, using an arithmetic mean (AM) of the dispersion force attractions to estimate the magnitude of the interaction between dissimilar materials, has found that the following equation yields some interesting results:

$$\gamma_{12} = \gamma_1 + \gamma_2 - (\gamma_1^d + \gamma_2^d). \quad [17]$$

The two terms by which the γ's are diminished are each taken to be $(\gamma_1^d + \gamma_2^d)/2$. Of course, the arithmetic mean (AM) has not as great a scientific basis for its use as does the geometric mean. However, the AM approach has been previously more successful for some interaction constants (26).

From interfacial tension data with hydrocarbons, $\gamma_{H_2O}^d$ is still found to be 22 ergs/cm² with the use of the arithmetic mean. However, γ_{Hg}^d is found to be only 108 ergs/cm² instead of 200 by this averaging technique. With the use of these $\gamma_{H_2O}^d$ and γ_{Hg}^d values in Eq. [17] to calculate γ_{Hg/H_2O}, a value of 426 dynes/cm is obtained. This interfacial free energy again agrees with the best experimental values and indicates that the arithmetic mean has the same internal consistency as does the geometric mean approach.

The arithmetic mean technique predicts an interaction of mercury with the pi-electron system of aromatic compounds at a magnitude of *ca.* 15 ergs/cm², as might be expected. Table II shows a comparison of the GM and AM approaches in calculating the

TABLE II

ENERGY OF MERCURY-ORGANIC LIQUID
INTERACTIONS IN EXCESS OF
DISPERSION FORCES

Liquid No. 2	Geometric mean $(ergs/cm^2)$	Arithmetic mean $(ergs/cm^2)$
Hexane	0	0
Benzene	0	13
Toluene	0	17
p-Xylene	0	15
Bromobenzene	0	25
Aniline	0	34
1,2-Dibromomethane	0	29
Cyclopentanol	−9	10
Methanol	−11	−8
n-Propanol	−8	−3

TABLE III

CALCULATED WORKS OF ADHESION—HEPTANE
ON IRON

$$W_A = \gamma_s + \gamma_{LV} - \gamma_{SL}$$

Geom. Mean	Arith. Mean
$W_A = 2\sqrt{\gamma_s^d \gamma_{LV}^d}$	$W_A = \gamma_s^d + \gamma_L^d$
$W_A = 2\sqrt{(108)(20.4)}$	$W_A = 73.4 + 20.4$
$W_A = 93.8\ ergs/cm^2$	$W_A = 93.8\ ergs/cm^2$

Experimental

$$W_A = \pi_e + 2\gamma_{LV}$$
$$W_A = 53 + 2(20.4) \qquad \text{(Harkins and Loesser (21))}$$
$$W_A = 93.8\ ergs/cm^2$$

polar interfacial attractions at the mercury-organic liquid interface. The AM calculation gives expected contributions of some of the ring and dipole compounds, whereas the GM calculation does not. It is interesting to note that both averaging techniques predict no polar interaction of short-chain alcohols with mercury. Negative values must be regarded as zero.

Table III shows the work of adhesion of heptane on "iron" as calculated by both averaging procedures. The calculated values are identical and agree with the value of 93.8 ergs/cm² measured by Harkins and Loesser (21). This agreement again shows the internal consistency of both averaging techniques.

THE DISPERSION PART OF THE SURFACE FREE ENERGY OF SOLIDS, γ_s^d

A table of γ_s^d values for different solids is of considerable value as can be seen from previous liquid/liquid discussions in predicting interactions with liquids whose γ^d's are also established. Three methods for estimating the γ_s^d's will be discussed:

1) From the π_e's of hydrocarbons obtained from the adsorption isotherms.

2) From contact angles for nonwetting systems.

3) From heats of immersion.
The last method is probably the least controversial. The first method will be discussed next.

From Adsorption Isotherms. Table IV shows a comparison of the geometric and arithmetic mean in calculating γ_s^d values from the free energy of adsorption of heptane vapors on the solids. The γ_{SL} is obtained from the measured isotherms; it is the same as the π_e of Eq. [12] when the adsorbed film reaches the liquid state at P_0. In all cases, the arithmetic mean predicts a lower contribution of the London dispersion forces to the surface free energy. It will be seen that we are pleased with the lower results, particularly for graphitic solids, when compared with other measurements to be presented later.

From Contact Angles. An important source of information concerning the chemical nature of surfaces is that of contact angles. Fowkes' scheme is very useful in this connection. The Young equation for the angle of a liquid L on a plane surface S is:

$$\gamma_{LV}\cos\theta = \gamma_s - \gamma_{SL} - \pi_e, \qquad [18]$$

where π_e is the equilibrium film pressure of adsorbed vapor on the solid surface. For solid-liquid systems interacting by dispersion forces from Eq. [16] and [18]:

$$\gamma_{LV}\cos\theta = -\gamma_{LV} + 2\sqrt{\gamma_L^d \gamma_s^d} - \pi_e. \qquad [19]$$

The expectation might be that for high contact angles, the value of π_e would be negligible; we shall see later that this is not always the case. Then from Eq. [19]:

$$\cos\theta = -1 + \frac{2\sqrt{\gamma_s^d}\sqrt{\gamma_L^d}v}{\gamma_{LV}} - \frac{\pi_e}{\gamma_{LV}}, \qquad [20]$$

TABLE IV

CALCULATION OF $\gamma_s{}^d$ VALUES FROM HEPTANE
AT 25°C

Solid	π_e (dynes/cm)	$\gamma_s{}^d$ Geom. Mean	$\gamma_s{}^d$ Arith. Mean
Copper	29	60	49.4
Silver	37	74	57.4
Silica	39	78	59.4
Anatase (TiO$_2$)	46	92	66.4
Lead	49	99	69.4
Tin	50	101	70.4
Iron	53	108	73.4
Ferric iron oxide	54	107	74.4
Stannic oxide	54	111	74.4
Graphite	56,58,63	115,120,132	76,78,83

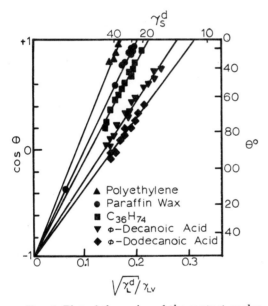

FIG. 6. Plot of the cosine of the contact angles of various liquids of differing surface tensions on five solids against the function $\sqrt{\gamma_L{}^d}/\gamma_{LV}$. This technique is used to obtain $\gamma_s{}^d$, the dispersional contribution to the free energy of the solid. Equation [20].

and a plot of contact angles vs. $\sqrt{\gamma_{LV}{}^d}/\gamma_{LV}$ should give a straight line with its origin at $\cos\theta = -1$ and with a slope of $2\sqrt{\gamma_s{}^d}$. In Fig. 6 such a plot of some of Zisman's extensive data (22) is shown. Note that since the origin is fixed, one contact angle value is theoretically sufficient to determine

$\gamma_s{}^d$. A similar equation yielding similar $\gamma_s{}^d$'s is obtained by the arithmetic mean approach.

Thelen (23) used Eq. [20] to calculate the surface dispersion energies on the noble metals from the contact angles with water measured by Erb after 3650 hours of continuous condensation of pure water on the metals. See Table V. Although most metals have oxide coatings and are wet by water, the noble metals under Erb's conditions are apparently hydrophobic and must be substantially free of oxide. Erb's value of the contact angle for water on gold is confirmed by White's (24) extremely careful work although the latter results are quite at odds with some measurements by Zisman and co-worker (25). The latter workers found gold was wet by water, but White concluded that the wetting was due to small quantities of residual hydrophilic abrasive used by Zisman in his cleaning procedure. Zisman's erroneous results are only one example of the extreme care which must be applied to any surface chemical measurement to ensure that the material being studied is perfectly clean and free from contaminants. Too much emphasis cannot be placed on this point especially where the study of very active surfaces such as those of metals is concerned.

Thelen was able to relate the dispersion energies to the cohesive energy densities, $\Delta E_v/V$, of the metals with a relationship developed by Hildebrand (26). Thelen got:

$$\Delta E_v/V = 166.7\gamma/V^{1/3}, \qquad [21]$$

where ΔE_v is the energy of vaporization of a gram atom, V is the gram-atomic volume in cm^3, and γ is the surface energy in ergs/cm^2. For metals, a straight line plot of $\gamma^d/V^{1/3}$ vs. cohesive energy density was obtained (Fig. 7) again clearly illustrating the operation of dispersion forces across an interface. Thelen did not measure the π_e's which are believed to contribute according to Eq. [20] to the determination of the γ^d's. The straight-line plots, as might be expected, suggest that the π_e decreases in an orderly fashion with the increase in contact angle. Also, since ΔE_v includes all the types of interactions, other

TABLE V

DISPERSION AND SURFACE ENERGIES OF NOBLE METALS

	Ag	Au	Pd	Pt
Contact angle of water, θ	79.5	65.5	62.5	40.0
Dispersion energy, ergs/cm², γ_s^d	84.95	121.63	129.86	189.56
Energy of vaporization, kcal, ΔE_v	67.4	88.8	92.7	136.9
Atomic volume, cm³, V	10.3	10.2	8.9	9.1

interactions besides the dispersion part of the total interaction in ΔE_v must also decrease approximately linearly with the contact angle.

From Heats of Immersion. The heat of immersion technique is one of the most direct methods for determining the γ^d's of solids. In this method, an evacuated solid is broken into a hydrocarbon liquid in a calorimeter as shown in Fig. 3. The heat liberated is measured by following the change in resistance of a thermistor. The free energy change per unit area is:

$$g_i = \gamma_{SL} - \gamma_S. \qquad [22]$$

Then from Eq. [16]:

$$g_i = \gamma_{LV} - 2\sqrt{\gamma_s^d \gamma_{LV}^d}. \qquad [23]$$

Since $h_i = g_i - T\dfrac{(dg_i)}{dT}$ (Eq. [9]), then:

$$
\begin{aligned}
h_i = {} & \gamma_{LV} - 2\sqrt{\gamma_s^d \gamma_{LV}^d} \\
& - T\left[\frac{(d\gamma_{LV})}{dT} - 2\sqrt{\gamma_{LV}^d}\,\frac{d\sqrt{\gamma_s^d}}{dT}\right. \\
& \left. - 2\sqrt{\gamma_s^d}\,\frac{d\sqrt{\gamma_{LV}^d}}{dT}\right]. \qquad [24]
\end{aligned}
$$

Values of γ_s^d can be calculated from Eq. [24]. As a rule γ_{LV}, γ_{LV}^d, and $d\gamma_{LV}/dT$ are known, but $d\sqrt{\gamma_{LV}^d}/dT$ and $d\sqrt{\gamma_s^d}/dT$ are rarely known. The temperature coefficients of the dispersion contributions can be estimated from the fourth power of the density. Since the temperature coefficients are small, errors arising from such estimates must be even smaller, but they are not necessarily insignificant.

A simple case is a graphitic solid like Graphon (a graphitized carbon black) where the interaction with liquids might often be

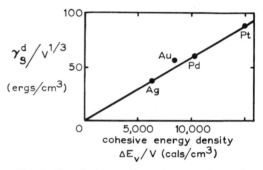

FIG. 7. Correlation of dispersion energies of the noble metals with cohesive energy densities. It is important that the plot is easily drawn through the zero point. ΔE_v includes all interactions so those other than dispersion energies apparently increase linearly along with the dispersion energies.

expected to be due only to dispersion forces. From a variety of sources, $\gamma_s^d = 70$ ergs/cm². (Fowkes (19) claimed 110 ergs/cm², but certain doubts have arisen with regard to the data being used. We shall find later that the areas of graphitic solids appear to be larger, and hence the h_i's smaller, than was earlier supposed.) Together with the liquid data, the results in Table VI have been calculated from Eq. [24]. The agreement between the measured and calculated values is good. Graphite or Graphon does not show any appreciable interaction with the liquid dipoles as was earlier expected (15).

For a hydrocarbon on a solid, the equation according to the AM rule to calculate γ_s^d from h_i is quite simple. From:

$$g_i = \gamma_{SL} - \gamma_S, \qquad [22]$$

$$h_i = g_i - T\frac{dg_i}{dT}, \qquad [9]$$

$$\gamma_{SL} = \gamma_S + \gamma_{LV} - (\gamma_s^d + \gamma_{LV}^d) \qquad [25]$$

TABLE VI
Heats of Immersion of Graphon at 25°C

| Liquid | $-h_i$ (meas.) | | $-h_i$ (calc.) | $\gamma_s{}^d$ (calc.) | |
	$N_2 = 16.2\ A$	$N_2 = 20.2\ A$		Geom. M.	Arith. M.
	(ergs/cm²)	(ergs/cm²)	(ergs/cm²)	(ergs/cm²)	(ergs/cm²)
Hexane	103	82		65	61
Heptane	112	90		68	69
Octane	127	102		78.5	81
Average				70	70
BuOH	114	91	91		
BuNH₂	106	85	91		
BuCl	106	85	91		
BuCOOH	115	92	91		
Average		88	91		
Water	32	26	25		

and therefore, when the liquid is a hydrocarbon so that $\gamma_{LV} = \gamma_{LV}^d$:

$$h_i = -\gamma_s{}^d + Td\frac{\gamma_s{}^d}{dT}. \qquad [26]$$

For low-energy surfaces, taking $d\gamma_s{}^d/dT$ to be about -0.07 from Schonhorn (27), we have at ordinary temperatures:

$$\gamma_s{}^d = -h_i - 21 \quad \text{(Low-Energy} \qquad [27]$$
$$\text{Solids into Hydrocarbons)},$$

obtained by measuring the heat of immersion into a hydrocarbon. Other investigators also agree with this magnitude for the temperature coefficient of $\gamma_s{}^d$ for low-energy polymers (28).

The heats of immersion of Graphon into hexane, heptane and octane have been used to calculate the $\gamma_s{}^d$'s by the GM and AM approaches as given in Table VI. Both approaches are only modestly successful in that the results are not constant, although the averages agree well. With the use of Eq. [27] and $\gamma_s{}^d = 70$, the h_i's for the butyl derivatives are -91 ergs/cm², in good agreement with the measured values.

Origin of Hydrophobicity

Before we leave the comparisons of the GM and the AM rules, let us introduce them in turn for γ_{SL} into Eq. [14] for the spreading

coefficient of water on a hydrophobic solid:

$$S = -2\gamma_W + 2\sqrt{\gamma_s{}^d\gamma_W{}^d} \quad \text{(GM)}; \quad [28]$$

$$S = -2\gamma_W + \gamma_s{}^d + \gamma_W{}^d \quad \text{(AM)}. \quad [29]$$

For simplicity, consider Eq. [29] for the water/Graphon system. With $\gamma_W = 72$, $\gamma_s{}^d = 70$, and $\gamma_W{}^d = 22$ ergs/cm², S is obviously negative and water will not spread as observation tells us. If it was not $\gamma_W{}^d$ interacting across the interface, but the total γ_W instead, then S would be very nearly positive and the contact angle would be very low or zero. The reason for the hydrophobic character of graphitic surfaces becomes clear!

Very simply, $S = \gamma_W{}^d - \gamma_0$ for oil spreading on water according to the AM rule. Both approaches agree, however, in that γ_{LV} must be less than 22 ergs/cm² for hydrocarbons, i.e., $<C_8$, for them to spread on water.

Contact Angles vs. Heats of Immersion

Any inconsistency between contact angles and heats of immersion could be checked. If we use the definition of $h_{i(SV)}$, which differs from h_i in that the initial state is not the bare solid but the solid immersed in the saturated vapor:

$$h_{i(SV)} = h_{SL} - h_{SV} \qquad [30]$$

and

$$g_{i(SV)} = \gamma_{SL} - \gamma_{SV} = -\gamma_{LV}\cos\theta, \quad [31]$$

together with the Young equation:

$$\gamma_{SL} = \frac{\gamma_s - \pi_e}{\gamma_{SV}} - \gamma_{LV} \cos \theta,$$

then:

$$h_{SL} - h_{SV} = \gamma_{SL} - \gamma_{SV} - T \frac{dg_i}{dT}$$

$$= -\gamma_{LV} \cos \theta$$

$$+ \left[T\gamma_{LV} \frac{d \cos \theta}{dT} + \cos \theta \frac{d\gamma_{LV}}{dT} \right. \quad [32]$$

$$\left. -\gamma_{LV} - \frac{Td\,\gamma_{LV}}{dT} \right] \cos \theta + \gamma_{LV}\, T \frac{d \cos \theta}{dT}.$$

Therefore:

$$-h_{i(SV)} = h_{LV} \cos\theta - \gamma_{LV} T \frac{d \cos \theta}{dT}. \quad [33]$$

Equation [33] allows direct comparisons to be made between the two methods, contact angle and heat of immersion. Impeding progress in testing Eq. [33] is that we need powders to get sensible adsorption isotherms and, contrariwise, flat platelets of greater than microscopic size for measurement of contact angles. Whalen and Wade (29) recently tackled the problem with Teflon powder (polytetrafluoroethylene). Even in this case it could be argued that when Allen and Roberts (30) pressed the powder into discs for the contact angle measurements (Whalen and Wade used their data), the surface was changed in some manner.

The C_{10}, C_{12} and C_{16} hydrocarbons on Teflon gave no detectable vapor adsorption and no π_e's. Some slight difference might have developed between powders and discs because the γ_s^d's calculated from contact angles, Eq. [20], and from the heats of immersion of the powders, Eq. [24], did not agree too well. The average from the first method was 19.2, from the second 16.1 ergs/cm^2. Whalen and Wade reported π_e's for hexane and octane on Teflon as listed in Table I. Correction for these π_e's in the calculation by means of contact angle increases γ_s^d to 22–25 ergs/cm^2, but then the agree-

ment with the γ_s^d's determined with the higher homologs is lost. The h_i values for C_6 and C_8 also gave 20–24 ergs/cm^2. The AM rule does just as poorly as the GM rule in this regard. More such comparisons with greater experimental precision are needed, however, before definite conclusions about the comparison of results from the two approaches can be made.

SPREADING PRESSURES ON HYDROPHOBIC SURFACES

Of course, the general problem of "what is the true contact angle for real surfaces?" remains. It will now be interesting to examine our present knowledge of π_e's for water and other substances on hydrophobic surfaces. Many times (19, 22, 23) π_e's have been neglected or considered negligible for contact angle systems in such equations as Eq. [20]. We shall find that this assumption is sometimes poor.

Derjaguin and Zorin (31) pointed out that in a system possessing a contact angle, the adsorption isotherm must cross the p_0 line and there must be an unstable region as depicted in Fig. 8. To the first crossing:

$$\gamma_s - \gamma_{SV} = \pi_e = kT \int_{\Gamma=0}^{\Gamma^0} \Gamma\, d\ln P, \quad [34]$$

and to the infinitely thick, duplex film:

$$\gamma_s - \gamma_{SL} - \gamma_{LV}$$

$$= kT \int_{\Gamma=0}^{\infty} \Gamma\, d\ln P = \pi_D \quad \text{or} \quad S_{SL}. \quad [35]$$

The difference between these two integrals is represented by the net shaded area in Fig. 8. Upon subtraction:

$$\pi_D - \pi_e = \gamma_{SV} - (\gamma_{SL} + \gamma_{LV})$$

$$= kT \int_{\Gamma=\Gamma^0}^{\infty} \Gamma\, d\ln P,$$

$$= S_{SVL} = \gamma_{LV}(\cos\theta - 1), \quad [36]$$

$$= S_{SL} - \pi_e,$$

where S_{SVL} is the spreading coefficient when the solid has the equilibrium adsorbed film

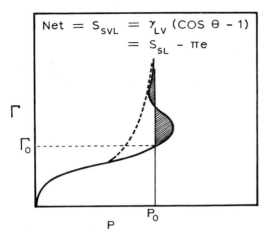

$$Net = S_{SVL} = \gamma_{LV}(\cos \theta - 1)$$
$$= S_{SL} - \pi e$$

FIG. 8. Adsorption isotherm for the case of equilibrium contact angle; nonduplex film at P_0.

on it. Adamson and Ling (32) developed a tentative analytical expression for Eq. [36]. The important conclusion is that in the contact angle situation, the equilibrium adsorbed film may be in quite a different state than the bulk liquid. Derjaguin (31) has reported an exceedingly strange state for water and other liquids ordered at the surfaces of solids. Such a different adsorbed liquid phase and the transition at the boundary might greatly affect our consideration of fluid flow through capillaries and porous media.

An obvious implication is that measurement of the adsorption isotherm could establish the presence or absence of a π_e. Of course, the amount adsorbed is often small so that it is difficult to obtain accurate results. This difficulty is compounded in attempting to work with water vapor, and rather few laboratories have reported water adsorption isotherms even on polar surfaces. From such measurements, Boyd and Livingston (33) reported a value of 59 ergs/cm^2 for π_e for water on graphite (Dixon 0708), and 19 for a different sample (34). These results seem unusually high and are probably largely due to impurity sites produced by the ash content and by oxides of carbon. Although the value may vary with the sample, a more expected magnitude of 6 ergs/cm^2 was estimated by us from the water vapor adsorption data of

Bassett (35) on the graphitized carbon black, Graphon. See Table I. Water adsorption on this substance (and on other hydrophobic solids—see later section) increases slightly with increasing temperature; therefore, the π_e does also. There are other interesting thermodynamic consequences of this experimental fact to be examined later.

Measurable values of π_e begin to occur for hydrocarbons on polytetrafluoroethylene (Teflon), as we have noted before, below 10 carbon atoms. Thus, for octane in Table I, where π_e is 3 ergs/cm^2, the term for π_e/γ_{LV} is 0.15, a sizable contribution to the $\cos \theta$ expression, Eq. [20].

ADSORPTION ON HYDROPHOBIC SURFACES

The discussion of the adsorbed "film" in equilibrium with the bulk liquid in the contact angle situation has already led us to consider adsorption over hydrophobic surfaces. At least some of these "films" are definitely not continuous, as we shall learn, but consist of clusters around specific sites. We shall examine the latter situation, which pertains particularly to polar adsorbates on residual or impurity polar sites in the hydrophobic matrix, after we examine the surface area determination of hydrophobic powders, and adsorption from vapors and solutions essentially over the entire surface.

AREA DETERMINATION

Pierce (36) was responsible, in three papers starting in 1958, for establishing techniques useful for the problem at hand. He applied the Frenkel-Halsey-Hill equation (FHH):

$$\left(\frac{v}{v_m}\right)^s = \frac{C}{\log P_0/P} \qquad [37]$$

to the nitrogen isotherms for many high surface energy powders to establish the "ideal isotherm." Shull (37) had done this earlier for a few isotherms. The C value is related to the energy of adsorption in the first layer and the s value to the diminution of forces emanating from the surface with successive layers in the adsorbed film. The "ideal isotherm" is given in Table VII and Fig. 9 in terms of the number of layers $n = v/v_m$ adsorbed at various relative pressures. The s value is 2.75 for polar surfaces as obtained from the $\log v/v_m$ versus $\log \log P_0/P$ plots; the C value will not be discussed here. Note that the s value is established in the multilayer region after the

TABLE VII
IDEAL ISOTHERMS FOR NITROGEN ON VARIOUS SUBSTRATES

P/P_0	Nature of the surface High-energy		Low-energy
	n	v/v_m	n
0.20	1.25		1.18
0.30	1.39		1.35
0.40	1.54		1.55
0.50	1.70		1.77
0.60	1.90		2.05
0.70	2.17		2.43
0.75	2.34		2.69
0.80	2.58		3.07
0.85	2.90		3.46
0.90	3.35		4.06
0.98	6.17		12.9
s Value	2.75		2.12

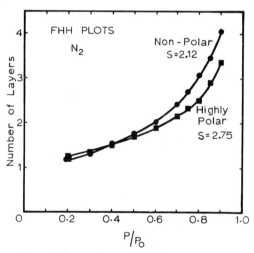

FIG. 9. The number of monolayers of nitrogen adsorbed as a function of the relative pressure for polar and nonpolar surfaces. The "ideal" isotherms for N_2.

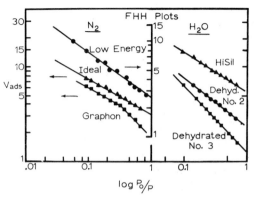

FIG. 10. FHH plots of the log volume adsorbed vs. the log log P_0/P for nitrogen and water on surfaces of varying polarity. For both adsorbates, lower s values are found for lower polarity surfaces. Higher slope at low P's for Graphon is ascribed to loose packing, lower than liquid-like in monolayer.

specific effects of the surface are dissipated. To avoid using data involving condensation between particles, values at high relative pressure were established from nitrogen adsorption on "metal" foil (38). Deviations above the line at high relative pressures indicates condensation in pores of one kind or another. Scores of isotherms gave the same s value, but these substances all possessed hydrophilic and high-energy surfaces.

Early work in this Laboratory (39) established the general utility of the BET approach to the determination of the surface areas of polyethylene, Nylon, Teflon, and collagen. When the FHH plots for nitrogen on the first three and on polypropylene (40) were recently plotted as shown in Fig. 10, excellent straight lines were found, but the s values were constant at the lower value of 2.1 as listed in Table VIII. Therefore, an "ideal isotherm" could be established for such low-energy surfaces, Table VII. Lower specificity of these surfaces is shown by the lower adsorption at the low relative pressures.

For graphitized black, Sterling MT (3100°), Pierce obtained two branches intersecting at 0.4 P/P_0. The branch at higher relative pressures had the same slope as the line for the "master" isotherm, but the slope

TABLE VIII

Slopes of FHH Plots

Adsorbent	s	Adsorbent	s
N_2 at $-195°C$		H_2O at $25°C$	
Teflon	2.12	Anatase	2.50
Polyethylene	2.12	α-Fe_2O_3	2.45
Polypropylene	2.12	α-FeOOH	2.45
Nylon	2.12	HiSil 233	2.49
Collagen	0.8	HiSil (#1 dehydrated)	1.88
"Ideal"	2.75	HiSil (#2 dehydrated)	1.92
HiSil 233	2.75	HiSil (#3 dehydrated)	1.40
HiSil (dehydrated)	2.48	Kaolinites (Li, Cu, Ba, Al)	1.80
Cabosil	2.20	AgI	1.3

of the line between 0.2 and 0.4 P/P_0 was steeper (similar to our plot for Graphon in Fig. 10). Therefore, the indication was that the packing of the nitrogen molecules was normal (close to liquid-like) in the multilayer region, but that much looser packing occurred in the first layer. Thus, the conclusion was reached that homogeneous graphitized blacks adsorb nitrogen in loose array with a co-area of about 20 Å^2 per molecule rather than the close-packed value of 16.2 Å^2 commonly used. The MT surface area was corrected from 7.65 to 9.4 m^2/gm.

Graphon, possessing a slightly less homogeneous surface than the MT (3100°), also yields the double branch FHH plot as shown in Fig. 10. The specific surface area of Graphon is much higher (now corrected to about 120 m^2/gm) than that for the MT (3100°), and so it has been frequently used in surface chemistry studies of all kinds. Many of these measurements, such as heats of immersion, should be corrected to the 25% higher area, or the energy values and amounts adsorbed per unit area should be reduced by 20%.

Examples are given in Table VI for the heats of immersion of Graphon into hexane, heptane, and octane showing our original results (15) and then the values corrected for the higher area. These values give the same average value of $\gamma_s{}^d$ by either the geometric mean or arithmetic mean approach. In turn, this estimated $\gamma_s{}^d$ was used to calcu-

late expected h_i's for four butyl derivatives. Good agreement with our earlier measurements is shown. On the other hand, this $\gamma_s{}^d$ and a π_e of 6 leads to a calculated contact angle of about 71° for water using Eq. [18] with the AM approach and 91° with the GM approach. The measured value from a pressed disc of Graphon was 82°, midway between these calculated values.

The only previous application of the FHH equation to water adsorption has been Halsey's plot (41) of the Harkins and Jura data for anatase. Halsey's s value of 2.50 is tabulated with several other new values for oxides in Table VIII, all being close to the average of 2.47. These oxides no doubt possess hydroxyl surfaces. On the other hand, the dehydrated silicas, obtained by heating at 650°C for several hours, also yielded lower s values of about 1.90. The kaolinites (42), which are suspected by a number of investigators to be partially hydrophobic, also gave the lower s value. Of course, the v_m's for water adsorption for the low s value surfaces are much lower than expected from the nitrogen surface areas. On the dehydrated silicas, the water adsorbs only on the residual hydroxyls. After the "nominal" first layer has formed, clusters of adsorbed molecules develop around the hydrophilic sites. We shall consider some practical implications of this situation in the section on Hydrophilic Sites in Hydrophobic Matrices.

ADSORPTION FROM SOLUTION

Only two aspects of adsorption from aqueous solutions onto hydrophobic surfaces will be discussed here. The one has to do with the adsorption of surfactants, the other with the adsorption of simple organic molecules to study model systems. Both are of continuing interest in our Laboratory. Graphon has been used as the model adsorbent in both cases.

Surfactants on Graphon. The mechanisms and the energetics of the adsorption of surfactants onto Graphon have proved fascinating. The hydrophobic character of this substrate surface, together with its homogeneity, makes it ideal to study packing in adsorbed films and heat of adsorption by heat of immersion techniques from aqueous solutions. These calorimetric techniques have been reviewed elsewhere (43).

Adsorption isotherms having double plateaus, as illustrated in Fig. 11, were obtained with sodium dodecyl sulfate, NaDS. The lower plateau was ascribed (44) to a lower packing than that which occurred in the higher plateau region. The co-areas in Table

ADSORPTION OF DODECYL SULFATE ANIONS AND SODIUM IONS
GRAPHON / SOLUTION INTERFACE

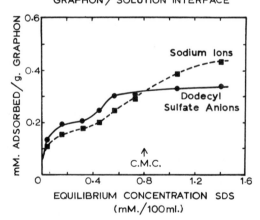

FIG. 11. Adsorption isotherms of dodecyl sulfate anions and sodium ions at the Graphon/solution interface exhibit double plateaus. The lower plateau has a lower packing of the DS⁻ than in the higher plateau region.

TABLE IX
SURFACTANT HEATS OF ADSORPTION ON GRAPHON

Surfactant	Area per anion (A^2)	ΔH_{ads} (kcal/mole)
NaDS	46	−7.6
	84	−9.5
NaDBS	60	−8.8
α-Sulfo esters	61–104	−3.7, −5.5

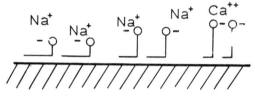

NaDS on GRAPHON

FIG. 12. The proposed configuration of the sodium dodecyl sulfate when adsorbed onto Graphon in the lower and higher plateau regions of the isotherm. The bridging of two dodecyl sulfate anions by a divalent calcium cation is also depicted.

IX have been corrected to 119 m²/gm for the surface area of the Graphon used instead of the 95 m²/gm previously reported. The heats of adsorption, determined from heats of immersion by correcting for small heats of dilution and of demicellization, were constant in the two plateau regions. The value decreased with increased packing (lower co-area) in the higher plateau region owing to increased repulsion between the negatively charged head groups. The two situations are depicted in Fig. 12.

Another curious phenomenon is shown by the sodium adsorption isotherm, followed with radiotagged sodium. It should be mentioned that no metal ions were adsorbed on the Graphon unless the organic ion was previously adsorbed. Of course, the adsorption of the counterion occurs in the diffuse double layer, but the amount taken up was less than the DS⁻ below the critical micelle concentration, and greater above the cmc, where the adsorbed DS⁻ film took up Na⁺ preferentially leaving the micelles deficient. This curious situation deserves further study.

The addition of Ca⁺⁺ to the NaDS/

Graphon system produced further interesting results (44). The Ca^{++} enhanced the adsorption of DS^- as shown in Fig. 13, schematically depicted in Fig. 12. The mole ratio Ca^{++}/DS^- adsorbed approached the stoichiometric ratio of 0.5 as shown in Fig. 14. Here, the packing tended toward 27.5 A^2/DS^- (corrected from 22 A^2 previously reported), as the closely held Ca^{++} reduced repulsion between the head groups. To achieve a ratio of one for Na^+/CA^{++} at the Graphon/solution interface required a solution concentration ratio of 500/1. Interesting results were also achieved in following the flocculation of the Graphon occurring on the addition of Ca^{++}, and in the adherence to cotton so produced.

Apparently, only the straight-chain surfactants adsorb in several distinct configura-

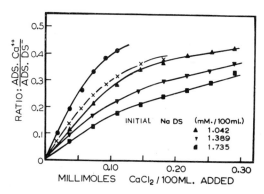

Fig. 14. The ratio of the adsorbed calcium ions to adsorbed dodecyl sulfate anions is plotted as a function of added calcium chloride. The ratio tends toward the stoichiometric value of 0.5.

tions on a homogeneous surface. Detailed studies have been made of sodium dodecylbenzenesulfonate, showing only one limiting co-area in Table IX, and of a series of the α-sulfoesters of differing R and R'. The heat of immersion of Graphon into sodium dodecylsulfate (Na lauryl sulfate) solutions remains the largest yet measured.

A few words about amino acid adsorption on Graphon might prove of interest. Even though highly important, the surface chemistry of these building blocks of proteins has been little studied. Beyond the adsorption and heat of immersion measurements in our Laboratory, the only other important measurements that have been found are the surface tensions of solutions of the simple ones reported thirty years ago (45). The first members of the series, glycine and alanine, behave like inorganic salts and are negatively adsorbed at the air/water interface. Much the same behavior is found at the Graphon/solution interface although alanine is adsorbed at the maximum on an average of one on each seven "benzene rings" on the graphite surface. On the other hand, the apparent monolayer capacity for leucine occurs at a coverage of one on each three surface rings.

It is interesting to contrast the adsorption of *dl*-phenylalanine with that of *dl*-tyrosine, the latter including a para hydroxyl group. See Table X. The adsorption isotherms and

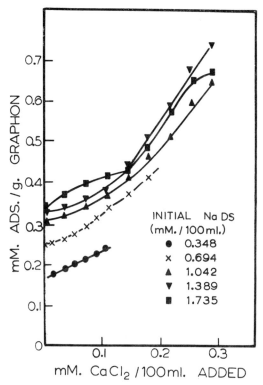

Fig. 13. The amount of DS^- adsorbed as a function of the concentration of added CA^{++} at several initial NaDS concentrations. The addition of Ca^{++} obviously enhances the DS^- adsorption. The model for this is given in Fig. 12.

TABLE X
ADSORPTION OF AMINO ACIDS ON GRAPHON

Amino acid	Monolayer (*millimoles/gm*)	Co-area (A^2/molecule)	Molec. model co-area (A^2)	Min. conc. for $\theta = 1$, (*millimoles/l*)
dl-Phenylalanine	0.18	98	104	36.
dl-Tyrosine	0.11	163	115	1.9

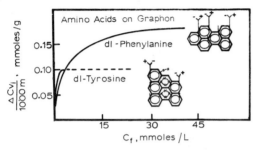

FIG. 15. The adsorption isotherms for *dl*-phenylalanine and *dl*-tyrosine onto Graphon. The proposed configurations of these molecules on the Graphon surface explain the 3/2 coverage for tyrosine as due to hydrogen bonding.

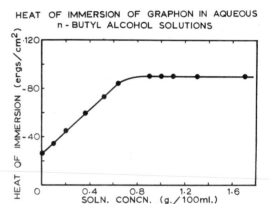

FIG. 16. Heat of immersion of Graphon in aqueous *n*-butyl alcohol solutions. The calculated line plot agrees entirely with the experimental points. The interfacial film behaves ideally.

thus the monolayer capacities were determined by depletion analysis with differential refractometry (46). Figure 15 gives the adsorption isotherms determined at the isoelectric point and the sketches suggest an explanation for the lower packing of the tyrosine. Tyrosine takes up about ³⁄₂ as much area as the phenylalanine. If each benzene ring at the amino acids sits on a graphite ring, then the lower packing of the tyrosine can be accounted for by hydrogen bonding across one of the graphite hexagons as depicted. In addition, the monolayer capacity of the phenylalanine has been confirmed by measurement of heats of immersion which reached a plateau value of 54 ergs/cm² at 0.18 mM adsorbed per gram of Graphon anywhere above 36 mM/l. For tyrosine, the plateau value was rather high at 127 ergs/cm² at 0.11 mM adsorbed per gram above 1.9 mM/l. concentration.

The fact that the tyrosine reached monolayer adsorption at a much lower concentration than the phenylalanine is in accord with its lower solubility limit: 1.44 versus 85.4 mM/l.

Heats of Immersion into Solutions. Rather little has been done on this important topic, but one study in this Laboratory (47) suggests that a fairly ideal type of behavior can occur. The system *n*-butanol/water/Graphon was chosen to avoid ions and to deal with species that could also be adsorbed from the vapor. Therefore, besides the heats of immersion of the bare surface into solutions of increasing butanol concentration, samples precoated from the vapor up to monolayer coverage could be measured. For the butanol-covered Graphon, the concentration of the solutions in the calorimeter had to be chosen so that no net adsorption or desorption would occur.

Here too, a monolayer of butanol was completed at a low concentration of butanol. The heat effects had to be corrected for the heat of dissolution upon adsorption; these values were directly determined by measuring heats of solution of butanol into water. Ideal behavior would be indicated if at each θ coverage, the prorated heat effects could be added to yield the measured heat of immersion. Figure 16 indicates that close agreement was obtained; these results are again corrected for the larger area of Graphon. Apparently, each portion of the adsorbed film, water and butanol, could be treated independently, and no special butanol-water interaction occurred in the interfacial region.

Another interesting study was concerned with the fatty acid solution/Graphon systems (48). Heats of immersion were determined at both low and high pH. At low pH, the un-ionized acid was adsorbed; at high pH, the double layer with Na^+ as the counterion was formed. Comparisons of the two at the same packing or co-area allowed the direct determination of the heat of formation of the double layer. It was about -1 kcal/mole adsorbate. From studies of different chain lengths, the contribution of each $-CH_2$ and $-COOH$ could be estimated.

Again, there is great need for additional measurements of a variety of solutions by the heat of immersion technique.

HETEROGENEITIES ON HYDROPHOBIC SURFACES

There are two kinds of heterogeneities we can examine. The one is geometric and is typified by the studies by Graham (49) of steps and recesses left in graphitized blacks. The other is the chemical or polar variety characterized by hydrophilic sites left behind on burning off oxides from carbon blacks in producing graphtic carbons such as Graphon. Or they could be produced by oxidation of the surface of a polymer like polyethylene, catalyzed by heat, light, etc. Or they could be due to "impurities" like surfactant molecules used in emulsion polymerization, as in the manufacture of polytetrafluorethylene (Teflon). This polar variety of heterogeneous sites will provide the main thrust of the following discussion. Minor attention will be paid to the geometric heterogeneities at the end of this review.

Hydrophilic Sites in a Hydrophobic Matrix.

There is no doubt that sometimes such hydrophilic sites are detrimental to the utility of the hydrophobic material. Take water repellency of a cloth: Hydrophilic areas could easily form the route to the passage of water. Surprisingly, at first glance, such sites can produce beneficial or useful effects. Take the lubricity of graphite: it is the water cluster around the hydrophilic sites that gives it utility in commutator brushes, and the like. Take the adhesion of organic coatings and printing ink to polyethylene packaging film: the surface of the film is treated so that polar sites are purposely produced by chemical action to enhance adhesion. Take ice nucleants such as nature provides to "seed" clouds and thus cause rainfall, or as might be artfully introduced: in "cloud seeding," as for the polar sites on graphitic solids, the first water molecules to adsorb go down on these impurity sites and then the next adsorbate molecules cluster around these. The embryo thus formed, probably after it grows past the critical radius at which the bulk free energy change can counteract the surface free energy increase, goes over to ice, the stable phase at the ambient temperatures of clouds. It is the curious thermodynamic properties of these polar site/adsorbate interactions on such otherwise hydrophobic surfaces which will be discussed here.

The surface concentration of these polar sites can vary enormously. On Graphon, about one in 1500 surface sites is polar and accepts highly polar adsorbates like water. On Teflon 6, about one in 200. On silver iodide, the surface concentration varies considerably with the preparation, but it is often about one in four surface sites. The work in this Laboratory seems to show that the contact angles of water on pressed discs of these powders agree with those obtained on the purest single crystals or film. At fairly high surface concentrations of polar sites as on treated polyethylene film, the contact angle of water was indeed found to decrease, but no relationship with the amount of treatment nor performance in adhesion could be found.

Adsorption and heat of immersion measurements, however, are successful in characterizing these sites.

Fifteen years ago, our Laboratory was studying (50) water adsorption on Graphon. In the early sixties, the earlier findings proved useful in studying and in producing synthetic ice nucleants (51). But let us go back to an earlier date. In the late forties, Vonnegut (52) and Langmuir chose silver iodide as a "cloud seeder" on the basis that the common basal planes matched ice closely, and considering that epitaxy was important to the nucleation process. Silver iodide, even though expensive, remains the most effective readily available nucleant and is easily dispersed by burning acetone solutions. Nevertheless, silver iodide is highly insoluble and the really pure solid surface would not be expected to adsorb water. Early reports (53) on the adsorption of water on silver iodide incorrectly claimed close-packed multilayer adsorption. Apparently, these workers used silver iodide prepared from ammonium iodide, and they got *ab*sorption as well as *ad*sorption. Indeed a trend toward lower water adsorption with purer samples has been reported (54), but a really pure surface as might be produced from silver and iodine under UHV has not been examined.

At least one of our substitute nucleants rivals silver iodide in efficiency tests in a cloud chamber (51). See Fig. 17. No exhaustive attempts to synthesize nucleants have been made. Instead, we have been interested in learning how nucleants work, and thus the required surface properties. Most of our synthetic nucleants have been made by hydrophobing silicas, that is, by proceeding in a direction which offhand would be expected to be opposite to that required. Besides the fact that silicas are inexpensive, their surface properties have been explored more completely than those of any other oxide. The Lehigh work will be reviewed here and some recent developments will be included.

Water adsorption isotherms at several

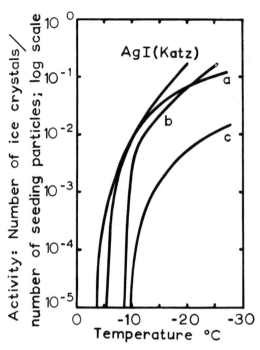

Fig. 17. Ice-nucleating ability of silver iodide and hydrophobed silicas. (*a*) A silica heated with AgNO₃ to 650°C compares favorably with the classical efficiency curve due to Katz for AgI; (*b*) a heat-treated silica made from (*c*) wet-precipitated silica. At low enough temperatures, all particles become nucleants.

temperatures on Graphon (55), AgI (56), and dehydrated HiSil No. 2 (57) are presented in Figs. 18–20. Such reversible isotherms displaying *increasing* amounts adsorbed with *increasing* temperatures were reported earlier by Lamb and Coolidge (58) for water on charcoal, but in this case the porosity complicates matters. This curious phenomenon leads to an interesting composite set of thermodynamic properties. The following results are required if reversible isotherms fall in this reverse order to that usually expected:

1) The isosteric heat of adsorption is less than the heat of liquefaction. Curves for the H₂O/Graphon system are shown in Fig. 21. Model isotherms which cross are shown in Fig. 22. We can write from the Clapey-

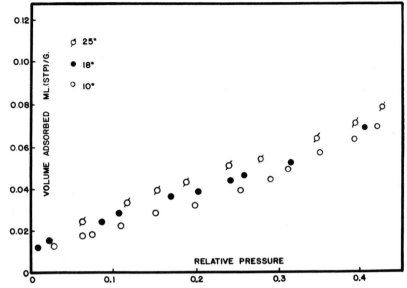

Fɪɢ. 18. Volume of water vapor adsorbed onto Graphon as a function of relative pressure at three temperatures. The isotherms are completely reversible. The increased adsorption with increased temperature was quite unusual when first reported.

WATER VAPOR ISOTHERMS ON AgI

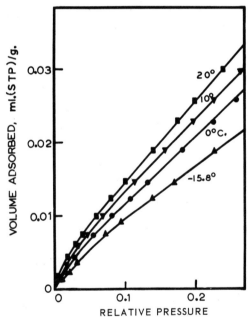

Fɪɢ. 19. Volume of water vapor adsorbed onto silver iodide as a function of relative pressure at low temperatures. Again, increased adsorption with increase in temperature. Impurity sites are thought to be responsible for the water adsorption by their water insoluble and hence hydrophobic compounds.

ron-Clausius equation:

$$\ln \frac{P_2}{P_1} = \ln \frac{(P_e/P_0)_2(P_0)_2}{(P_e/P_0)_1(P_0)_1}$$
$$= \frac{q_{st}}{R}\left(\frac{T_2 - T_1}{T_1 T_2}\right). \qquad [38]$$

Therefore, at the crossover point, q_{st} is just the heat of liquefaction. Below this point, q_{st} is less than the heat of liquefaction; above, it is greater as is usual.

2) The entropy of the adsorbed phase is high as shown for the water/Graphon system in Fig. 23. To calculate these entropies all the changes were ascribed to the adsorbate; this is a reasonable assumption for this weak adsorption. The "Hill hypothesis" has it that the differential entropy should cross the minimum in the integral entropy at the monolayer value (59). Interestingly, the BET value for water, only 1/1500 the nitrogen monolayer capacity, agrees well with the estimate from the entropies. For AgI, this agreement was not found (56).

The low-energy adsorption and the high entropy both indicate high mobility of the adsorbed water molecules. It is interesting to take the entropy values for the "nominal" first layer on Graphon and on a particular sample of AgI, which happens to be 29.3 eu in both cases. See Table XI. Following Graham (60), the translational part can be estimated by subtracting out the other con-

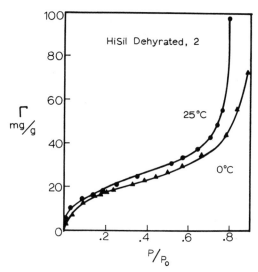

FIG. 20. Volume of water vapor adsorbed onto dehydrated HiSil as a function of relative pressure at two temperatures. Here is a "synthetic" nucleant which shows the same water adsorption behavior as do AGI and Graphon.

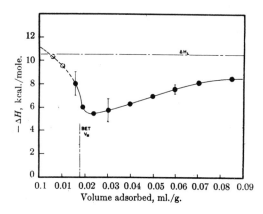

The difference in the internal energy E_L, per mole

FIG. 21. The variation in the isosteric heat calculated from the Clapeyron-Classius equation with increasing coverage for adsorption of water on Graphon. Note that the value of the heat of adsorption is lower than the heat of liquefaction at all coverages.

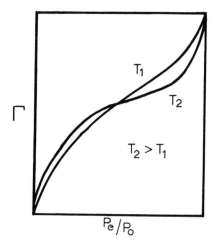

$$\ln \frac{P_2}{P_1} = \ln \frac{(P_e/P_o)_2 (P_o)_2}{(P_e/P_o)_1 (P_o)_1} = \frac{q_{st}}{R}\left[\frac{T_2 - T_1}{T_1 T_2}\right]$$

$$\Delta h_{ads} = \int_0^\Gamma q_{st} \, d\Gamma = \left[h_{i(SL)} - h_{i(SfL)}\right] + \Gamma \Delta H_L$$

FIG. 22. Hypothetical "cross-over" isotherm used to show the thermodynamic significance of the temperature dependence of adsorption isotherms. Below the "cross-over", the q_{st} is lower than the heat of liquefaction, and the heat of immersion rises with precoverage.

tributions. Then, the two-dimensional partition function can be used to calculate the surface area over which the adsorbed molecule migrates on the average. The result is 120 A^2 in both cases. For Graphon, the polar site density is so small that there is no lateral interaction. The AgI sample studied, on the other hand, has the maximum surface density at which lateral interaction would be expected to set in. A large concentration of sites, still isolated sufficiently for cluster formation, might be desirable for efficiency in ice nucleation.

3) The heat of immersion rises with precoverage (50) as shown in Fig. 24. For Graphon into water, this curve starts at 26 ergs/cm^2 and is still far from 118.5 ergs/cm^2, which is the surface enthalpy of liquid water, at 0.99 relative pressure. This unusual requirement of a rising heat of

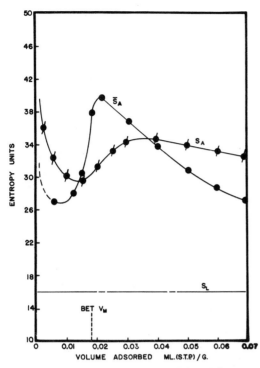

FIG. 23. Plots of the differential and integral entropies, $\bar{S}_s$ and S_s, respectively, as a function of the volume of water adsorbed.

TABLE XI

ENTROPY OF WATER ON HYDROPHOBIC SURFACES

$$Q_t = \frac{2\pi mkT}{h^2} a$$

$$S_t = R\left[1 + \ln\frac{2\pi mkT}{h^2} a\right]$$

	Graphon	AgI
Observed S, eu	29.3	29.3
Config.	1.0	
Rot.	8.4	
Vib.	2.6	
	12.0	12.0
Residual, trans.	17.3	17.3
	$a = 120$ A²	$a = 120$ A²
Available:	16,000	100

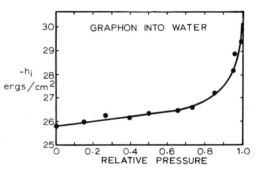

FIG. 24. Variation of the heat of immersion of Graphon in water with different precoverages of the Graphon surface with water vapor. Note that the heat of immersion increases as the surface of the Graphon adsorbs more water vapor. This shows that the bare surface of the Graphon has less affinity for water than does the adsorbed water.

immersion isotherm is indicated by the equation:

$$h_{ads} = (h_{i(SL)} - h_{i(SfL)}) + h_L. \qquad [39]$$

All these terms are minus quantities. This integral heat of adsorption from the vapor is less than the heat of liquefaction when $h_{i(SfL)}$ is greater in magnitude than $h_{i(SL)}$. Enthalpywise, then, water prefers the partially precovered surface to the bare surface.

These facts reveal much about the nature of adsorption of water on the polar heterogeneities on a hydrophobic surface. Silver iodide falls into this same category. So it became obvious that this knowledge should point the way to the development of what might be called "synthetic" ice nucleants or cloud seeders.

Ice Nucleants

It appears that for effective nucleants one should proceed oppositely to what might

be expected at first glance, as mentioned previously. If a substance like a precipitated silica is chosen, then its original surface will be highly hydrophilic and covered with —OH groups. Even heating to an elevated temperature to drive off water through the condensation of hydroxyls should develop nucleation ability if nature desires a surface similar to that of AgI. The silica should not be dead-burned; some of the —OH groups should be allowed to remain to provide sites around which clusters of water molecules might form.

TABLE XII
Surface Properties of Ice Nucleants

Substrate	Nucleation capability	Ar(−196°C) (σ = 16.6) (m^2/gm)	H₂O(25°C) (σ = 10.5) (m^2/gm)	$\dfrac{\Sigma_{H_2O}}{\Sigma_{Ar}}$	$\dfrac{\Sigma_{i\text{-propanol}}}{\Sigma_{Ar}}$	$\dfrac{V_m\, i\text{-propanol}}{V_{m_{H_2O}}}$
Cabosil	Poor	187	42.9	0.23	0.79	1.09
Dehydrated HiSil, 1	Good	113	57.8	0.51	0.82	0.51
Dehydrated HiSil, 2	Good	104	51.6	0.50	0.69	0.44

Figure 17 gives the classic nucleation efficiency curve of AgI due to Katz (51). The onset temperature is −4°, and the curve rises steeply as the temperature in the cloud chamber used for testing is lowered. At low enough temperatures almost any solid particle becomes a nucleant. Thus, at very low temperature, at the far right, even a wet precipitated silica, HiSil 233, becomes a nucleant. But importantly, simply heating HiSil at an elevated temperature produced a nucleant with much improved efficiency as indicated by the second curve from the right. When the silica was heated in the presence of soluble salts like NaCl or Ag NO₃, even better results were achieved. The best sample rivaled that of AgI as shown by the curve which coincides over part of the temperature range in Fig. 17. The quest for understanding of what further detailed surface structure is needed for efficient nucleation continues to prove fascinating.

That the better "synthetic" nucleants have similar adsorptive properties to AgI is indicated by Table XII. Higher adsorption at higher temperature and the resulting similar thermodynamic properties are indicated by what has been presented before. Curiously, however, numerous silica samples dehydrated to the same degree varied markedly in nucleating ability. Some additional difference in surface properties had to be sought.

New-neighbor or patchy sites. It was recalled that sterically hindered amines had been used to discover the closeness of the spacing between acid sites on cracking catalysts (61). Why not try isopropanol as a probe for the spacing between the residual

hydroxyls? The answer obtained for numerous samples is summarized for three examples in Table XII. Cabosil would appear to possess a proper surface concentration of —OH groups as manufactured. For this flame-hydrolyzed product, an apparently appropriate hydrophilic/hydrophobic ratio was found from the ratio of apparent water area/total area. Contrariwise, it is a poor nucleant. The isopropanol results suggest the cause. The better silica nucleants have neighboring hydroxyls, on the average two together, so the isopropanol gage of hydroxyl population gives roughly half that measured by the much smaller water molecule. In other words, isolated hydroxyls do not furnish efficient nucleation sites on silica surfaces.

Infrared Studies. Infrared studies both of the hydroxyls and of water adsorption on the silica nucleants support the conclusions reached from the isopropanol adsorption measurements. The work was performed by Hamilton and Katsanis in our Laboratory (62). The IR beam was passed directly through pressed discs made from the powders. The discs were held in a cell so that the relative humidity over the samples could be varied at will. Figure 25 displays several interesting spectra carefully developed in the—OH stretch region each at the same initial intensity.

The contrast between the IR spectra for the poor nucleant and the good nucleant is striking. The flame-hydrolyzed Cabosil has a hydrophilic/hydrophobic site ratio of about 1/4, similar to the usual silver iodide surface, whereas the dehydrated HiSil has a ratio of about 1/1.7. The Cabosil is also more

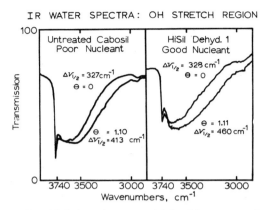

IR WATER SPECTRA: OH STRETCH REGION

FIG. 25. Infrared spectra obtained from pressed discs of Cabosil and dehydrated HiSil 1 under high vacuum, $\theta = 0$, and at $\theta = 1.1$ monolayers. Of particular significance are the relative intensity of the broad peak between 3700 and 3200 cm^{-1} of the dehydrated HiSil 1 corresponding to hydrogen-bonded, clustered surface hydroxyl groups. The isolated hydroxyl peak at 3740 cm^{-1} on the Cabosil fails to show any interaction with the water up to a coverage of four monolayers.

transparent to IR radiation than is the HiSil, as is also apparent in the visible light region, but that is not the main concern here. Most important is the fact that the poor nucleant has a much stronger absorption peak at 3740 cm^{-1}, representing isolated hydroxyls, than the broader neighboring band, representing interacting hydroxyls. The opposite is the case for the good nucleant, for which the interacting hydroxyls are more populous than the isolated ones.

In each case, curves are also given at 1.1 θ coverage with adsorbed water. The isolated hydroxyl peak for the poor nucleant (Cabosil) was not altered in accord with Kiselev's finding that water does not prefer these isolated species. On the other hand, the good nucleant (dehydrated HiSil) yielded a diminution in the isolated hydroxyl peak upon adsorption of water. The hydroxyl surface concentration is so great in this case that water adsorption leads to an interaction even with the previously isolated hydroxyls; a kind of overlapping sets in.

The IR technique indeed confirms the findings from the isopropanol adsorption studies. More importantly, the IR technique is more informative, more quantitative, and simpler to perform.

FHH plots—Strong Interactions in Multilayers.—One more finding has been added to the story of nature's efforts to cause rainfall by the study of the s values from the FHH plots. It would be expected that adsorption above the first layer would be most important since the water vapor is supersaturated with respect to ice under nucleation conditions. Here is just where the FHH plots are effective. We have already touched on what has been found. Consider Table VIII again. If s is about 1.9 for the water isotherm, good nucleation results are obtained. At 1.4 and at the other extreme of 2.5, poor results are obtained. Neither isolated sites nor a completely nonspecific surface will provide effective surfaces. Heterogeneous surfaces that are effective, however, extend their force fields to deeper layers. Nature arranges that the nucleant has weak forces at low coverages but extends its interaction more forcibly into the multilayer region. Cluster formation around groups of sites otherwise isolated from other groups appears to be established as an important part of the mechanism by which the embryo grows over the critical size. Weak adsorption of highly entropic water molecules is an additional criterion.

These silica nucleants are essentially amorphous so that there is little chance that epitaxy plays a role in contributing to their activity. It should be mentioned that some trend toward crystallinity has been detected (35), but the temperatures and times to which the silicas were exposed were higher than optimum, and in its basal planes the cristobalite structure developed doesn't match ice well. On the other hand, the importance of epitaxy in the nucleation efficiency of silver iodide should not be ruled out. Different nucleants may operate in different ways.

Geometric Heterogeneities

The graphitized blacks provide quite homogeneous surfaces, as we have seen, for the study of interfacial phenomena at hydrophobic surfaces. They have enjoyed wide use in the past and no doubt their use will increase. We have seen that the two most popular, Graphon and P-33, differ widely in surface area: 120 versus 10 m^2/gm. The hydrophilic sites on both are so infrequent that adsorption over the matrix, when it occurs, is not affected. A look at the surface concentration of physical heterogeneities is necessary to complete the current appraisal of the situation.

Graham (63) studied nitrogen adsorption on two graphitized blacks, Graphon and P-33, at low relative pressure and at several temperatures. A linear or Henry's law region was found in each case, but with a knee or Type II character at quite low pressures. This knee was attributed to strong geometric sites: 1.25% of the total sites on Graphon and 0.10% on P-33. The isosteric heat of adsorption for these sites was found to be about 4 kcal/mole, or almost twice that on the predominant sites. There was no indication of more than one type of strong site. The suggestion arose quite naturally, then, that the strong sites are composed of positions such as steps or regions of close approach between particles where interaction of one molecule to two planes can occur.

Electron micrographs of P-33 supported the latter contention at first glance. The particles are made up of doubly truncated polygonal bipyramids. One possibility was that the intersection of the planes in the surface was responsible for the strong sites. The fraction of strong sites was far too low, however, to be expained away by this effect. In accord with the earlier suggestion of Kmetko (64), the planes are apparently joined to each other by valence bonds so that they do not provide the double interaction.

The surface nonuniformity at levels of 1% or less might not be expected to produce interesting effects. On the contrary, the adsorption of CF_4 at 90.5°K shows a two-dimensional condensation on P-33, but not on the less uniform Graphon. On the other hand, ethane at the same temperature displays two-dimensional condensation on both graphitic blacks; the lateral interaction between the ethane molecules is greater than for the CF_4 molecules.

ACKNOWLEDGMENTS

Very helpful discussions were held with Dr. W. C. Hamilton and J. Lavelle of this Laboratory. Dr. Hamilton and the author had many enjoyable sessions studying current knowledge of hydrophobic surfaces, and particularly the nature of the FHH plots of adsorption on such surfaces.

Some of this work was supported by NSF Grant G.A. 560X.

REFERENCES

1. Young, Thomas, *Phil Trans. Roy. Soc.* p. 84 (1805).
2. Johnson, R. E., Jr., *J. Phys. Chem.* **63**, 1655 (1959).
3. Test Method of American Assoc. of Textile Chemists and Colorists.
4. Wenzel, R. N., *Ind. Eng. Chem.* **28**, 988 (1936).
5. Wenzel, R. N., *J. Phys. & Colloid Chem.* **53**, 1466 (1949).
6. Baxter, S., and Cassie, A. B. D., *J. Textile Inst.* **36**, T67 (1945); Cassie, A. B. D., and Baxter, S., *Trans. Faraday Soc.* **40**, 546 (1944).
7. Dettre, R. H., and Johnson, R. E., Jr., *Symp. Contact Angle, Bristol, 1966.*
8. Ellison, A. H., and Zisman, W. A., *J. Phys. Chem.* **58**, 260 (1954).
9. Johnson, R. E., Jr., and Dettre, R. H., *J. Phys. Chem.* **68**, 1744 (1964).
10. Dettre, R. H., and Johnson, R. E., Jr., *J. Phys. Chem.* **69**, 1507 (1965).
11. Shafrin, E. G., and Zisman, W. A., *J. Phys. Chem.* **64**, 519 (1960).
12. Leslie, J., *Tilloch's Phil. Mag.* **14**, 201 (1802).
13. Parks, G. J., *Phil. Mag.* **4**, 240 (1902).
14. Brunauer, S., Emmett, P. H., and Teller, E., *J. Am. Chem. Soc.* **60**, 309 (1938).
15. Chessick, J. J., Zettlemoyer, A. C., Healey, F. H., and Young, G. J., *Can. J. Chem.* **33**, 251 (1955).
16. Chessick, J. J., and Zettlemoyer, A. C., *Advan. Catalysis* **11**, 263 (1959).
17. Bangham, D. H., and Razouk, R. I., *Trans. Faraday Soc.* **33**, 1459 (1937).

18. HARKINS, W. D., "The Physical Chemistry of Surface Films," Chapt. 2. Reinhold, New York, 1952.

19. FOWKES, F. M., *Ind. Eng. Chem.* **56**, [12], 40 (1964). Marchessault, R. H., and Skaar, S., "Surface and Coatings Related to Paper and Wood," Chapt. 5. Syracuse University Press, 1967.

20. GRIFALCO, I. A., AND GOOD, R. J., *J. Phys. Chem.* **61**, 904 (1957).

21. HARKINS, W., AND LOESSER, E. H., *J. Chem. Phys.* **18**, 556 (1950).

22. ZISMAN, W. A., *Ind. Eng. Chem.* **55**, 19 (1963).

23. THELEN, E., *J. Phys. Chem.* **71**, 1946 (1967).

24. WHITE, M. L., AND DROBEK, J., *J. Phys. Chem.* **70**, 3432 (1966).

25. BEWIG, K. W., AND ZISMAN, W. A., *J. Phys. Chem.* **69**, 4238 (1965).

26. HILDEBRAND, J. H., AND SCOTT, R. L., "The Solubility of Non-electrolytes," 3rd ed. Reinhold, New York, 1950.

27. SCHONHORN, H., *J. Phys. Chem.* **70**, 4086 (1966); *Nature* **210**, 896 (1966).

28. DETTRE, R. H., AND JOHNSON, R. E., JR., *J. Colloid and Interface Sci.* **21**, 367 (1966); *J. Phys. Chem.* **71**, 1529 (1967); Starkweather, H. W., *SPE Trans.*, p. 5 (Jan., 1965).

29. WHALEN, J. W., AND WADE, W. H., Paper presented at the 41st National Colloid Symposium, Buffalo, New York, June 1967.

30. ALLEN, A. J. G., AND ROBERTS, R., *J. Polymer Sci.* **39**, 1 (1967).

31. DERJAGUIN, B. V., AND ZORIN, A. M., *Proc. Intern. Congr. Surface Activity 2nd London, 1957*, **II**, 145.

32. ADAMSON, A. W., AND LING, I., *Advan. Chem. Ser.* **43**, (1964)

33. BOYD, G. E., AND LIVINGSTON, H. K., *J. Am. Chem. Soc.* **64**, 2383 (1942).

34. HARKINS, W. D., "The Physical Chemistry of Surface Films," p. 290. Reinhold, New York, 1952.

35. BASSETT, D. R., Ph.D. Thesis, Lehigh University, 1967.

36. PIERCE, C., *J. Phys. Chem.* **63**, 1076 (1959); Pierce, C., *J. Phys. Chem.* **64**, 1184 (1960); Pierce, C., and Ewing, B., *J. Am. Chem. Soc.* **84**, 4070 (1962).

37. SHULL, C. G., *J. Am. Chem. Soc.* **70**, 1405 (1948).

38. BOWERS, R., *Phil Mag.* [7] **44**, 4676 (1953).

39. ZETTLEMOYER, A. C., CHAND, AMIR, AND GAMBLE, ERNEST, *J. Am. Chem. Soc.* **72**, 2752 (1950).

40. GRAHAM, D., *J. Phys. Chem.* **68**, 2788 (1964).

41. HALSEY, G., *J Chem. Phys.* **16**, 931 (1948).

42. JURINAK, J., *J. Soil Sci. Soc. Am. Proc.* **27**, 269 (1963).

43. SKEWIS, J. D., AND ZETTLEMOYER, A. C., *Proc. 3rd Intern. Congr. Surface Activity*, **IIB**, p. 401 (1960).

44. ZETTLEMOYER, A. C., SKEWIS, J. D., AND CHESSICK, J., *J. Am. Oil Chemists Soc.* **39**, 280 (1962).

45. PAPPENHEIMER, J. R., LEPIE, M. P., AND WYMAN, J., *J. Am. Chem. Soc.* **58**, 1851 (1936).

46. KATSANIS, E. P., Undergraduate thesis, Lehigh University, Center for Surface and Coatings Research, 1967.

47. YOUNG, G. J., CHESSICK, J. J., AND HEALEY, F. H., *J. Phys. Chem.* **60**, 394 (1956).

48. IYER, S. R., ZETTLEMOYER, A. C., AND NARAYAN, K. S., *J. Phys. Chem.* **67**, 2112 (1963).

49. GRAHAM, D., *J. Phys. Chem.* **61**, 1210 (1957).

50. ZETTLEMOYER, A. C., YOUNG, C. J., CHESSICK, J. J., AND HEALEY, F. H., *J. Phys. Chem.* **57**, 649 (1953).

51. ZETTLEMOYER, A. C., TCHEUREKDJIAN, N., AND HOSLER, C. L., *J. Appl. Math. Phys.* (Zamp), **14**, 496 (1963).

52. VONNEGUT, B., *J. Appl. Phys.* **18**, 593 (1947).

53. BIRSTEIN, S. J., *J. Meteorol.* **12**, 324 (1954).

54. CORRIN, M. L., EDWARDS, H. W., AND NELSON, J. A., *J. Atmospheric Sci.* **21**, 565 (1964).

55. BASSETT, D. R., Ph.D. Thesis, Lehigh University, Center for Surface and Coatings Research, Bethlehem, Pennsylvania, 1967.

56. TCHEUREKDJIAN, N., ZETTLEMOYER, A. C., AND CHESSICK, J. J., *J. Phys. Chem.* **68**, 773 (1964).

57. HAMILTON, W. C., Private communication.

58. COOLIDGE, A. S., *J. Am. Chem. Soc.* **49**, 708 (1927).

59. HILL, T. L., EMMETT, P. H., AND JOYNER, L. G., *J. Am. Chem. Soc.* **73**, 5102 (1951).

60. GRAHAM, D., *J. Phys. Chem.* **60**, 1022 (1956).

61. ZETTLEMOYER, A. C., AND CHESSICK, J. J., *J. Phys. Chem.* **64**, 1131 (1960).

62. HAMILTON, W. C., KATSANIS, E. P., AND ZETTLEMOYER, A. C., *Proc. 1st Natl. Conf. Weather Modification, Albany, N.Y., April 1968.*

63. (a) GRAHAM, DONALD, *J. Phys. Chem.* **61**, 1310 (1957); (b) GRAHAM, DONALD, *J. Phys. Chem.* **62**, 1210 (1958); (c) GRAHAM, DONALD, AND KAY, WALTER S., *J. Colloid Sci.* **16**, 182 (1961).

64. KMETKO, E. A., *Proc. Conf. Carbon 1st and Buffalo 1953 1955*, p. 21 (1956).

1969

TERRELL L. HILL

UNIVERSITY OF CALIFORNIA, SANTA CRUZ
SANTA CRUZ, CA.

. . . for his penetrating contributions to the rigorous molecular theory of surface phenomena, multilayer gas adsorption, and the structure of interfaces; for his profound monographs on statistical mechanics which have guided his colleagues to the past and the future; and for his steady judgment of institutions and men.

The following article, selected by Prof. Hill as very similar to his Kendall Award address, is reproduced by permission from *Proceedings of the National Academy of Sciences,* **61**, 98 (1968)

ON THE SLIDING-FILAMENT MODEL OF MUSCULAR CONTRACTION, II*

By Terrell L. Hill

DIVISION OF NATURAL SCIENCES, UNIVERSITY OF CALIFORNIA (SANTA CRUZ)

Communicated July 15, 1968

In recent years, a number of physical chemists have suggested that muscular contraction could be understood at the molecular level in terms, essentially, of the equilibrium properties of a macroscopic elastic system of one kind or another.[1-8] This line of thought might be said to have reached its logical conclusion in a very recent note by Hill[9] (part I), in which it is shown that the sliding-filament model itself, viewed as an equilibrium macroscopic polymeric system, exhibits elasticity and, possesses the very desirable feature of a phase transition.[7, 8]

However, the recent paper by Gordon, Huxley, and Julian[10] and the earlier work of Ramsey and Street[10] make it quite clear that an equilibrium phase transition cannot be the basic feature in muscular contraction. For it is observed experimentally and unambiguously that in the region of simple actin-myosin overlap, the isometric force that can be generated by the actin-myosin filament system *decreases* linearly with increasing length (the force is proportional to the extent of overlap). In contrast, an equilibrium phase transition system would show a free energy that increases linearly with length and a *constant force*. Further comments on a related topic will be contained in the appendix to part IV.[11]

A second very convincing point is that a phase transition system would show considerable and random fluctuations[9] in the lengths of the sarcomeres of a single myofibril, whereas these lengths are observed experimentally to be very uniform.[10]

Thus, although a macroscopic sliding-filament system would show a phase transition as part of its hypothetical *equilibrium* behavior, this is a property of little concern in the physiological problem of muscular contraction. In active isotonic or isometric contraction, the actin-myosin filament structure seems to be, in essence, a cyclic steady-state system which cannot be analyzed by equilibrium methods.

Outline of the Study.—The general type of mechanism suggested by A. F. Huxley[12] requires that the myosin cross-bridge experience an improbable fluctuation in configuration in order to be able to exert a significant force *after* becoming attached to actin. The main purpose of a subsequent paper with G. M. White (ref. 11, part III) will be to examine this question and suggest that the necessary fluctuations do *not* imply an impossibly low rate of attachment of myosin to actin.

A second significant result, to be emphasized here, is that the *free energy* (or partition function) of various cross-bridge substates, the *force* exerted in some of these states, and the interstate *rate constants* are parameters that are not independent of one another. Self-consistency must be imposed.

The third principal conclusion to be reached is that although the *average* isometric force exerted by individual cross-bridges at *equilibrium* is zero, this quantity would, in general, *not* be zero at steady state. Furthermore, it will be suggested that at a steady state far from equilibrium, the force exerted by the individual cross-bridges should far exceed the macroscopic (phase-transition) force.

In effect, then, this paper and its sequels (III and IV)[11] will be concerned with a reexamination of some aspects of A. F. Huxley's proposed type of cyclic mechanism.[12] Relatively general considerations will be presented here; numerical examples will be reserved for parts III and IV. For simplicity, following Huxley, we shall assume in these papers that it suffices to consider one cross-bridge and one actin site at a time. (This could prove to be a very restrictive assumption.)

The molecular details at a resolution of less than about 25–50 Å are not known with any certainty at present, so it will be necessary throughout to discuss exemplary possibilities rather than the "true" mechanism. But many general principles introduced below are independent of the particular mechanism.

Biochemical Cycles.—Although the biochemical details are not yet clear, Figures 1 and 2 include many of the possible cyclic sequences that might be involved in active isometric or isotonic contraction. For example, the particular cycle used (in effect) by A. F. Huxley[12] is designated by solid arrows in Figure 2. A cycle we shall consider below, as another example, is indicated by the solid arrows in Figure 1. In both figures, dashed lines show other possible transitions, which are assumed to be relatively unimportant. However, there is no difficulty, in principle, in including every transition explicitly with its corresponding rate constant (forward and backward).

A myosin cross-bridge (M) contributes to the force on the actin site (A) only while M and A are combined (top three states in each figure).

In isometric contraction, for a given position of the actin site relative to the origin of the myosin cross-bridge, the steady-state population of each state in Figure 1 or Figure 2 and the steady-state flux between any two states may be expressed in terms of the rate constants by means of a diagram method.[13] Then

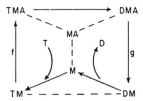

Fig. 1.—States and transitions for a single myosin cross-bridge and actin site.

(A) Actin site; (M) myosin cross-bridge; (T) adenosine 5′-triphosphate (ATP); and (D) adenosine 5′-diphosphate (ADP).

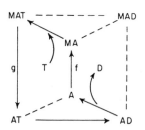

Fig. 2.—States and transitions for a single myosin cross-bridge and actin site. Symbols are as in Fig. 1.

240

to obtain gross properties, an average must be taken of the different relative actin-site positions. The isotonic contraction case is rather more complicated because the actin site is moving relative to the cross-bridge (ref. 11, part IV; ref. 12). Only the isometric case will be considered in the present paper.

A. F. Huxley[12] assumed implicitly, in the solid cycle of Figure 2, that the two steps labeled with rate constants f and g are rate-determining (i.e., relatively slow). The corresponding steps (i.e., attachment and detachment of M and A) are similarly labeled in Figure 1. The formal treatment of these two cases is the same.

Qualitatively, in using the solid cycle in Figure 1, we are assuming (a) that M does not bind readily to A unless T is previously bound, and (b) that M does not act as an enzyme for $T \rightarrow D$ until combined with A. Incidentally, in rigor (absence of T), the step $M \rightarrow MA$ would become significant as an alternate pathway.

It should be noted that the use of the cyclic sequences in Figures 1–3 implies that even under nonequilibrium conditions (time-dependent or steady-state), no further subdivision of states or rate constants is necessary in order to be able to describe the kinetic behavior of the system. That is, each state is assumed to be in "internal equilibrium" even though the different states are not in equilibrium with each other. This, in turn, implies that "internal" processes are fast compared to those shown in the figures. If for a given set of states we do not in fact have internal equilibrium (see ref. 11, part III), then the rate "constants" cannot be strictly constant.

Mean Isometric Force.—General considerations: Our object in this section is to use the cycle shown in Figure 3 to illustrate the principles involved in calculating the mean steady-state isometric force $\bar{F}$ contributed by a single myosin cross-bridge. Simple special cases will be considered in part IV of this series.[11]

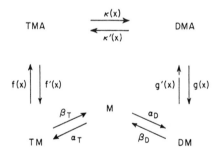

FIG. 3.—States and rate constants for a single myosin cross-bridge and actin site, at a specified value of x. Symbols are as in Fig. 1.

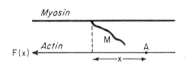

FIG. 4.—Schematic: myosin and actin filaments.

M, unattached cross-bridge; A, actin site; $F(x)$, force.

The Z-line is to the right of the figure.

Following A. F. Huxley,[12] we use the variable x to locate the position of the actin site A (Fig. 4) relative to a convenient origin on the myosin filament. The myosin may be considered fixed. The force is reckoned as positive in the direction indicated by the arrow in the figure. This would be the direction of motion of the actin filament relative to myosin in isotonic contraction. Force is exerted only when the cross-bridge M is attached to A; M spends only a fraction of its time so attached. $\bar{F}$ is the mean value of $F(x)$ averaged over an appropriate range (the distance between neighboring actin sites) of possible positions of A relative to the origin $x = 0$.

Although Figure 4, as drawn, implies that the dominant effect is a *pull* on A at positive values of x, it could just as well be a *push* at negative values of x. The fact that heavy meromyosin "arrowheads" on actin are observed to point away from the Z-line (which is to the right in Fig. 4) suggests a push, of course. However, no commitment one way or the other is needed here. Also, it is not necessary in this paper[14] to speculate on the possible molecular origins of this force (e.g., helix-random coil extension or compression of the cross-bridge, other kinds of configurational change in the cross-bridge, bending of the angle of attachment of the cross-bridge at either its myosin end or its actin end, distortion of ATP in the TMA state, etc.).

Figure 3 indicates that the cross-bridge M may exist in and make transitions among five possible states. The rate constants, or transition probabilities, are introduced in the figure. The α's and β's (adsorption or desorption of T or D onto or off M) are independent of x, but the other rate constants would, in the most general case, be functions of x.

Because of detailed balance between neighboring states at equilibrium, the rate constants are related to equilibrium partition functions. We therefore start with a digression on the equilibrium situation and return later to the steady-state case of primary interest.

For simplicity of notation, we shall treat the $T \rightleftarrows D$ reaction as if it were isomeric. For example, at equilibrium, we shall write $\lambda_T = \lambda_D$ ($\lambda = e^{\mu/kT}$ = absolute activity). Actually, λ_D includes a constant factor λ_P, and partition functions for D-states include a factor q_P (P = inorganic phosphate).

We denote the partition functions of the five states in Figure 3 by Q_M, Q_{TM}, $Q_{TMA}(x)$, $Q_{DMA}(x)$, and Q_{DM}. At equilibrium, with $\lambda_T = \lambda_D = \lambda$, the relative probabilities[15] of the five states are Q_M, $Q_{TM}\lambda$, $Q_{TMA}(x)\lambda$, $Q_{DMA}(x)\lambda$, and $Q_{DM}\lambda$. Thus, for example, the equilibrium probability of state M is

$$p_M{}^e(x) = Q_M/\xi(x), \tag{1}$$

where

$$\xi(x) = Q_M + [Q_{TM} + Q_{TMA}(x) + Q_{DMA}(x) + Q_{DM}]\lambda. \tag{2}$$

Q_M is the partition function of an unattached cross-bridge with neither T nor D bound to it. M is a small system[16] with length of order 150 Å. The length and other configurational properties of M fluctuate, of course (under zero force).

The partition function Q_{TM} includes a molecule of T bound on M. In the simplest case $Q_{TM} = Q_M q_T$, where q_T refers to the bound molecule only, but includes the energy (or free energy) of binding. But in general, we should expect the configurational states of M to be perturbed considerably by the binding of T. Then such a simple relationship would not hold. As in the case of Q_M, Q_{TM} would include configurations with different lengths, shapes, etc., fluctuating under zero force.

$Q_{TMA}(x)$ refers to a cross-bridge with T bound, which is itself attached to an actin site at x and is presumably held to the site by electrostatic, hydrogen-bond, or van der Waals' forces, or a combination of these. Because of the attachment at x, only those configurations of TM that are consistent with the given value of x are included in $Q_{TMA}(x)$. These configurations comprise a subclass of the states available to TM. This feature by itself would make $Q_{TMA}(x) < Q_{TM}$. On the other hand, TMA is stabilized relative to TM by the M-A bond energy (or free energy).

TM can attach to A at x to form TMA only when TM has fluctuated (under zero force) into the subclass of states consistent with x. The rather crucial kinetic problem implied here will be examined in part III of this series. The state of the T molecule in TMA may be quite different than in TM.

Rather analogous remarks could be made about Q_{DM} and $Q_{DMA}(x)$.

It is reasonable to expect (in order for the proposed mechanism to work well) rather weak binding of D to M; that is, for biologically important values of λ_D, $Q_M \gg Q_{DM}\lambda_D$. As is well known,[17] D is a much more stable molecule than T. For given λ_D, if T in solution were in equilibrium with D ($\lambda_T = \lambda_D$), the concentration of T would be about 10^{-5} times that of D. Hence, *at equilibrium*, the amount of binding of T would probably be much smaller even than that of D. Typically, at equilibrium we might expect, then, for relative probabilities [equations (1) and (2)],

$$Q_M \gg Q_{DM}\lambda, Q_{DMA}(x)\lambda \gg Q_{TM}\lambda, Q_{TMA}(x)\lambda. \qquad (3)$$

These inequalities would hold for easily accessible values of x. Of course, for extreme values of x, because of the improbable cross-bridge configurations required, $Q_{TMA}(x)$ and $Q_{DMA}(x)$ would approach zero.

Under biological steady-state conditions, the concentration of T could be as much as 10^7 or 10^8 times its equilibrium concentration.[17] This might well effectively saturate the myosin site with T. The order of relative steady-state probabilities would then be quite different than in (3) (see below).

When the cross-bridge is in the state $TMA(x)$ at equilibrium, the force exerted on the actin filament (positive in the direction shown in Fig. 4) is

$$F_{TMA}(x) = -kT \frac{\partial \ln Q_{TMA}(x)}{\partial x} = \frac{\partial A_{TMA}(x)}{\partial x}, \qquad (4)$$

where A_{TMA} is the corresponding Helmholtz free energy. This force may be positive or negative, depending on the value of x. There is, of course, a similar expression for $F_{DMA}(x)$. No force is exerted on the actin filament in the other states (M, TM, DM) because the corresponding Q's are independent of x. For a given value of x, the statistically averaged force (over the five states)[18] is then

$$F(x) = p_{TMA}{}^e(x) F_{TMA}(x) + p_{DMA}{}^e(x) F_{DMA}(x). \qquad (5)$$

For extreme values of x, the F's on the right-hand side of equation (5) will become large, but the p's will approach zero even more rapidly. Hence, $F(x) \rightarrow 0$. For example, suppose that

$$A_{TMA}(x) = A_0 + \frac{1}{2}\gamma(x - x_0)^2, \qquad F_{TMA}(x) = \gamma(x - x_0),$$

$$Q_{TMA}(x) = Q_0 \exp\left[-\frac{\gamma(x-x_0)^2}{2kT}\right], \tag{6}$$

where x_0 is the point of minimal free energy. Then

$$p_{TMA}{}^e(x)\, F_{TMA}(x) = Q_0\lambda\gamma\xi(x)^{-1}(x-x_0)\exp\left[-\frac{\gamma(x-x_0)^2}{2kT}\right] \to 0$$

$$\text{as } |x-x_0| \to \infty. \tag{7}$$

Equations (6) and (7) are practically thermodynamic in nature; no particular assumption about the cross-bridge configuration problem need be made.

Next, we note that $-kT \ln \xi(x)$ is the potential of the average force $F(x)$. That is, from equations (2) and (5),

$$F(x) = -kT \frac{\partial \ln \xi(x)}{\partial x}. \tag{8}$$

This has the following important consequence. Since every position x for the actin site is equally likely, $\bar{F}$, the mean value of $F(x)$, averaged over x, is proportional to $\int F(x)\, dx$, where the integral is taken over the entire range of values of x, say $x_1 < x < x_2$, in which $F(x)$ is nonzero. This integral is also a "work." But since

$$Q_{TMA}(x_1) = Q_{TMA}(x_2) = Q_{DMA}(x_1) = Q_{DMA}(x_2) = 0 \tag{9}$$

and hence

$$\xi(x_2) = \xi(x_1) = Q_M + (Q_{TM} + Q_{DM})\lambda,$$

we have

$$\int_{x_1}^{x_2} F(x)\, dx = -kT[\ln \xi(x_2) - \ln \xi(x_1)] = 0. \tag{10}$$

Therefore $\bar{F} = 0$: the positive and negative contributions to $\bar{F}$ cancel at equilibrium. Formally, this is a consequence of the existence of a potential function (with the same value at $x = \pm\infty$).

Thus, the cross-bridges, *acting individually*, would make no net contribution to the (equilibrium) thermodynamic force in a macroscopic system of actin and myosin filaments. This is what one might have expected intuitively since there is no way at equilibrium to get a "handle" or control (*inside* the system) on the individual cross-bridges that would induce them to pull or push, on the average, one way or the other.

The conventional thermodynamic force may, of course, be controlled by means of an equal and opposite *external* force applied at a *boundary* of the macroscopic system. This nonzero thermodynamic force is a macroscopic (phase transition) property of the whole system. In fact, this force would be (for the present model), per myosin filament, from equation (12) of reference 9,[19]

$$Fa/kT = \overline{\ln j_0} - \overline{\ln j_n} = \overline{\ln \xi} - \ln(Q_M + Q_{TM}\lambda + Q_{DM}\lambda), \tag{11}$$

where a is the distance along a myosin filament per cross-bridge (about 71 Å) and ln ξ is an unweighted average over x (range of integration equals distance between actin sites).

As we shall see, at steady state, the *individual* cross-bridges *can* generate a nonzero average force, $\bar{F}$. The "handle" in this case is the drive of the far-from-equilibrium $T \to D$ reaction operating through rate constants with suitably asymmetric x dependences.[20] Furthermore, the steady-state force generated *per cross-bridge*, far from equilibrium, may well be (see below) of the same order of magnitude as F in equation (11), which is a force *per myosin filament* (there are approximately 110 cross-bridges per one-half sarcomere per filament). Thus, the steady-state analogue of the macroscopic force F in equation (11) could be negligible, as seems to be the case experimentally.[10]

A numerical check: In view of the size of a cross-bridge and of the magnitude of the MA bond energy (6 kcal per mole maximum?), the right-hand side of equation (11) is probably of order 10. Then it is easily calculated that the force per cm² from this equation is too small (compared with experiment) by a factor of order 40.

To summarize: at equilibrium, those cross-bridges that are in the myosin-actin overlap region make a *direct* (additive) contribution to the appropriate free energy but not to the force. The effect on the force comes about *indirectly* through a change in the number of "overlapping" cross-bridges with a change in the length of the system. In effect, in a contraction of length a, only the *one* cross-bridge per myosin filament, which newly overlaps with actin (see equation (11)), contributes to the work and force. On the other hand, at steady state, *all* overlapping cross-bridges are *directly* involved in the force (though there will, of course, still be some positive-negative cancellation in $\bar{F}$ from different x values).

Next, we write the relations between the rate constants in Figure 3 imposed by the statistical-mechanical Q's already introduced. These relations follow from a consideration of detailed balance at equilibrium, but they apply as well under nonequilibrium conditions (recall the discussion of "internal equilibrium" above):

$$\alpha_T = Q_{TM} \beta_T \lambda_T / Q_M, \qquad \alpha_D = Q_{DM} \beta_D \lambda_D / Q_M \tag{12}$$

$$f(x) = Q_{TMA}(x)f'(x)/Q_{TM}, \qquad g'(x) = Q_{DMA}(x)g(x)/Q_{DM} \tag{13}$$

$$\kappa(x)/\kappa'(x) = Q_{DMA}(x)/Q_{TMA}(x). \tag{14}$$

Thus, for example, if some assumption is made about $f'(x)$ (e.g., $f' =$ constant) and the function $Q_{TMA}(x)/Q_{TM}$ is based on some macromolecular model,[11] then no arbitrariness is left in the assignment of $f(x)$.

Furthermore, even under nonequilibrium conditions (because of the "internal equilibrium" assumption), the forces $F_{TMA}(x)$ and $F_{DMA}(x)$ are still related to $Q_{TMA}(x)$ and $Q_{DMA}(x)$ as in equation (4). Thus, the Q's completely determine the forces associated with the different states and partially determine the rate constants.

Because there is only one cycle in Figure 3, it is a very simple matter, by using the diagram method,[13] to obtain explicit expressions for the steady-state prob-

245

abilities (and for the flux) as functions of the rate constants. We shall omit these equations, but we will examine special cases in part IV of this series. In terms of these steady-state probabilities, the steady-state force at x is

$$F(x) = p_{TMA}(x)\, F_{TMA}(x) + p_{DMA}(x)\, F_{DMA}(x). \tag{15}$$

The essential difference between this equation and equation (5) is that, in general, even after advantage is taken of equations (13) and (14), the x-dependent quantities appearing in the p's in equation (15) include not only $Q_{TMA}(x)$ and $Q_{DMA}(x)$ but also one or more x-dependent rate constants. Thus, in general, we must expect that $F(x)$ does not have a potential analogous to $-kT \ln \xi$ and that $\bar{F} \neq 0$.

As to the magnitude of the steady-state $\bar{F}$ from equation (15): complete cancellation gives $\bar{F} = 0$ at equilibrium, and we must therefore expect almost complete cancellation near equilibrium; but far from equilibrium, which is the case of biological interest ($\lambda_T \gg \lambda_D$), $\bar{F}$ (per cross-bridge) should be, according to equation (4), of roughly the same magnitude as F (per myosin filament) in equation (11). Two numerical examples will be included in part IV.

I am indebted to Drs. J. Gergely, A. Litan, W. Mommaerts, M. Morales, and R. Podolsky for helpful comments on this work.

* This work was supported in part by research grants from the General Medical Sciences Institute of the U.S. Public Health Service and from the National Science Foundation.

[1] Meyer, K. H., *Biochem. Z.*, **214**, 253 (1929).

[2] Kuhn, W., and A. Katchalsky, *Nature*, **165**, 514 (1950).

[3] Katchalsky, A., in *Progress in Biophysics*, ed. J. A. V. Butler and J. T. Randall (London: Pergamon Press, 1954), vol. 4.

[4] Riseman, J., and J. G. Kirkwood, *J. Am. Chem. Soc.*, **70**, 2820 (1948).

[5] Polissar, M. J., *Am. J. Physiol.*, **168**, 766, 782, 793, 805 (1952).

[6] Morales, M. F., and J. Botts, *Discussions Faraday Soc.*, **13**, 125 (1953).

[7] Hill, T. L., *J. Chem. Phys.*, **20**, 1259 (1952); Hill, T. L., *Discussions Faraday Soc.*, **13**, 132 (1952).

[8] Flory, P. J., *Science*, **124**, 53 (1956).

[9] Hill, T. L., these PROCEEDINGS, **59**, 1194 (1968) (part I in this series).

[10] Gordon, A. M., A. F. Huxley, and F. J. Julian, *J. Physiol.*, **184**, 170 (1966); Ramsey, R. W., and S. F. Street, *J. Cellular Comp. Physiol.*, **15**, 11 (1940).

[11] Hill, T. L., and G. M. White, these PROCEEDINGS, in press (parts III and IV in this series).

[12] Huxley, A. F., in *Progress in Biophysics*, ed. J. A. V. Butler and B. Katz (New York: Pergamon Press, 1957), vol. 7.

[13] Hill, T. L., *J. Theoret. Biol.*, **10**, 442 (1966); Hill, T. L., *Thermodynamics for Chemists and Biologists* (Reading, Mass.: Addison-Wesley, 1968), chap. 7.

[14] To be discussed in part III of this series.

[15] Hill, T. L., *Introduction to Statistical Thermodynamics* (Reading, Mass.: Addison-Wesley, 1960), pp. 140–143.

[16] Hill, T. L., *Thermodynamics of Small Systems, Part I* (New York: W. A. Benjamin, 1963).

[17] Lehninger, A. L., *Bioenergetics* (New York: W. A. Benjamin, 1965), pp. 54 and 55.

[18] Hill, T. L., *J. Am. Chem. Soc.*, **78**, 3330 (1956), Section II; see also ref. 15, p. 358.

[19] In equation (12) of ref. 9, let $x = 1$ (since $y = x^C$) to obtain the force at the phase transition. Averaging was implicit in part I, but is explicit here. Both notations are shown in equation (11) ($j_0 \equiv \xi$, etc.).

[20] This is reminiscent of active transport where $T \rightarrow D$ (again with directional asymmetry) drives the Na^+ transport across a membrane. See ref.13; and also Hill, T. L., and O. Kedem, *J. Theoret. Biol.*, **10**, 399 (1966), especially pp. 432–441.

1970

JEROME VINOGRAD

CALIFORNIA INSTITUTE OF TECHNOLOGY
PASADENA, CA.

. . . for his penetrating contributions to the theory and practice of equilibrium buoyant density ultracentrifugation, and the advances in our knowledge of deoxyribonucleic acids to which this had led; in particular to his studies of circular and catenated forms of these macromolecules, and in recognition of his help to other investigators.

The following article, selected by Prof. Vinograd as very similar to his Kendall Award address, is reproduced by permission from *Engineering and Science Magazine,* November 1968. The references were selected by Prof. Vinograd for this volume.

CENTRIFUGES, CIRCLES, AND CANCER

By JEROME VINOGRAD

Molecular biology has among its goals the understanding of the molecular events that occur in living cells. Centrifugation, circular DNA, and cancer have played major roles in this exciting new science.

Centrifugation is one of the principal tools of the molecular biologist. It is used to isolate and characterize cellular macromolecules and organelles. Circular DNA, because of its well-defined size and the absence of ends, sharply tests our knowledge of the cell's machinery for replicating DNA and expressing its information. Although the problems of cancer—the causes and cures—appear to be awaiting new discoveries in the molecular biology of normal cells, we already know that there are abnormalities in the chromosomes of patients having certain types of leukemia. We also know that when tumor viruses such as *polyoma* transform normal mouse cells into malignant cells, the originally circular viral DNA becomes an integral part of a long strand of nuclear DNA.

About ten years ago, three years after Watson and Crick described the duplex (double-stranded) structure of DNA, Matthew Meselson, then a graduate student in chemistry, approached me with a problem. Could light and heavy DNA molecules, differing in mass by 1 to 2 percent, be separated in a centrifugation experiment? If so, could a hybrid molecule of DNA containing one light progeny strand and one heavy parental strand also be distinguished? The presence of such hybrid molecules in the DNA of an organism initially grown in a medi-

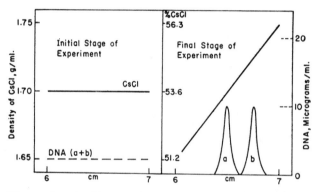

The results of a buoyant density experiment show how two DNA's (a and b), when centrifuged in a concentrated cesium chloride solution, move to regions of neutral buoyancy and form bands in the solution.

um containing heavy isotopes and then transferred for further growth to a medium containing normal isotopes would provide further direct evidence for the proposed duplex structure of DNA and for the Watson-Crick proposal that a DNA molecule replicates semi-conservatively (replicates, that is, to form two daughter molecules, each of which contains one of the original strands).

I replied that the resolving power of sedimentation velocity analyses as then practiced was inadequate, but that infinite resolution could be attained if one DNA species were held stationary while the other moved. I suggested that such a condition could be achieved by performing the sedimentation in concentrated salt solutions of high liquid density.

249

Each experiment begins with a change in DNA. With delicate specificity

the machinery of the cell amplifies and channels each change

until it reaches its functional expression and its harsh trial,

subject to the verdict of survival and reproduction or failure and extinction.

Thus in the cumulative laboratory of evolution has arisen the whole intricate pattern of life

which the mind of man now attempts to unravel.

Some probe the DNA itself.

Our first experiment—in the autumn of 1956 in the subbasement of Church Laboratory—reminded us that the salt itself, redistributing in the centrifugal field, would form a significant density gradient. In density gradients of this sort, DNA species move to regions of neutral buoyancy and there form bands. Dense DNA's form bands in the denser salt solution, and "light" DNA's form bands in the less dense salt solution. The transfer experiment, now thought of in terms of separating dense and less dense species as opposed to heavy and light masses, could clearly be done and was carried out by Meselson and Franklin W. Stahl in their now classic experimental validation of the Watson-Crick hypothesis.

The theory and practice of buoyant density centrifugation has, since its inception, been vigorously investigated in the Caltech laboratories. The density of the macromolecular complex at band center, a quantity now known as the buoyant density, is numerically equal to the density of the solution at band center and is readily measured with high precision and accuracy. Proteins, carbohydrates, DNA, and RNA exhibit widely different buoyant densities and are therefore very easily separated by this procedure. Nucleoproteins such as viruses normally form bands at densities that correspond to the weight fraction of the nucleic acid. DNA's of differing base composition have different buoyant densities. The buoyant method has become a favorite procedure for the analysis of the base composition

of DNA. The *Handbook of Biochemistry* lists the buoyant densities of some 300 DNA's from various organisms and viruses.

Our interest in circular duplex DNA arose in 1963 when Roger Weil and I and Renato Dulbecco and Marguerite Vogt discovered that the DNA in the tumor-inducing virus *polyoma* occurred in the form of a new DNA structure—a closed circular duplex without ends. The experiments that led to this con-

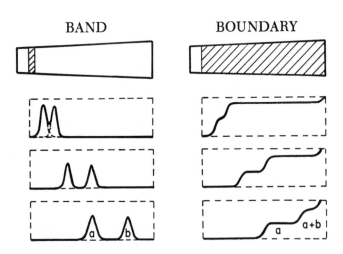

A comparison of the results of two methods of performing sedimentation velocity experiments shows the advantages of Caltech's new band method (left) over the boundary method (right). A smaller amount of DNA (indicated by striped lines) can be used. As sedimentation progresses (top to bottom frames), the concentration distributions of the DNA can be seen.

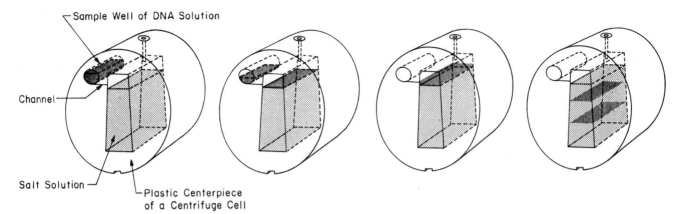

In a sedimentation velocity experiment a thin layer of a solution of DNA, stored in a sample well, flows out through a narrow channel under the influence of the centrifugal field and spreads out as a thin layer onto the surface of a salt solution. Diffusion of water from that layer into the slightly denser salt solution occurs rapidly and forms the density gradient necessary for stabilizing the band of DNA as it sediments through the salt solution. Both the DNA sedimentation and the necessary gradient formation occur almost simultaneously. The last drawing shows the position of two bands of DNA during the experiment.

clusion were performed by a new method that we had just evolved for performing sedimentation velocity experiments with very small amounts of DNA. The new method, called *band sedimentation in self-generating density gradients,* had two significant advantages over the classical boundary sedimentation experiments. Smaller amounts of DNA could be used, and fairly large quantities of contaminants did not interfere. The procedures underlying this tricky method are illustrated above, and a typical set of results is shown below.

A band sedimentation experiment performed with a mixture of two variants of λ virus. Photographs of the rotating centrifuge cell taken at four-minute intervals show bands which contain about five billion viruses each.

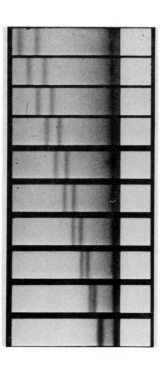

Closed circular duplex DNA consists of two, covalently closed, single-stranded DNA molecules that are interwound. The two strands share no atoms in primary bonding, yet they cannot be separated from each other without breaking a covalent bond. Such systems are said to contain a *topological bond.* Molecules containing interlocked circular submolecules are called catenanes. Organic chemists have enjoyed contemplating the various possibilities for stereoisomerism in catenanes. One catenane containing two rings of 30 carbons has been prepared and isolated. The DNA from *polyoma* virus contains approximately 5,000 nucleotides in each ring. The covalent backbone chains are interwound about 500 times to assume the normal DNA structure.

When closed duplex DNA is dissolved in common solvents, it assumes a structure in which the potential energy of the ensemble of atoms is at a minimum. The duplex winding number (the number of times one strand crosses over the other) in *polyoma* DNA, as in all other naturally occurring closed circular DNA's, turns out to be somewhat larger than the topological winding number obtained by counting the number of crossings when the helix axis lies in a plane. Although it is impossible to change the duplex winding number in a planar system, it can be changed if the helix axis is itself allowed to become helical so as to compensate exactly for the change in the duplex winding number. Closed circular DNA forms interwound superhelices, twisted molecules, in common solvents. The

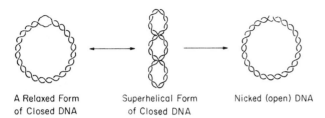

A Relaxed Form
of Closed DNA

Superhelical Form
of Closed DNA

Nicked (open) DNA

In these three forms of circular duplex DNA, the lines represent single strands of DNA. Closed circular DNA forms interwound superhelices (twisted molecules) in common solvents. The handedness of the superhelix allows us to conclude that the duplex was slightly underwound in the cell at the moment the last bond was formed.

handedness (the direction, clockwise or counterclockwise, in which the superhelix turns) of the superhelix allows us to conclude that the duplex was slightly underwound in the cell at the moment the last bond was formed (above).

Twisting of the molecule obviously requires an expenditure of energy which is partly stored in the molecule. The twisted molecule is like a spring, ready to unwind if the restraining forces are released. Untwisting occurs spontaneously if any one of 10,000 phosphate ester bonds in *polyoma* DNA is hydrolyzed. The duplex then spins around the rotatable chemical bonds in the intact strand opposite the single-strand break (the nick). The untwisted or relaxed molecule has a very different conformation. It moves through a solvent in band sedimentation experiments at a much slower rate

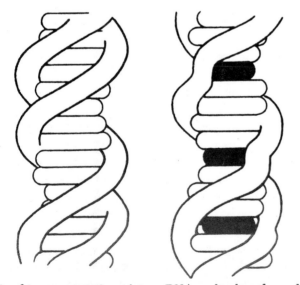

In this representation of two DNA molecules, the molecule on the right shows how drug (dye) molecules (black bands) are intercalated in the DNA molecule.

than the superhelical molecule. This effect is a useful tool in the study of the cell's machinery for "nicking" DNA—a process which must obviously occur if a closed circular molecule is to be replicated semiconservatively.

The duplex DNA can be unwound in a controlled way by using any one of a set of intercalating dyes —dyes that slip in between base pairs of DNA. These intercalators force the base pairs of DNA apart and unwind the duplex slightly. The effects of the controlled addition of the drug ethidium bromide are exactly as predicted from our understand-

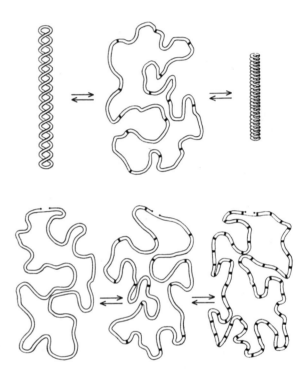

The effects of intercalating the drug ethidium bromide (black bands) into closed circular DNA are shown in the upper half of the diagram. First the drug relaxes the DNA, then it winds it up tighter. Binding the drug to nicked (open) circular DNA (lower part of the diagram) causes a rotation of the duplex at the site of the nick (the gap in the line). Each line represents a single strand of DNA.

ing of the topological properties of closed, double-stranded molecules. William Bauer, a Caltech graduate student in chemistry, recently showed that small amounts of this compound unwind and relax the intact molecule; larger amounts wind in superhelical turns that are left-handed instead of right-handed as in the native molecule. However, the nicked molecule merely spins at its rotation site, called a swivel, while binding the ethidium. It will

We are now faced with a large number of unsolved problems. Centrifugation, the chemistry of circular DNA, electron microscopy, hard work, and inspiration will all be needed to obtain solutions.

become saturated with the drug when one drug molecule is bound for every two base pairs. The relaxed DNA-ethidium complex will then have the appearance of a stack of sandwiches strung together in a circle, whereas closed molecules will wind up into a very tight spring with the free energy of spring formation opposing the binding of the ethidium. In consequence, less ethidium is bound by the closed molecules than by the nicked molecules.

The combination of the drug-binding by closed circular DNA and the buoyant density centrifugation method provides a very easy means to fish closed circular DNA out of mixtures which contain much larger amounts of linear DNA. The drug molecules are light and act like balloons attached to the denser DNA molecule. They cause the complex to move up in the gradient column to a region of lower density. The closed molecule, which takes up less drug, then has a higher buoyant density than do the

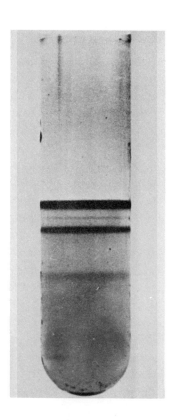

Three species of HeLa mitochondrial DNA are shown resolved in a cesium chloride-ethidium bromide density gradient. The dark upper band contains nicked (open) DNA, the narrow middle band contains singly nicked catenanes, the bottom band, closed DNA. The pale band in the lower part of the tube contains carbohydrate.

nicked or linear molecules. The bands shown in the test tube (below left) are easily separated into containers for further study.

This simple preparative method made it possible for us to investigate the occurrence and the properties of a type of closed circular DNA contained in almost all animal cells, beginning with protozoa and continuing to man. This particular DNA is contained in a cytoplasmic organelle called the mitochondrion. Mitochondria are responsible for the transfer of energy in oxidizable compounds to ATP (adenosine triphosphate), a key source of metabolic energy in the cell. The mitochondria are therefore often referred to as the cell's power plants. We do not yet understand why mitochondria have their own DNA genetic systems, spatially separated from the chromosomal DNA in the nucleus. Nor do we know the identity of the proteins specified by the information in the mitochondrial DNA. It is known, however, that certain genetic traits in yeast and molds are inherited through mitochondria.

When the mitochondrial DNA from the lower band in the test tube was examined in the electron microscope, Roger Radloff, formerly a Caltech graduate student in biology, made a surprising observation. Not only were there circles five microns in contour length (as had been reported by other researchers earlier), but there were also double, triple, and quadruple length circles in small amounts. Careful measurement of photographs of these molecules revealed crosspoints which divided the multiple length molecules into five micron subunits. Bruce Hudson, a graduate student in chemistry, showed that the double length molecules were catenated or interlocked pairs of closed circular duplexes. Such molecules are properly called catenated catenanes. If one submolecule is nicked and the other is closed, the restriction on drug-binding will be only half as large as in a simple closed molecule. The middle band in the test tube was highly enriched in singly nicked catenated dimers.

These structural studies were performed with

mitochondrial DNA obtained from HeLa cells, a line of human cancer cells that have been grown in tissue culture for more than 25 years. The biological implications of the interlocking of two or more sets of mitochondrial genes were puzzling. It was obvious, however, that we should try to find whether these structures were unique to HeLa cells or whether they occurred also in the mitochondria of normal and malignant tissues that had not been grown in tissue culture.

The first phase of this population survey is now almost complete and has taken a surprising turn. David Clayton, a graduate student in biology, has found that a large fraction of the mitochondrial DNA in the circulating white cells of patients with granulocytic leukemia are in the form of double length molecules that appear to have exactly twice the contour length of monomeric mitochondrial DNA (right). We call this kind of molecule a circular dimer, and tentatively we think of it as a villain. Chemotherapy appears to reduce the frequency of the circular dimer by a factor of about five. My collaborators, David Clayton, John Jordan, Charles A. Smith, Marlyn Teplitz, and Eric Wickstrom, have searched diligently but without success for a circular dimer among thousands of mitochondrial DNA molecules from various organs of healthy rabbits, guinea pigs, and rats. The circular dimer is also absent in mitochondrial DNA from immature, circulating white cells of patients with nonmalignant maladies that give rise to high white blood cell counts. In the course of this search we have, however, found catenanes in varying frequencies from 3 to 9 percent in every one of the mitochon-

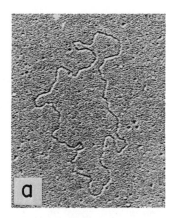

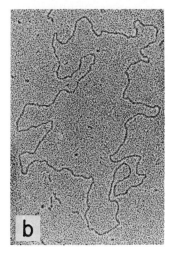

Electron micrographs of mitochondrial DNA from human leukemic white cells show (upper) a circular monomer five microns in contour length and (lower) a ten-micron circular dimer.

drial DNA samples. It may safely be said that catenanes are normal constituents of our cells.

We are now faced with a large number of unsolved problems. What is the mitochondrial DNA distribution in other kinds of leukemia and in solid tumors? How are the catenanes and circular dimers formed in the cell? How does the cell control their frequency? Do these molecules represent precise duplications of the mitochondrial DNA genes? If so, are excess gene products (proteins) formed? Are abnormal gene products formed? Finally, are we any closer to understanding the cancer problem? Is the correlation that we have so far observed between the occurrence of the circular dimer and granulocytic leukemia trivial or meaningful? Is the change in the size of the molecule an early event in an undifferentiated cell which gives rise after many cell divisions to white cells that do not mature properly and do not go about their job in an orderly and controlled way? Centrifugation, the chemistry of circular DNA, electron microscopy, hard work, and inspiration will all be needed to obtain answers to these questions.

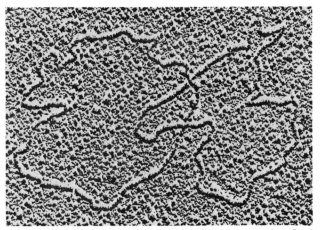

This electron micrograph shows a catenated mitochondrial DNA molecule from a human cell. The submolecules are connected like closed links in a chain.

ADDENDUM 1972

A list of references to the original articles is given for readers who may wish to investigate the material presented here in more detail.

Equilibrium Sedimentation of Macromolecules in Density Gradients
Matthew Meselson, Franklin W. Stahl, and Jerome Vinograd
Proc. Nat. Acad. Sci. USA, **43** (7), 581-588 (1957)

Equilibrium Sedimentation of Macromolecules and Viruses in a Density Gradient
Jerome Vinograd and John E. Hearst
Fortsch. Chem. Org. Naturs. (Progress in the Chemistry of Organic Natural Products), **XX**, 372-422 (1962)

Band-Centrifugation of Macromolecules and Viruses in Self-Generating Density Gradients.
Jerome Vinograd, Robert Brunner, Rebecca Kent, and Jean Weigle.
Proc. Nat. Acad. Sci. USA, **49** (6), 902-910 (1963)

The Cyclic Helix and Cyclic Coil Forms of Polyoma Viral DNA
Roger Weil and Jerome Vinograd
Proc. Nat. Acad. Sci. USA, **50** (4), 730-738 (1963)

The Twisted Circular Form of Polyoma Viral DNA
Jerome Vinograd, J. Lebowitz, R. Radloff, R. Watson, and P. Laipis
Proc. Nat. Acad. Sci. USA, **53** (5), 1104-1111 (1965)

A Dye-Buoyant Density Method for the Detection and Isolation of Closed Circular Duplex DNA: The Closed Circular DNA in HeLa Cells
Roger Radloff, William Bauer, and Jerome Vinograd
Proc. Nat. Acad. Sci. USA, **57** (5), 1514-1521 (1967)

The Interaction of Closed Circular DNA with Intercalative Dyes. I. The Superhelix Density of SV40 DNA in the Presence and absence of Dye
William Bauer and Jerome Vinograd
J. Mol. Biol., **33** (1), 141-172 (1968)

Complex Mitochondrial DNA in Leukemic and Normal Human Myeloid Cells
David A. Clayton and Jerome Vinograd
Proc. Nat. Acad. Sci. USA, **62** (4), 1077-1084 (1969)

1971

MILTON KERKER

CLARKSON COLLEGE OF TECHNOLOGY
POTSDAM, N. Y.

. . . for his experimental and theoretical studies of light scattering, in particular for coated spheres and cylinders, having great impact on meteorology for particle size analysis of colloids, and for multiple scattering to explain radiative transfer in the atmosphere and space, the hiding power of pigments and opalescence in fluids.

The following Kendall Award Address is reproduced by permission from *The Journal of Colloid and Interface Science,* **39,** 2 (1972). Copyright © 1972 by Academic Press, Inc.

Some Optical and Dynamical Properties of Aerosols[1,2]

MILTON KERKER

Department of Chemistry and Institute of Colloid and Surface Science, Clarkson College of Technology, Potsdam, New York 13676

Hover through fog and filthy air

William Shakespeare (1606)

1. INTRODUCTION

Interest in the atmosphere has had a most pervasive influence on the direction of aerosol research. Consider the following passage by Leonardo da Vinci (*ca.* 1500), which may be the first description of a laboratory aerosol:

> I say that the blueness we see in the atmosphere is not intrinsic color, but is caused by warm vapor evaporated in minute and insensible atoms on which the solar rays fall, rendering them luminous against the infinite darkness of the fiery sphere which lies beyond and includes it. . . . If you produce a small quantity of smoke from dry wood and the rays of the sun fall on this smoke and if you place [behind it] a piece of black velvet on which the sun does not fall, you will see that the black stuff will appear as a beautiful blue color. . . . Water violently ejected in a fine spray and in a dark chamber where the sun beams are admitted produces then blue rays. . . . Hence it follows, as I say, that the atmosphere assumes this azure hue by reason of the particles of moisture which catch the rays of the sun.

John Tyndall's celebrated experiments with aerosols formed by photochemical decomposition of organic vapors may mark the origin of aerosol science. The motivation for these experiments is clearly derived from an interest in the atmosphere for he has stated:

> . . . the blue color of the sky, and the polarization of skylight . . . constitute, in the opinion of our most eminent authorities, the two great standing enigmas of meteorology. Indeed it was the interest manifested in them by Sir John Herschel, in a letter of singular speculative power, that caused me to enter upon the consideration of these questions so soon. . . .
>
> Mixed air and nitrite-of-butyl vapour was permitted to enter [the tube and] the condensed beam of the electric light passed for some time in darkness through this mixture. . . . Soon, a superbly blue cloud was formed along the track of the beam, and it continued sufficiently long to permit of its thorough examination. . . . Now the instance cited here is *representative*. In all cases, and with all substances, the cloud formed at the commencement, when the precipitated particles are sufficiently fine is *blue*, and it can be made to display a colour rivalling that of the purest Italian sky. (*Phil. Mag.* 37, 384–394 (1869).)

Such examples can be multiplied many times over.

Our own interest has been in the elucidation of some optical and dynamical properties of aerosols, and certainly there is no pretense at tackling the real problems of meteorology; yet we hope to make clear the implications of our own work for under-

[1] Presented on the occasion of the 1971 American Chemical Society Award for Colloid or Surface Chemistry, sponsored by the Kendall Company, at the ACS Meeting, March 30, 1971, Los Angeles, CA.

[2] This investigation has been supported by the U. S. Atomic Energy Commission and the U. S. Public Health Service, Grants AP-00048 and AP-00743 by the National Air Pollution Control Administration.

259

standing the role of aerosols in the atmosphere. We will review three recent experiments: a study of Brownian coagulation (1), a study of the scavenging of aerosol particles by a freely falling water droplet (2), and a study of multiple scattering (3, 4). Each of these is treated at greater length in the cited references.

There are two experimental techniques utilized in this work which will be reviewed first in Parts 2 and 3. These are the laboratory preparation of aerosols, both liquid and solid, having a narrow distribution of sizes in the tenth micron range and the particle size analysis of such aerosols by light scattering. Then the above-mentioned experimental studies will be described in Parts 4, 5, and 6.

There is a common thread running through these studies which goes beyond their applicability to atmospheric phenomena. In each case the main physical processes can be identified in terms of clear concepts and they can be articulated quantitatively. However, the equations can be solved numerically only for the simplest models. For these, experiments should be able to discriminate among the possibilities. Unhappily, the phenomena are not always so simple.

The theorists have had their heyday with these problems over the past three to seven decades and have done the job well. On the other hand, there has been a dearth of experimental work and there is a long-standing need to redress the balance in favor of experiments. Our present efforts, which are still in progress, are in that direction. They can serve to elucidate the models and, when nature is particularly intractable, can provide at least a natural history and possibly some new insights.

2. PREPARATION OF SUBMICRON AEROSOLS BY CONDENSATION

The Sinclair–La Mer (5) aerosol generator made possible the laboratory preparation of aerosols consisting of a narrow distribution of liquid droplets with radii in the tenth micron range and opened up a new era of experimental work. Aerosol is formed upon cooling a mixture of inert

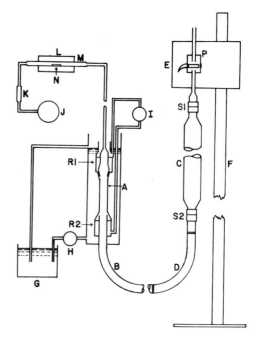

FIG. 1 Liquid aerosol generator, coagulation tube, and light-scattering photometer. Helium source J, flowmeter K, combustion furnace L, combustion tube M, combustion boat N, boiler tube A, reservoirs of dibutylphthalate R1, R2, tube pump I, constant-temperature oil bath G, circulating pump H, exit tube B, recondenser D, joints S1, S2, coagulation chamber C, light-scattering photometer E, column for photometer F, light-scattering cell P.

carrier gas, condensation nuclei, and condensable vapor in a flowing system. Cooling occurs sufficiently slowly that the supersaturation is relieved by condensation upon the nuclei and subsequent diffusion of vapor to the growing particles, rather than by successive bursts of homogeneous nucleation. The particle concentration is controlled by the concentration of nuclei, and the particle size by the ratio of condensable vapor to nuclei.

Figure 1 depicts our liquid aerosol generator (6, 7) which has evolved over a period of many years from the Sinclair La Mer generator. This generator formed the aerosol used in the coagulation experiments to be described in Part 4. There are three sections: the nucleator, the boiler, and the reheater.

In the nucleator, vapors of a salt, such as sodium chloride, are picked up by a stream of filtered helium from a reservoir of the salt contained in a boat (N) which is located in a combustion tube (M) maintained at 500–800°C in a combustion furnace (L). These vapors condense upon emergence from the nucleator to form an aerosol consisting of very small salt particles.

The helium stream, now laden with these nuclei, enters the boiler which is immersed in a constant temperature oil bath maintained at 85–120°C. The present generator has been operated mainly with dibutyl phthalate (DBP). This runs down the inner wall of tube A from reservoir R1 to reservoir R2, whence it is recycled through a variable-speed tubing pump I. The flow is controlled by a fluted ground-glass joint at the entrance to tube A. The DBP vapor condenses on the NaCl nuclei upon emergence from the heated region into tube B. This leads into tube D, the reheater, which is maintained at 150°C, by a strip of heating tape. This temperature is sufficient to evaporate the DBP, which then recondenses downstream in the cooler zone. The remainder of the apparatus as depicted here is set up for coagulation studies and will be discussed in Part 4.

The entire system is devoid of constrictions and sharp bends. Typical flow rates range from 1–2 liters/min. The flow is laminar with Reynolds numbers of the order of 10. The stability and reproducibility of this generator are excellent. The size distribution of the aerosol remains quite constant over the course of an experiment and when the generator is shut down and then started up on another occasion the same aerosol is obtained for the same operating conditions. The standard deviation of the particle size distribution ranges from 7–10 % of the mean size compared to 12–20 % for the best of the earlier generators.

The mean size increases with increasing boiler temperature since this gives a higher DBP vapor concentration. It decreases with furnace temperature since this in-

FIG. 2. Photograph of paraboloidal-shaped condensation zone. Illuminated region represents scattering of aerosol by light beam at $\theta = 90°$ in a plane parallel to the flow.

creases the nuclei concentration. It also decreases with flow rate. This is due in part to the decreasing mass transfer in the boiler from the flowing liquid to the vapor with a corresponding lower degree of saturation.[3] Also there is a higher concentration of nuclei at higher flow rates resulting

[3] A recent analysis of the convective diffusion has elucidated the concentration profile within the boiler as a function of position within the tube, flow rates of gas stream and liquid film, physical properties of the aerosol material and carrier gas, and temperature. Further analysis in the region of tube B follows the radial redistribution of the vapor to a uniform mixture (Davis, E. J. and Nicolaon, G., *J. Colloid Interface Sci.*, **37,** 768 (1971)).

possibly from the greater cooling rate upon emergence from the nucleator.

Under illumination, condensation downstream from the reheater can be seen to occur along the surface of a paraboloidal-shaped region as illustrated in Fig. 2. The effect is quite striking. The region upstream of the condensation zone appears optically void, whereas downstream the higher order Tyndall spectra characteristic of an aerosol of narrow size distribution is seen. The line of demarcation along the paraboloidal surface is very sharp, indicating that condensation occurs rapidly and that there is little further growth beyond the condensation zone.

The temperature profile, both radially and axially, has been measured downstream from the reheater and has also been calculated from heat-transfer theory. The measurements and theoretical results agree precisely with each other and demonstrate that condensation in the paraboloidal-shaped region occurs along an isotherm.

An important conclusion that can be drawn from this is that this is primarily an equilibrium process rather than one controlled by diffusion of vapor to the nuclei and the rate of heat transfer associated with condensation on the nuclei. If diffusion-controlled growth were a leading factor, the large residence time of nuclei in the region near the wall would lead to formation of aerosol at a different temperature than in the more rapidly moving center of the stream.

Presumably, condensation takes place when the temperature falls to a value where the supersaturation is appropriate to the particular nuclei. Assuming that the nuclei behave as smooth perfectly wetted spheres, it is possible to estimate their size from a knowledge of the supersaturation along the isotherm. Values of 25–40 Å in radius have been obtained, and preliminary electron microscope observations indicate that the nuclei are primarily in this size range. Indeed, this suggests the possibility of using this generator to study heterogeneous nucleation in much the same way that a diffusion cloud chamber is used to study homogeneous nucleation, and we plan to explore this more fully.

The silver chloride aerosol generator used in the scavenging experiments (2), which will be discussed in Part 5, is represented in Fig. 3. Three combustion furnaces were used to produce the aerosol in stages. The first served as a generator for

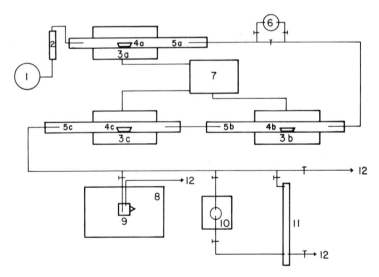

Fig. 3. Solid aerosol generator. Helium source 1, flowmeter 2, combustion furnaces 3a, 3b, 3c, combustion boats, 4a, 4b, 4c, combustion tubes 5a, 5b, 5c, viewing chamber 6, potentiometer 7, light-scattering photometer 8, light-scattering cell 9, thermal precipitator 10, filter tube 11, vent 12.

NaCl nuclei, the second produces a partially grown AgCl aerosol, and the third supplies additional AgCl vapor from which the final aerosol of narrow size distribution is formed. The salts were contained in combustion boats centered in the combustion tubes. The first furnace was usually maintained at about 700°C, the second furnace from 775 to 850°C, and the third furnace at a somewhat higher temperature from 775 to 900°C. The helium flow rate ranged from 1 to 3 liters/min.

The three furnaces in the generator play a role similar to the nucleator, boiler, and reheater in the liquid aerosol generator and the performance is similar. The particle size decreases with increasing flow rate or with increasing concentration of nuclei. It increases when the temperatures of the second and third furnaces are increased. Just as for the liquid generator, the size is controlled by the ratio of the concentration of condensable vapor (AgCl) to concentration of nuclei (NaCl). Stable aerosols can be obtained only with careful

voltage control, which in turn maintains constant furnace temperatures.

The aerosol particles, illustrated by the electron micrograph in Fig. 4, are spherical. They consist either of exceedingly small microcrystals or of a glassy material. The standard deviation of the particle size distribution ranged from 10 to 25% of the mean. Sodium chloride aerosols (8) can also be obtained with this generator. Coated aerosols consisting of AgCl cores coated with a concentric spherical shell of a liquid material (9) can be obtained by a hybrid of the solid and the liquid generators. Aerosols of vanadium pentoxide (10) were formed with a somewhat different system.

3. PARTICLE SIZE ANALYSIS BY LIGHT SCATTERING

We now turn to the determination of the distribution of particle sizes by light scattering (11). The object is to find a distribution function $p(\alpha)$ such that

$$P(\alpha) = N \int_{\alpha}^{\alpha+\Delta\alpha} p(\alpha)\, d\alpha \qquad [1]$$

gives the number of particles per unit volume with sizes between α and $\alpha + \Delta\alpha$. N is the total number of particles per unit volume. The function which we have used to describe the distribution of sizes is

$$p(\alpha) = \frac{\exp - [(\ln \alpha - \ln \alpha_M)^2/2\sigma_0{}^2]}{\sqrt{2\pi}\sigma_0 \alpha_M \exp(\alpha_0{}^2/2)}, \quad [2]$$

where α_M is the modal value of the size parameter and σ_0 is a parameter which measures the width of the distribution. The size parameter α is the ratio of the particle circumference ($2\pi a$) to the wavelength λ of the radiation.

The light-scattering quantities which are measured are the polarized components of the Rayleigh ratio $V_v(\theta)$ and $H_h(\theta)$. These denote the radiances scattered by a unit volume of aerosol in a particular direction θ for unit irradiance when the incident light is linearly polarized perpendicular and parallel, respectively, to the scattering plane. The scattering plane is defined by the incident and scattered directions. The scattering angle θ is between the transmitted and the

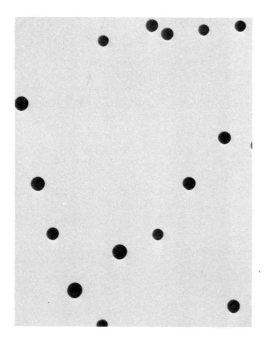

FIG. 4. Electron micrograph of silver chloride aerosol; modal radius $a_M = 0.183$ μm, width parameter $\sigma_0 = 0.08$ (see Eq. [2]).

scattered directions. Then

$$V_v(\theta) = \frac{\lambda^2 N}{4\pi^2} \int_0^\infty i_1 p(\alpha)\, d\alpha \qquad [3]$$

$$H_h(\theta) = \frac{\lambda^2 N}{4\pi^2} \int_0^\infty i_2 p(\alpha)\, d\alpha, \qquad [4]$$

where i_1 and i_2 are angular intensity functions which depend upon α, θ and the refractive index of the aerosol material. We also define the polarization ratio

$$\rho(\theta) = \frac{H_h(\theta)}{V_v(\theta)} = \left(\int_0^\infty i_2 p(\alpha)\, d\alpha \right) \bigg/ \left(\int_0^\infty i_1 p(\alpha)\, d\alpha \right). \qquad [5]$$

The problem is to determine α_M and σ_0. This is done by comparison of $\rho(\theta)$ obtained by measurement of the Rayleigh ratios at various angles with theoretical calculations of the expression on the right-hand side of Eq. [5] corresponding to various combinations of α_M and σ_0 for the appropriate refractive index. It is hardly an overstatement to note that this is a very long computation.

Figure 5 shows the effect upon $\rho(\theta)$ of varying σ_0 from 0.100 to 0.300 while keeping $\alpha_M = 5.0$ and the refractive index $m = 1.43$. The two extreme distribution curves are plotted in the inset. The radii plotted along the abscissa of the inset correspond to light of wavelength 0.546 μm ($a = \lambda\alpha/2\pi$).

For $\sigma_0 = 0.100$, the curve shows the typical oscillations characteristic of systems of rather narrow size distribution. However, this structured character is obliterated as the distribution becomes broader. It is the structure in these curves, acting as a kind of fingerprint, that permits the precise determination of the size distribution from the light-scattering data. Accordingly, as the distribution becomes increasingly broader, the precision with which a distribution can be fitted to the data becomes increasingly poorer.

One example using a DBP aerosol (12) is shown in Fig. 6, where the experimental values of $\rho(\theta)$ plotted as circles are compared with a theoretical curve which best agrees with them and which corresponds to $\alpha_M = 2.67$ and $\sigma_0 = 0.15$. These data obtained at

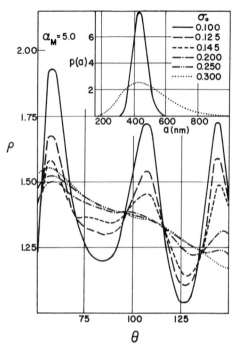

FIG. 5. Polarization ratio ρ versus scattering angle θ for $\alpha_M = 5.0$, refractive index $m = 1.43$ and $\sigma_0 = 0.100$, 0.125, 0.145, 0.200, 0.250, 0.300. Size distributions in the inset correspond to $\sigma_0 = 0.100$ and 0.300 for $\lambda = 0.546$ μm.

$\lambda = 0.546$ μm illustrate typically the ability to find a size distribution corresponding to the experimental results.

The internal consistency of the technique is illustrated by Fig. 7 which gives the experimental results for a particular aerosol at three wavelengths, $\lambda = 0.436$, 0.546, and 0.578 μm. The corresponding size distributions obtained from these data are shown in Fig. 8. Although the scattering curves are quite different, the size distributions are in excellent agreement, as they must be.

For any matching technique, such as used here, one must raise the question of the uniqueness of the solution. Granted that there is excellent concordance between experimental and calculated results, and also that there is internal consistency among results obtained at several wavelengths, the question still arises as to whether there might not be other size distributions that would fit these data equally well.

The error contour map in Fig. 9 illustrates

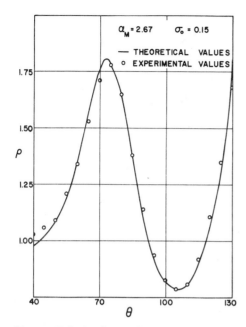

FIG. 6. Polarization ratio ρ versus scattering angle θ for a DBP aerosol. Smooth curve is from theory for $\alpha_M = 2.67$, $\sigma_0 = 0.15$, $\lambda = 0.546$ μm.

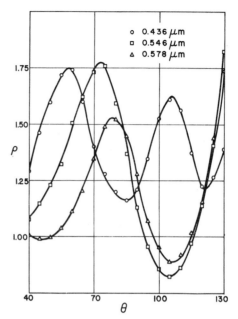

FIG. 7. Polarization ratio ρ versus scattering angle θ for DBP aerosol at $\lambda = 0.436$, 0.546, and 0.578 μm.

this point. The contour lines in the $\alpha_M - \sigma_0$ domain represent loci of equal values of the mean square of the deviation between experiments and calculations. The topography of the contour map shows a single deep well. The size distribution chosen to represent this aerosol is given by the center of the bottom of the well. Because this is the only well in the entire domain, this solution is unique. Figure 9 represents only a small section of the $\alpha_M - \sigma_0$ domain which was searched from $\alpha_M = 0.5$ to 9.9 and $\sigma_0 = 0.01$ to 0.19.

4. BROWNIAN COAGULATION

Colloidal dispersions are intrinsically unstable, an aspect which makes experimentation with them so difficult. An important dynamical process by which this instability may be relieved is coagulation, a process in which those particles not bouncing apart upon collision accrete or coalesce. The "aging" of the primary particles entering the atmosphere to give the aerosol normally encountered may involve coagulation.

Although Smoluchowski's theory of Brownian coagulation was published more than half a century ago (13), numerical solu-

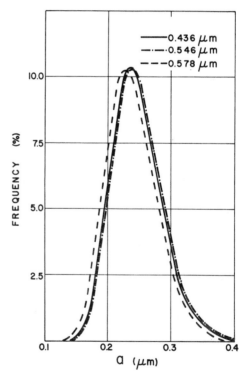

FIG. 8. Size distributions corresponding to scattering data in Fig. 7.

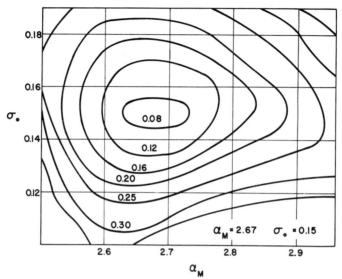

FIG. 9. Error contour map for results of Fig. 6.

tions to his nonlinear integrodifferential equation have only been obtained recently with the advent of electronic digital computers. Virtually all earlier work had been restricted to the initial rate of coagulation of a monodisperse system.

Experimental studies have lagged even further. Indeed, we do not believe there has been a definitive experiment in which the particle size distribution of a coagulating system has been followed and compared with Smoluchowski's theory. We would now like to describe such an experiment (1).

The apparatus is shown schematically in Fig. 1. Aerosol is formed in the generator which has been described in Part 2. Tube C is the coagulation chamber. It serves to hold up the aerosol sufficiently long for appreciable coagulation to occur. The hold-up time is varied by using tubes of different volumes.

The light-scattering photometer (E) and cell (P) were constructed according to our design. The main feature for this work is the ability to adjust its vertical position by moving it along column F. With aerosol flowing through the apparatus, the radiance of each of the polarized components of the scattered light was measured as a function of scattering angle prior to entry and after emergence from the coagulation chamber.

The strategy of the experiment is to calcu-late the particle size distribution of the aerosol from the light-scattering data obtained prior to entry into the coagulation chamber. This must occur early enough in the life history of the aerosol so that the distribution is sufficiently narrow and unimodal. The size distribution of this initial aerosol is then calculated as a function of time using Smoluchowski's theory. Theoretical light-scattering results corresponding to the distribution for the coagulated system can then be calculated with Eqs. [3]–[5], and these are compared with the experimental light-scattering data for the coagulated aerosol.

Presumably, if the calculated result can be fit to the experimental data, the coagulation mechanism proceeds in accordance with Smoluchowski's model. Furthermore, if the experimental and theoretical time scales agree, there is no potential barrier to coalescence upon collision. If the experimental time is greater, the collision efficiency is less than unity and the potential barrier can be calculated. If the experimental time is less than the calculated time, the aerosol is coagulating faster than predicted by Brownian diffusion so that other mechanisms must also be involved.

Clearly, the requirements for the success of this experiment are (a) the ability to prepare aerosols of narrow size distribution and

266

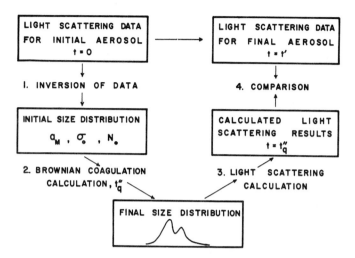

FIG. 10. Flow chart for coagulation calculation.

to handle them in such a way that Brownian coagulation is the only process occurring, and (b) the ability to determine *in situ* the size distribution of the aerosol at the outset and to follow its evolution. The new aerosol generator was designed to meet the first requirement by minimizing the possibility of gradient coagulation, turbulent coagulation, or wall loss (14). Light scattering provides the ability to meet the second requirement.

The calculation is outlined in Fig. 10. The light-scattering data obtained prior to entry and after exit from the coagulating chamber are depicted in the two boxes at the top of the figure. The first step is the inversion of the data in the left-hand box to obtain the size distribution parameters α_M and σ_0 with the aid of Eq. [5]. The particle concentration was obtained from the mass concentration m and the density of DBP ρ_L using

$$N = \frac{m}{\frac{4}{3}\pi\rho_L \int_0^\infty a^3 p(a)\,da}\qquad[6]$$

rather than directly from the light-scattering data using Eqs. [3] or [4]. The value of m is determined by direct weighing of a sample collected by thermal precipitation.

The second step is the calculation of the size distribution that evolves after the initial aerosol has undergone coagulation. The rate of coagulation is

$$\begin{aligned}
\frac{dn_h}{dt} =\ & \frac{kT}{3\eta}\int_{a_o}^{a_{h-o}} (a_i + a_{h-i}) \\
& \cdot \left[\frac{1}{a_i}\left(1 + \frac{A_i\ell}{a_i}\right)\right. \\
& \left. + \frac{1}{a_{h-i}}\left(1 + \frac{A_{h-i}\ell}{a_{h-i}}\right)\right] \\
& \cdot \left(\frac{a_h}{a_{h-i}}\right) n_i n_{h-i}\,da_i \qquad[7] \\
& - \frac{2kT}{3\eta}\,n_h\int_{a_o}^{a_f}(a_h + a_i) \\
& \cdot \left[\frac{1}{a_h}\left(1 + \frac{A_h\ell}{a_h}\right)\right. \\
& \left. + \frac{1}{a_i}\left(1 + \frac{A_i\ell}{a_i}\right)\right]n_i\,da_i
\end{aligned}$$

where n_h is the concentration of particles of size $a_{h'}$ k is the Boltzmann constant, T is the Kelvin temperature, ℓ is the mean free path of the gas molecules, η is the viscosity, and A is the Cunningham correction to the Stokes drag.

The first integral above describes the rate of formation of particles of radius a_h by coagulation of particles of radius a_i with those of radius a_{h-i}. The second term gives the rate of disappearance of particles of radius a_h by coagulation with other particles. The evolution of the size distribution was followed by solving this integrodifferential equation by finite differences using time increments of 3 sec.

The angular distribution of the polarization ratio corresponding to the size distribution of the coagulated aerosol is calculated next as step 3 using Eqs. [3]–[5]. This result is compared with the light-scattering data measured for the coagulated aerosol in step 4. The value of the hold-up time t'' is altered until a best fit is obtained between the calculated and measured light-scattering results. Finally, the experimental hold-up time t' is compared with the calculated time t''.

There is a problem in selecting an appropriate time for the coagulation calculation. The experimental hold-up time is given by

$$t' = V/F \qquad [8]$$

where F is the volumetric flow rate and V is the volume of the coagulation chamber. However, this is only appropriate for a coagulation calculation in a static system or in a flowing system having a zero velocity gradient. Then the aerosol in every part of the system would undergo coagulation for precisely the same time. We refer to such a calculation carried out for a single time t_q'' as the quasistatic calculation.

Actually, the aerosol is in Poiseuille flow with a parabolic velocity profile. The particles near the wall are moving much more slowly than those along the axis of the tube and therefore undergo coagulation for a longer time. We have carried out the calculation by dividing the tube into 20 annular regions and solving Smoluchowski's equation separately for the hold-up time appropriate to each such region. The result is a size distribution for each annulus, and these are then appropriately averaged to give the overall distribution emerging from the coagulation chamber.

The following assumptions are implicit in this model.

1. There is no appreciable diffusion of particles from one annular region to the next. Since coagulation itself takes place by such diffusion, this assumes that collisions occur primarily between neighboring rather than between distant particles.

2. There is perfect mixing of the aerosol from the various annular regions in the con-

stricted part of the tubing where the aerosol enters the light-scattering chamber.

If there were perfect mixing within the coagulation chamber by diffusion of particles across the annular boundaries, this would correspond to the quasistatic calculation. Partial mixing would give intermediate results.

Since the time that the aerosol spends in the outermost annulus is much greater than the average hold-up time, the Poiseuille flow calculation may become exceedingly long. However, we have made the following observation which leads to a short cut. The quasistatic calculation goes through a sequence of size distributions which is very similar to the Poiseuille flow calculation except that the latter proceeds more slowly. This permits preparation of a calibration curve shown in Fig. 11, which is a plot of the hold-up time t_p'' calculated for Poiseuille flow against t_q'' calculated for quasi-static flow to achieve the same angular variation of the polarization ratio.

A modified procedure is used to carry out the calculation. Assuming quasistatic flow, t_q'' is varied until the calculated light-scattering results best fit the experimental data.

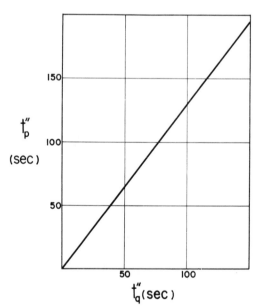

FIG. 11. Quasistatic hold-up time t_q'' versus Poiseuille hold-up time t_p'' for same angular distribution of polarization ratio.

TABLE I

COMPARISON OF EXPERIMENTAL HOLD-UP TIME t' WITH THOSE CALCULATED FOR QUASISTATIC FLOW t_q'' AND POISEUILLE FLOW t_p''

Run	Modal radius a_M (μ)	Breadth parameter σ_0	Exp t' (sec)	Quasistatic t_q'' (sec)	Poiseuille t_p'' (sec)	Standard deviation
1132	0.240	0.11	41	30	36	
1134	0.240	0.11	41	21	24	
1147	0.245	0.11	41	27	33	
1155	0.237	0.10	41	27	33	
1156	0.237	0.10	41	27	33	
1170	0.234	0.09	41	27	33	
Average			41	27	33	±2
1076	0.237	0.09	56	33	41	
1077	0.237	0.09	56	30	36	
1128	0.240	0.11	56	30	36	
1129	0.240	0.11	56	36	45	
1130	0.240	0.11	56	36	45	
Average			56	33	41	±5
1079	0.237	0.09	78	48	61	
1080	0.237	0.09	78	36	45	
1081	0.237	0.09	78	33	41	
1082	0.237	0.09	78	48	61	
1157	0.237	0.10	78	57	72	
1168	0.236	0.09	78	42	53	
Average			78	44	56	±11
1084	0.237	0.09	110	84	107	
1085	0.237	0.09	110	66	84	
1086	0.237	0.09	110	84	107	
1087	0.237	0.09	110	90	117	
1088	0.237	0.09	110	72	92	
1154	0.237	0.10	110	69	90	
1169	0.234	0.09	110	81	103	
Average			110	78	100	±12

Then t_p'' is obtained from the appropriate value of t_q'' with the aid of Fig. 11, and this is finally compared with the experimental value t'.

The results are presented in Table I. Four coagulation tubes were used having volumes of 1.37, 1.87, 2.60, and 3.64 liters with hold-up times $t' = 41, 56, 78,$ and 110 sec, respectively. These are listed in the fourth column. Columns 2 and 3 contain the size distribution parameters a_M and σ_0. The hold-up time calculated for quasi-static flow t_q'' is in the next column, and the final column contains t_p'', the value for Poiseuille flow.

Figure 12 represents two typical results. The experimental values of the polarization ratio $\rho(\theta)$ are plotted for an initial aerosol with $\alpha_M = 2.72$, $a_M = 0.237$ μm, $\sigma_0 = 0.10$ and $N = 1.2 \times 10^7$ cm^{-3} and at coagulation times $t' = 41$ and 110 sec. The theoretical values calculated for Poiseuille flow which best fit these data are plotted as curves. The corresponding theoretical coagulation times are $t_p'' = 33$ and 90 sec. The agreement between the theoretical and experimental light-scattering curves is excellent and well within the experimental error in the light-scattering measurements. The theoretical coagulation times are shorter than the experimental values. These examples correspond to runs 1156 and 1154 of Table I. In this case, the full Poiseuille flow calculation was used instead of the modified procedure. The computed size distributions at $t_p'' = 0, 33,$ and 90 sec from which the theoretical polarization ratio curves are obtained are plotted on Fig. 13. One can see the development of the second mode at the radius for the doublet.

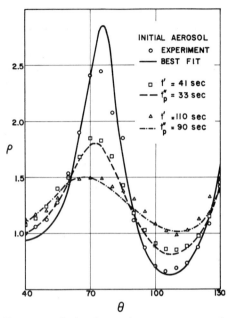

FIG. 12. Polarization ratio ρ versus scattering angle θ for initial aerosol and at later times. Curves are calculated; points are measured.

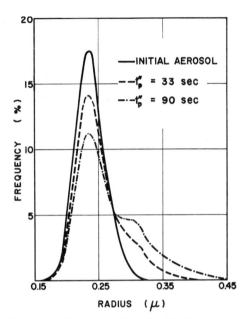

FIG. 13. Computed size distributions corresponding to aerosols in Fig. 12.

We now return to a consideration of Table I. The correction of the quasistatic calculation brings the values of the experimental and theoretical coagulation times closer together and corroborates the model of Poiseuille flow.

The best agreement is obtained at the longest time ($t' = 110$ sec), but this may be fortuitous since the interpretation of the light-scattering data becomes precarious at longer times; indeed, it was not possible to interpret any data beyond about $t' = 2$ min. The reason is that the polarization ratio curves for broad distributions become quite flat, as can be seen in Fig. 12. These curves then hardly change with increasing polydispersity; accordingly, light scattering is no longer a sufficiently sensitive tool for monitoring the coagulation. The theoretical coagulation times are about 20% lower than the experimental values. This might be due to a collision efficiency of less than unity as suggested earlier. On the other hand, the experimental uncertainties, particularly the value of the number concentration, easily suffice to account for this difference. The errors for the initial size distribution may be as high as 10% in N, 4% in $\alpha_{M'}$ and 10% in σ_0. The discrepancies between t' and t_p'' are about of the magnitude that would be expected for experimental errors of this magnitude. Indeed, by a judicious adjustment of the value of N within the uncertainty of the error, it is possible to bring the values into excellent agreement. Accordingly, it appears that within experimental error the coagulation proceeds in agreement with Smoluchowski's theory.

5. SCAVENGING OF AEROSOL PARTICLES BY A FALLING WATER DROPLET

The mechanism of removal of atmospheric aerosol by rain is still hardly understood. The primary event is either (a) washout, in which the aerosol particle attaches itself to a cloud droplet by diffusion or by acting as a condensation nucleus, or (b) scavenging, in which the aerosol particle attaches itself to a falling raindrop. For uncharged systems, at least four microphysical events may be involved in the scavenging process: Brownian diffusion, thermophoresis, diffusiophoresis, and aerodynamic capture. These events of course only lead to impact. They do not take into consideration the possibility that the droplet and the particle may bounce apart upon collision.

Aerodynamic capture predominates for particles larger than a micron, and theoretical efforts have been mainly directed to it. The problem is appallingly difficult. It is necessary to characterize the high Reynolds number fluid flow around the droplet and then to estimate the trajectory of the smaller particle upon which both inertial and fluid forces act. The result is expressed as the collection efficiency E, which is that fraction of particles contained in the track of the droplet that is actually captured. Both droplet and particle are assumed to be spherical.

Langmuir (15) has discussed the limiting cases for very low and very high Reynolds numbers, and his treatment has hardly been advanced (16). The viscous or creeping flow pattern which applies to low Reynolds numbers is not realistic for freely falling raindrops at their terminal velocities. For high Reynolds numbers, the potential flow approximation provides a description of the flow field near the forward surface of the droplet, but neglects the wake on the backside. This wake, which may be quite dramatic, provides an additional mechanism for capture. Further approximations are required for intermediate Reynolds numbers.

There is a dimensionless group in Langmuir's theory called the Stokes number or impaction parameter P, which is the ratio of the stopping distance of the particle to the radius of the droplet

$$P = \frac{2}{9}\frac{a^2\rho}{\eta}\frac{U}{A} = \frac{1}{9}Re\left(\frac{a}{A}\right)^2 \qquad [9]$$

where a is the particle radius, A is the droplet radius, ρ is the fluid density, η is the fluid viscosity, U is the velocity of the droplet relative to the fluid, and Re is the Reynolds number. The collection efficiency is usually presented as a function of the impaction parameter under the assumption of dynamical similarity between particle trajectories having the same value of P when launched at corresponding points in dynamically similar flows.

As the particle size moves into the submicron range, the inertial effects become less important and Brownian motion becomes a

significant mechanism for diffusion through the streamlines and impaction on the droplet. The case of Brownian diffusion has been treated by Friedlander (17). His study is restricted to creeping flow ($Re \rightarrow 0$) and to large values of the Péclet number defined by

$$Pé = 2AU/D \qquad [10]$$

where D is the particle diffusion coefficient. Zebel (18) has proposed that the results can be correlated by

$$E = 3.18\,Pé^{-2/3}. \qquad [11]$$

However, as pointed out above, creeping flow is hardly realistic for a freely falling droplet.

It would seem that in such a complicated system the role of theory must be mainly to shed light on the important features of the process, but that definitive quantitative results must come from experiments. Yet there have been only a small number of experimental investigations. Langmuir's work was stimulated by his interest in the growth of a raindrop upon traversing a water cloud, and almost all of the experimental work has been with droplets having radii greater than 1 μm, mainly with cloud-sized droplets.

The lacuna with respect to smaller aerosol particles is understandable when one considers the experimental difficulties. The formation, reproducibly, of submicron aerosols of uniform size and the subsequent particle size and mass analyses are still formidable experimental feats. Furthermore, the collection efficiency for submicron particles is so small that extremely sensitive analytical techniques must be utilized and extreme care must be exercised to avoid even the slightest contamination which can mask the results. Indeed, it was not until the latter obstacles were resolved after a long effort that we were able to consummate the experiments described below.

Yet the tenth-micron range of the atmospheric aerosol is exceedingly important. Hidy and Brock (19) have estimated that, although the mass of aerosol is concentrated among the larger particles, some 25% of the mass may lie below a radius of 0.5 μm, and Quenzel (20) estimates that the maximum in the size distribution of maritime aerosol particles is about 0.5 μm in radius.

These particles are mainly responsible for the degradation of visibility in the atmosphere. They also reduce the solar radiation reaching the earth's surface and enhance that reflected back into space. Furthermore, heat absorption by small particles is closely related to scattering. Thus, this constituent of the atmospheric aerosol, particularly for materials which absorb radiant energy, may have an important effect on the earth's heat balance and climate.

Washout of the aerosol cleanses the atmosphere and thereby enhances visibility, but it also results in radioactive fallout. The radioactive nuclides adsorb onto the aerosol particles, and to this process there has been added the direct injection of radioactive particles into the troposphere and stratosphere by nuclear explosions. It is likely that most of the atmospheric radioactivity is associated with particles whose radii are 0.5 μm or smaller (21).

Clearly, a laboratory study of the collection efficiency of a water droplet for aerosol particles in the tenth-micron range is in order. In our work we have measured the collection efficiency of water droplets (0.71–2.54 mm radius) freely falling through a silver chloride aerosol consisting of a relatively uniform distribution of spherical particles with radii in the range of several tenths of a micron. The particle size distribution of the aerosol was determined by light scattering. The amount of AgCl scavenged by the droplet was analyzed by a very sensitive colorimetric technique. Collection efficiencies measured were extremely low (5 × 10⁻⁶–3 × 10⁻⁴).

A schematic diagram of the scavenging apparatus is depicted in Fig. 14. The aerosol chamber H (inner diameter 4.0 cm, length 134 cm) is bounded on the top and bottom by sliding valves D1 and D2, so that while filling with aerosol the upper and lower portions may be sealed off. The chamber is filled through tube K, which can be closed with slide valve D3. Valves E1 and E2 are used to pass gas through the chamber in order to flush out the aerosol.

The two photographic shutters, C1 and

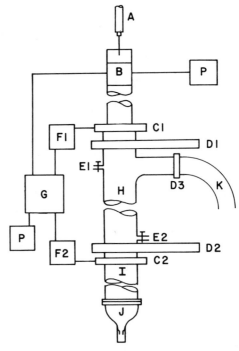

Fig. 14. Scavenging apparatus. Dropping device A, acceleration tube B, photographic shutters C1, C2, slide valves D1, D2, D3, valves E1, E2, solenoids F1, F2, timing circuit G, power supply P, scavenging tube H, lower tube I, collection assembly J, filling tube K.

C2, enclose the aerosol chamber H when the slide valves D1 and D2 are open. The shutters are operated by solenoids F1 and F2, which are actuated by a timing circuit so that the shutter C1 may be opened to allow the falling droplet to enter the chamber, and C2 opened subsequently to allow the droplet to exit.

In the upper chamber, or acceleration tube B, there is a cylinder containing a light beam which shines through the tube upon a detector. As a droplet falls, it interrupts the beam and starts the timing circuit which opens the shutters. This mechanism was used only during the initial stages of the work since we found that a set of 20–80 droplets could be run through the apparatus in less than 4 min and that during this time there was no apparent leakage of aerosol out of chamber H, even with the shutters open. The lower tube I is mounted in order to

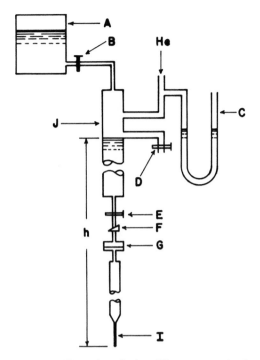

FIG. 15. Dropping device. Water reservoir A, stopcocks B, D, E, manometer C, filter G, constant head tube J, hypodermic needle or capillary I.

lengthen the path below H and so to minimize any contamination by aerosol particles settling out of H.

A simple device was built for the generation of water droplets as shown in Fig. 15. This maintains a constant head of water to which an overpressure of helium can be added when necessary to maintain a convenient dropping rate. A membrane filter (Millipore HAWP 0.45 μm) was used to avoid clogging the needle or capillary by impurities. Hypodermic-type needles and glass capillaries were used to give the various-sized droplets. The radii were determined by weighing groups of 10–30 droplets in order to get a sufficient mass for accurate weighing on a semimicroanalytical balance. There was no measurable evaporation, as evidenced by weighing droplets collected directly under the syringe, compared with those collected after falling through the system.

The silver chloride aerosol was generated and the size distribution parameters α_M and σ_0 were determined by light scattering as discussed in Parts 2 and 3, respectively. The particle concentration was calculated from Eq. [6]. The mass of aerosol in the scavenging tube was measured by flushing it with helium through a thermal precipitator where it was collected upon a sheet of aluminum foil and weighed.

The procedure was to allow a set of 20–80 droplets to fall through the system. The aerosol was flushed out of the scavenging tube for weighing and the scavenging tube refilled with aerosol after each such set. A complete run was comprised of a number of sets sufficient to give a measurable amount of AgCl. The amount of silver chloride collected by a single droplet ranged from about 5×10^{-12} to 5×10^{-9} gm, corresponding to between about 10 to 10^4 aerosol particles. A sufficient number of droplets was utilized in each run to bring the total mass of AgCl collected to between 5×10^{-9} and 2×10^{-7} gm. In some cases, this required as many as 1400 droplets.

Obviously, a very sensitive analytical technique is necessary to detect such small quantities. We utilized a colorimetric chemical analysis based upon the catalytic activity of ionic silver in the persulfate oxidation of manganous ion to permanganate

$$2Mn^{2+} + 5S_2O_8^{2-} + 8H_2O \xrightarrow{Ag^+}$$
$$2MnO_4^- + 10SO_4^{2-} + 16H^+.$$

The principle involves measurement of the rate of the catalytic reaction to determine the concentration of the catalyst. In this case, the rate is followed colorimetrically by the development of the permanganate color. The reaction does not proceed appreciably at room temperature so that it is necessary to heat the solution for a period of time in order to develop an appropriate coloration and then to quench the reaction by rapid cooling. The variables which must be controlled are the concentrations of acid, of manganous ion and persulfate, as well as the temperature and time of heating. A procedure originally developed by Underwood et al. (22) was modified for our purposes. The sensitivity of the method is about 2×10^{-9} gm. This method was at least as sensitive as, and more precise than, neutron-activation analysis.

The collection efficiency is obtained from

$$E = (m/n)/[(A/R)^2\bar{M}], \qquad [12]$$

where

m = total mass of AgCl collected
n = total number of droplets
A = radius of droplet
R = radius of scavenging tube
$\bar{M}$ = average mass of AgCl in scavenging tube

The results are given as a plot of the efficiency versus the droplet radius in Fig. 16. The collection efficiency drops off sharply from a value of 3.1×10^{-4} at $A = 2.54$ mm to a minimum of 6.3×10^{-6} at $A = 0.94$ mm, and then it appears to rise again. There is some variation in the radius of aerosol particles (0.2–0.5 μm) within each set of runs for each droplet radius. However, there do not appear to be any trends with particle radius over this range. The spread of the values of the efficiency for a given droplet size reflects the experimental error, and this appears to mask any effects of the particle radius.

We were unsuccessful in preparing droplets smaller than 0.071 cm by the present method. Although smaller droplets might be obtained by blowing a pendant droplet off the capillary, this would have perturbed our aerosol system. Also, vibratory methods

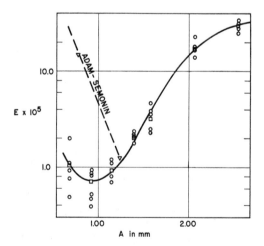

FIG. 16. Collection efficiency E versus droplet radius A.

would have released a train of droplets, which was undesirable. The upper limit of size was selected to avoid distortion from sphericity upon falling freely.

The droplets do not attain their terminal velocities in this system and are still accelerating while falling through the scavenging tube. Accordingly, the drop velocity has been calculated at the entrance and the exit of the scavenging tube after having fallen freely through the helium atmosphere for 30 cm and 160 cm, respectively. With this information, Stokes number P and the Péclet number $P\acute{e}$ can be calculated.

The Péclet number correlates the data quite well for the three smallest drops ($A = 0.71$, 0.94, and 1.14 mm) giving

$$E = 1.68 \, P\acute{e}^{-2/3},$$

where the coefficient 1.68 compares with the value 3.18 in Eq. [11] given by Zebel (18). The correlation coefficient of the values is 0.94. This agreement suggests that for these droplets the collection mechanism may be Brownian diffusion.

For larger droplets, the efficiency rises sharply even though for the conditions of our experiments the Péclet number remains the same. Therefore, a different mechanism must be involved. However, the results did not correlate properly with the impaction parameter as required by the theory of aerodynamic capture. The efficiency decreases rather than increases with increasing values of P, and this effect is particularly striking for the smaller values of P. Interestingly, in these experiments the smaller values of P actually correspond to the larger droplet sizes even though the Reynolds numbers are greater for the larger droplets. However, this is more than compensated by the smaller values of $(a/A)^2$ due both to the larger values of A as well as the fact that the aerosol particle size was fortuitously smaller in those experiments with the larger drops. The failure of the Péclet number to correlate the data also occurred for the larger droplets.

The clue to this failure may lie in the greater velocities and correspondingly greater Reynolds numbers of the large droplets. It may be that at these higher Reynolds numbers collection occurs by the particles being caught up by the eddies formed in the wake of the droplet with subsequent collection on the backside. If this is the case, it should be possible to increase the collection efficiency of the 1.16-mm droplet by increasing the length of the acceleration tube so that it attains its terminal velocity prior to entry into the scavenging tube. Similarly, the efficiencies of the larger droplets can be reduced by decreasing the pathlength over which they may accelerate. We plan experiments over a much greater range of variables in order to resolve these questions.

There is the matter of comparison of our results with those of Sood and Jackson (23) and Adam and Semonin (24), who carried out experiments using an aerosol of *Bacillus subtilis* spores and water droplets falling freely in air. In the latter case, the droplets entered the aerosol chamber at terminal velocity; in the former case, the droplets entered the aerosol chamber at low speeds, and for most drop sizes accelerated to terminal velocity within the chamber. For the overlapping ragne of drop sizes ($A = 0.5$–1.1 mm), the collection efficiencies reported by Sood and Jackson were larger than those of Adam and Semonin by as much as two orders of magnitude. Since the major experimental difficulty appears to arise from contamination, this suggests that Sood and Jackson's results may be in error, particularly at the smaller droplet sizes where the collection efficiency is so low.

The two results of Adam and Semonin which overlap with our results are plotted on Fig. 16. Their value for $A = 1.22$ mm falls precisely on our curve. The efficiency for smaller droplets rises very steeply. Of course, there are differences between our experiments. The spores used by Adam and Semonin are larger than our droplets and they are nonspherical. Their medium was air, ours was helium. Furthermore, their droplets had attained terminal velocity.

Further work will be required to determine whether these discrepancies are due to the different conditions of the experiments or to experimental problems such as those associated with the analysis or with contamination. We are redesigning our apparatus so that the velocity of the droplets upon entry into the scavenging tube can be varied from rest up to values close to the terminal velocity. We will explore a considerably greater range of particle sizes as well as other aerosol particle media and gaseous media. Indeed, the substitution of N_2 or air for He will significantly change the physical conditions. These experiments will provide a much broader range of impaction parameter and Péclet number. The data obtained with droplets falling near the terminal velocity may provide the information necessary to calculate scavenging rates in rainstorms.

6. MULTIPLE SCATTERING

The scattering of light by the atmospheric aerosol and by air itself affects both visibility in the atmosphere and the planetary heat balance. Because of the great extent of the atmosphere, photons undergo numerous successive scattering encounters before reaching the earth's surface or escaping into space. These higher-order encounters which constitute multiple scattering become predominant as the aerosol content becomes greater, such as in a hazy, foggy, polluted, etc., atmosphere. This affects the brightness of the sky, the angular distribution and polarization of the skylight, and the planetary albedo, which is the ratio of light reflected by the planet to that received.

Although the body of literature devoted to the theory of multiple scattering and to the problem of extracting numerical results from this theory is vast and longstanding (25–28), there are hardly any laboratory experiments. This is surprising because the geometrical configuration usually employed is a simple one; viz., the "infinite" plane-parallel slab, and more especially because the only two parameters which enter into the problem lend themselves readily to laboratory simulation. These are the phase function and the optical depth. They can be scaled in the laboratory to give conditions which simulate those found in the atmosphere or even in interstellar space.

In our work we have combined an experimental study with detailed numerical calculations (3, 4). The purpose has been to establish for a simple system concordance between the results of experiments and theory. Then one can devise systems which simulate real atmospheres, but which may be quite intractable to computation. Laboratory experiments with such systems can serve as an analog computer.

Multiple-scattering theory is based on the equation of transfer

$$dI/ds = -KI + KJ, \qquad [13]$$

where I is the irradiance, s is the pathlength, K is the extinction coefficient, and J is the source function. With only the first term on the right-hand side, this reduces to the familiar Bouguer–Lambert law. It gives the rate at which radiant energy is singly scattered out of the incident beam. The second term represents the multiply scattered energy which reappears in the incident direction.

Single-scattering theory may be expressed as a phase function which describes the angular distribution of the singly scattered energy. Both the extinction coefficient and the source function can be derived from the phase function which, in turn, is determined by the distribution of particle size, shape, and refractive index. For an infinite plane-parallel slab, the multiple scattering depends only upon the phase function and the optical depth which is given by

$$\tau = K\ell = \mathrm{N}C\ell, \qquad [14]$$

where ℓ is the pathlength, C is the extinction coefficient for single scattering, and N is the particle concentration.

The basis for the simulation in the laboratory of scattering by the atmospheric aerosol is now apparent. The atmospheric aerosol combines low particle concentrations with long pathlengths. Laboratory pathlengths are much shorter, but it is possible to obtain considerably higher particle concentrations so that the values of $N\ell$ may become comparable.

We have used a narrow size-distribution polystyrene latex suspension (Dow LS-057A) as the model system. Multiple-scattering systems of various optical depths were obtained by dilution of the concentrated stock solution.

The single-scattering properties of this system have been studied intensively. The ex-

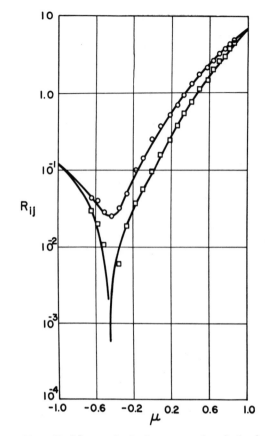

FIG. 17. Measured single scattered polarized radiances R_{ij} versus angular parameter ($\mu = \cos\theta$) for polystyrene latex suspension (Dow LS-057A). Circles R_{11}, squares R_{22}.

perimental values of R_{11} and R_{22} are plotted against μ in Fig. 17. These are defined by

$$R_{11} = (\lambda^2/\pi C)i_2(\theta) \qquad [15]$$

and

$$R_{22} = (\lambda^2/\pi C)i_1(\theta), \qquad [16]$$

where λ is the wavelength, C is the extinction coefficient, and $i_1(\theta)$, $i_2(\theta)$ are the polarized angular intensity functions (11). The parameter $\mu = \cos\theta$ where θ is the scattering angle. The phase function is given by

$$p(\theta) = (\lambda^2/2\pi C)[i_1(\theta) + i_2(\theta)]. \qquad [17]$$

The single scattering experimental values in Fig. 17 correspond to a distribution with $a_M = 0.122\ \mu\mathrm{m}$, $\sigma_0 = 0.07$. This is comparable to small atmospheric aerosol particles and simulates a rarefied atmosphere having only an aerosol component. Of course, a real aerosol would have a higher refractive index than polystyrene immersed in water.

The plane-parallel slab geometry is shown in Fig. 18 where z is the axial direction and θ and ϕ are the polar angles. The incident direction is denoted by (μ_0, ϕ_0) and the direction of the scattered beam by (μ, ϕ). Downward directions have negative values of μ.

Querfeld (3) has developed algorithms for numerical solution of the equation of transfer as well as a Monte Carlo method, and has obtained numerical results for conditions corresponding to the experiments. The calculations are extremely long. The algorithms have been adapted to the particular experimental situation since they include reflections at both boundaries of the slab which arise from the glass surfaces of the scattering cell. This is a novel but necessary feature of the calculation.

The results are expressed as the Stokes vector (11) which describes the polarization completely. We will limit this highly abbreviated discussion to the first component of the Stokes vector which is the radiance itself.

The light-scattering cell was 1 mm thick $\times$ 10 cm square, and, since the theory requires illumination of the entire infinite slab, the instrument was designed with a beam diameter of 95 mm in order to accommodate this, using a lens from an aerial camera. The instrument is shown in Fig. 19 and in schematic form in Fig. 20. Figure 21 depicts the filled cell under illumination. The foam at the top of the cell is to heighten the photographic contrast, but is normally absent.

The optical train consists of lenses, stops, and filters to give a polarized parallel monochromatic beam from the high-pressure mercury arc. The instrument goniometer is an ellipsometer with its collimator and telescope assemblies removed. The ellipsometer sample table supports the scattering cell, which can be rotated with respect to the incident beam. The detector assembly contains a Soleil compensator and a Glan–Thompson prism,

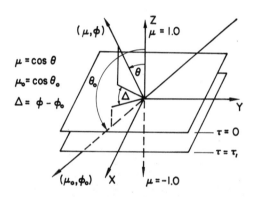

FIG. 18. Plane-parallel slab geometry. Incident direction (μ_0, ϕ_0); scattered direction (μ, ϕ).

FIG. 19. Multiple-scattering instrument.

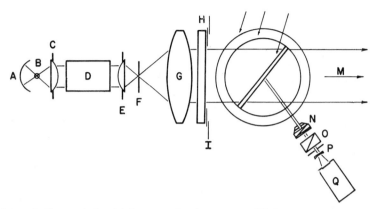

FIG. 20. Schematic diagram of multiple-scattering instrument. High-pressure mercury arc with retro-reflector A, B, filter pack D, lenses C, E, G, P, stops F, I, polarizer H, ellipsometer table J, cell holder K, cell L, Soleil compensator N, Glan–Thompson prism O, photomultiplier Q.

FIG. 21. Light-scattering cell.

as well as the photomultiplier assembly. The instrument is unenclosed, so it must be operated in a dark, dust-free room. It was calibrated against the incident beam irradiance by measurement of the radiance emerging from an opal glass diffuser of known transmittance.

A very extensive set of measurements was

obtained. Four incident tilt angles were used, $\mu_0 = -1.0$, -0.9947, -0.9328, and -0.8089, and three polarizer settings were made for each of the above tilt angles. Scattered radiances were measured at six angles in the lower half space ($\mu = -0.7290$, -0.8089, -0.8877, -0.9328, -0.9723, and -0.9947). A smaller number of measurements were made in the upper half space. Each of the above measurements was obtained for various settings of the Soleil compensator and Glan–Thompson prism to permit determination of the four Stokes parameters. Ten optical depths were used. Five ($\tau = 0.1, 0.4, 0.6, 1.0$, and 1.2) were made with the mercury violet line ($\lambda = 0.436$ μm) and five ($\tau = 0.05, 0.2, 0.3, 0.5$, and 1.0) with the green line ($\lambda = 0.546$ μm).

Obviously, this represents a vast amount of data, and only a small fraction of this has been discussed (3, 4). Here we will further limit the discussion to only three examples which are presented in Figs. 22–24 for opti-

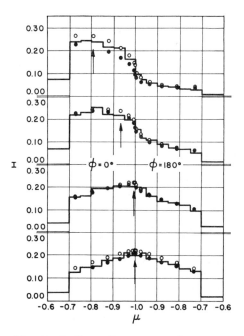

FIG. 23. Radiance I versus angular parameter ($\mu = \cos \theta$) for $\lambda = 0.436$ μm, $\tau = 0.6$. Histogram is for Monte Carlo calculation, open circles are for radiative transfer theory, closed circles are experimental. The arrow gives the incident direction. $\phi = 0°$ and $180°$ represent the two quadrants on the underside of the slab.

cal paths $\tau = 0.1, 0.6$, and 1.0, respectively. They are for violet light ($\lambda = 0.436$ μm) with the electric vector of the linearly polarized incident beam oriented $30°$ from the scattering plane.

The figures represent plots of the radiance versus μ for the downwelling radiance. The left side is for $\phi = 0°$; the right side for $\phi = 180°$. These give the radiances in the two quadrants on the underside of the slab. The direction of incidence μ_0 is shown by the arrow. The closed circles are experimental points, the open circles are obtained from solution of the equation of transfer, and the histogram represents the results of the Monte Carlo calculation. This latter technique counts photons emanating from the slab, which are collected by "buckets" extended over the indicated angular intervals. Accordingly, a histogram is obtained.

In making comparison of the experiments and calculations, it must be remembered

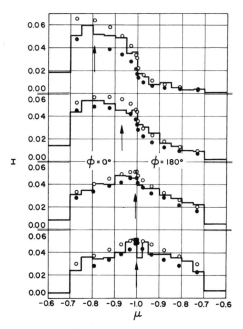

FIG. 22. Radiance I versus angular parameter ($\mu = \cos \theta$) for $\lambda = 0.436$ μm, $\tau = 0.1$. Histogram is for Monte Carlo calculation, open circles are for radiative transfer theory, closed circles are experimental. The arrow gives the incident direction. $\phi = 0°$ and $180°$ represent the two quadrants on the underside of the slab.

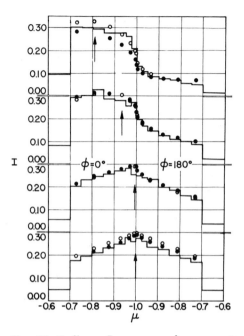

FIG. 24. Radiance I versus angular parameter ($\mu = \cos\theta$) for $\lambda = 0.436\ \mu m$, $\tau = 1.0$. Histogram is for Monte Carlo calculation, open circles are for radiative transfer theory, closed circles are experimental. The arrow gives the incident direction. $\phi = 0°$ and $180°$ represent the two quadrants on the underside of the slab.

that these experiments are absolute measurements and that whereas the angular trends may be readily established, agreement on an absolute scale is usually exceedingly difficult to attain. These three sets of results—experiment, Monte Carlo, and radiative transfer—agree remarkably well. The experimental values tend to be slightly low, but this is partly accounted for since neither the Monte Carlo nor radiative transfer calculations includes the effect of the reflection loss incurred by the incident beam at the entrance to the cell.

The statistics which comprise a portion of the Monte Carlo data keep track of the degree to which single scattering accounts for the radiance. Thus, for the smallest optical depth calculated ($\tau = 0.05$), there is very little multiple scattering present, and the system for this and lower optical depths can be treated by single-scattering theory. At $\mu_0 = -1.0$, the downward radiance is 95%

single scattering for $\tau = 0.1$, 73% for $\tau = 0.6$, and 50% for $\tau = 1.0$. The amount of multiple scattering is considerably greater at large scattering angles and also for the up-welling radiation.

The transmitted radiances increase as the optical depth increases, but at a somewhat smaller rate. The peak radiances in the $\mu_0 = -1.0$ curves increase from 0.05 to 0.22 and then to 0.29 as τ increases from 0.1 to 0.6 to 1.0. The attenuation in the slab thus begins to become evident. Also, the curves become flatter with increasing optical depth. This, of course, is expected since the radiance field will eventually become Lambertian.

The main point is that the radiance components measured in the experiments agree very well with those computed by the radiative transfer and the Monte Carlo calculations. Accordingly, carefully done experiments can simulate the multiple-scattered radiance fields in good accord with radiative transfer theory. However, the experiments can be effected with less labor. This offers the possibility of exploring with this device more complex systems such as appropriate mixtures of Rayleigh scatterers with colloidal particles in order to simulate a hazy or dusty atmosphere. Colloidal silica can simulate the gaseous or Rayleigh component so that a suitable mixture of colloidal silica and polystyrene latex can serve as a real atmosphere. Furthermore, by utilizing appropriately absorptive materials, the effects of absorption by either the air or the aerosol component can be explored.

ACKNOWLEDGMENTS

Although a perusal of the list of references will immediately indicate that this has been a cooperative endeavor, it can hardly suggest the central roles played by Doctors D. Cooke, V. Hampl, C. Huang, G. Nicolaon, and C. Querfeld, and by Professors E. Matijević and J. Kratohvil. This is their work at least as much as it is mine. The idea of using a falling film was conceived by Professor F. C. Goodrich, and our aerosol generator is patterned after one developed by him and Messrs. A. Sarmiento and S. Shahriari. Also, I want to record my continuing debt to the late Professor Victor K. La Mer in whose laboratory I first learned about these things.

REFERENCES

1. NICOLAON, G., KERKER, M., COOKE, D. D., AND MATIJEVIC, E., *J. Colloid. Interface Sci.* **38**, 460 (1972).
2. HAMPL, V., KERKER, M., COOKE, D. D., AND MATIJEVIC, E., *J. Atm. Sci.* **28**, 1211 (1971).
3. QUERFELD, C. W., Ph.D. Thesis, Clarkson College of Technology, Potsdam, New York 13676, 1970.
4. QUERFELD, C. W., KERKER, M., AND KRATOHVIL, J. P., *J. Colloid Interface Sci.*, in press.
5. SINCLAIR, D., AND LA MER, V. K., *Chem. Rev.* **44**, 245 (1949).
6. NICOLAON, G., COOKE, D. D., KERKER, M., AND MATIJEVIC, E., *J. Colloid Interface Sci.* **34**, 534 (1970).
7. NICOLAON, G., COOKE, D. D., DAVIS, E. J., KERKER, M., AND MATIJEVIC, E., *J. Colloid Interface Sci.* **35**, 490 (1971).
8. ESPENSCHEID, W. F., MATIJEVIC, E., AND KERKER, M., *J. Phys. Chem.* **68**, 2831 (1964).
9. ESPENSCHEID, W. F., WILLIS, E., MATIJEVIC, E., AND KERKER, M., *J. Colloid Sci.* **20**, 501 (1965).
10. JACOBSEN, R. T., KERKER, M., AND MATIJEVIC, E., *J. Phys. Chem.* **71**, 514 (1967).
11. KERKER, M., "The Scattering of Light and Other Electromagnetic Radiation." Academic Press, New York, 1969.
12. KERKER, M., MATIJEVIC, E., NICOLAON, G., AND COOKE, D. D., *in* "Assessment of Airborne Particles" (T. T. Mercer, ed.). Thomas, Springfield, Ill., 1972, p. 153–168.
13. SMOLUCHOWSKI, M., *Z. Phys. Chem.* Leipzig **92**, 129 (1917).
14. HUANG, C. M., KERKER, M., AND MATIJEVIC, E., *J. Colloid Interface Sci.* **33**, 529 (1970).
15. LANGMUIR, I., *J. Met.* **5**, 175 (1948).
16. RICHARDSON, E. G., ed., "Aerodynamic Capture of Particles." Pergamon, New York, 1960.
17. FRIEDLANDER, S. K., *J. Colloid Interface Sci.* **23**, 157 (1967).
18. ZEBEL, G., *J. Colloid Interface Sci.* **27**, 294 (1968).
19. HIDY, G. M., AND BROCK, J. R., "Proceedings of the 2nd Clean Air Congress," International Union of Air Pollution Prevention Associates, Washington, D. C., December 1970.
20. QUENZEL, H., *J. Geophys. Res.* **75**, 2915 (1970).
21. CADLE, R. D., "Particles in the Atmosphere and Space." Reinhold Publishing Co., New York, 1966.
22. UNDERWOOD, A. L., BURRILL, A. M., AND ROGERS, L. B., *Anal. Chem.* **24**, 1597 (1952).
23. SOOD, S. K., AND JACKSON, M. R., Report No. IIIRI-C6105-9, IIT Research Institute, Chicago, Ill. 60616, 50 pp., 1969.
24. ADAM, J. R., AND SEMONIN, R. G., "Proceedings of the Precipitation Scavenging Meetings" (W. G. Slinn, Ed.). U. S. AEC, Washington, D. C. 1960.
25. CHANDRASEKHAR, S., "Radiative Transfer." Oxford Univ. Press, London, 1950.
26. KOURGANOFF, V., "Basic Methods of Transfer Problems." Oxford Univ. Press, London, 1952.
27. BUSBRIDGE, I. W., "The Mathematics of Radiative Transfer." Cambridge Univ. Press, London, 1960.
28. PREISENFORFER, R. W., "Radiative Transfer in Discrete Spheres." Pergamon, Oxford, 1965.

1972

EGON MATIJEVIĆ

CLARKSON COLLEGE OF TECHNOLOGY
POTSDAM, N. Y.

. . . for his studies of the stability of colloidal dispersions, for bridging the idealized theoretical models of colloid stability and real systems by elucidating the role of counterion hydrolysis, complexing, and ion exchange in coagulation, reversal of charge, and adsorption, having an important impact on a multitude of practical technological problems.

Colloid Stability and Complex Chemistry

Egon Matijević

Institute of Colloid and Surface Science and
Department of Chemistry
Clarkson College of Technology
Potsdam, New York 13676

> "If a professor is obliged to discuss this unsatisfactory con-
> dition of the theory of coagulation for thirty or more years,
> in every term of the academic year, then it may easily happen
> that he becomes more and more impatient. Either he be-
> comes resigned or he commences to curse. The latter course
> is in general more fruitful."
>
> –Wolfgang Ostwald
> *J. Phys. Chem.*, **42**, 982 (1938)

Introduction

One of the essential tasks of colloid science is to study the formation and stability of colloidal sols. The world abounds in these systems as they are continuously produced (and eliminated) by myriads of natural phenomena. In addition, countless colloidal dispersions are of anthropogenic origin, either intentionally created for a variety of uses or arising as by-products of technological processes. The fact that the colloidal sols are in some cases highly desirable and in some other cases undesirable states of matter makes it crucial that we know how to prepare or destroy such dispersions and that we understand the reasons for their stability.

It is no surprise that for almost a century works on colloid stability have been reported in the scientific literature. For example, as early as 1857 Michael Faraday gave a detailed description of the properties of gold sols including their behavior towards electrolytes[1]. Unfortunately this area of science has been plagued by inadequate experiments and theory. The reason for this is not difficult to comprehend. In every colloidal sol, the particles are suspended in a liquid medium containing dissolved species. In hydrosols the solutes are, as a rule, electrolytes consisting frequently of a mixture of ions of various degrees of complexity. A solution of electrolytes is itself a complicated system whose properties are not always easily interpreted. When charged colloidal particles of different sizes, shapes, and surface characteristics are immersed in such an electrolyte medium the system becomes considerably more involved, especially because of the formation of distinctly structured solid/solution interfacial regions.

In most of the early attempts to explain the stability of hydrosols only the surface characteristics of the dispersed particles were considered, whereas the properties of the electrolyte medium were essentially disregarded. This was mainly due to the fact that the charge of colloidal particles (particularly in lyophobic systems) was recognized as the primary cause for sol stability. Thus, the idea of a coagulation mechanism based on the neutralization of charge by the adsorption of an equivalent amount of counterion seemed most reasonable[2]. Freundlich, who at first believed in this idea, was later on led by experimental observations to repudiate it[3]. It should be mentioned that in *some* cases the adsorption of solutes may indeed lead to charge neutralization and consequently to coagulation; however, this does not apply to simple counterions. Other theories based essentially on the properties of the particles, such as the one using critical electrokinetic potential as the determining parameter of stability[4], were eventually also found to be untenable.

Discouraged by these failures, Wolfgang Ostwald directed his attention to the electrolyte medium and proposed the "activity-coefficient theory of coagulation"[5]. This assumed that the interionic forces played a predominant role in coagulation, and that the activity coefficient of the counterions, being a measure of these forces, could serve as a criterion of stability. It is not unexpected that this theory was found inadequate. Firstly, it completely disregards the properties of the particles. Secondly, the activity coefficient is a rather crude way of characterizing an electrolyte medium, particularly when the limiting equation of Debye and Hückel is applied to calculate this quantity for asymmetrical electrolytes containing highly charged counterions.

Another approach to colloid stability which also emphasized the electrolyte medium was developed by Težak[6]. This was based on Bjerrum's ion association theory, as applied to the stabilizing ion-counterion pair, and resulted in an essentially empirical expression for the Schulze-Hardy rule for counterion-charge effects in coagulation[7,8].

A milestone in the development of the understanding of colloid stability was the double layer theory, which was independently presented by Derjaguin and Landau[9] and Verwey and Overbeek[10] (and is therefore called "the DLVO theory"). The coagulation power of electrolytes was shown to depend on the electric potential of the surface of the colloidal particles, on the charge of the counterions, and on the magnitude of the van der Waals attractive forces. In its simplest form it gave a quantitative formulation of the Schulze-Hardy rule[7,8]. This theory has had a twofold effect on colloid science. It resurrected the feeling among scientists that a reasonable physical model permits a quantitative attack even on such complicated phenomena as sol stability. It also gave a tremendous impetus to the renewal of interest in research in colloid science, which has led to the study of better defined systems and the development of more sophisticated experimental techniques.

The DLVO theory has also had negative effects. The situation is so complex that only the simplest systems can be handled. The electrolyte solution part is treated in terms of the Debye-Hückel theory of strong electrolytes and suffers from the limitations thereof. It is interesting, yet not surprising, that the basic premise of the Ostwald "activity coefficient" coagulation rule was actually shown to follow from the DLVO theory[11]. These limitations have led scientists to attempt to test the predictions of the theory by matching the colloidal sols studied experimentally to the theoretical model as closely as they could. As a result the studies have been carried out mostly with systems seldom encountered in nature or in practical applications. Lyklema et al.[12] recently stated: "Numerous experiments have shown that the ideas formulated in the DLVO theory for colloid stability are basically correct. On the other hand, the application of this theory to real systems is not always straightforward...". All attempts to refine the theory of the electrical double layer have made its application to colloid stability even less manageable. For example, the extension to include the so-called discreteness-of-charge effect proposed by Levine and Bell[13] requires the knowledge of four different dielectric constants, of which none is amenable to experimental measurement.

When it comes to "real" systems one ordinarily deals with electrolyte media which contain a variety of solute species. This is particularly so if polyvalent ions are present, as these tend to form complexes with ions of opposite charge or even with nonionic solutes. The complex species usually have charges different from those of the original ions, which in itself may dramatically change their coagulation ability. More importantly, these complexes quite frequently adsorb much more strongly on particle surface than the uncomplexed species. As a result the particle charge characteristics are altered, and the stability of the sol may either decrease or increase as a result. When the adsorption of counterions sufficiently reduces or neutralizes the charge, the sols are destabilized. However, the adsorption under certain conditions could be so strong as to cause a reversal of the sign of the charge of the colloidal particles. This recharging may produce sols of great stability.

It is obvious that, if one is to understand sol stability, it is of the utmost importance to consider complex reactions in the medium and at the solid/solution interface. Highly charged complexes have tremendous effects upon the rate of coagulation, but may constitute only a negligible fraction of the total electrolyte content of the dispersion medium. Failure to recognize this could lead to erroneous conclusions. In the past, systems subject to complexing were avoided or changes in the electrolyte solutions were disregarded. The latter happened much too often, as will be shown by a few examples.

It may be hard to believe, but even in the latest books on colloid science[14-19] the coagulation concentrations of various counterions for different sols are given in tables reproduced from Freundlich's own work published at the beginning of the century, although this fact has been omitted or misquoted in some of the texts[14,15,18]. Many old coagulation data were obtained with ill-defined sols under poorly controlled conditions, yet they are used to demonstrate the effect of counterion charge upon the critical coagulation concentrations (The Schulze-Hardy rule). Some of these tables list HCl, NaOH, and $Ba(OH)_2$ as coagulants for a hydrous ferric oxide sol despite the fact that H^+ and OH^- are potential-determining ions and as such will exercise much larger effects upon sol stability than other counterions. Also, in the case of polyvalent metal ions no information on pH is given; thus, neither their state nor charge in solution is known, making their coagulation concentrations a poor test of the charge rule.

A more flagrant example of complete disregard for chemical changes in the study of colloid stability can be found in work by Wenning[20] in which coagulation concentrations of various salts for a negatively charged latex are reported as functions of pH. In an attempt to interpret his results in terms of the Wo. Ostwald theory, the author calculates the activity coefficients of aluminum chloride at pH values up to 12 assuming $3\text{-}1_3$ electrolyte. Needless to say at these high pH values no cationic aluminum species exist and for that matter no electrolyte of the $3\text{-}1_3$ type is present in the medium.

Over a period of years we have carried out systematic studies of the stabilities of various hydrosols with special emphasis on the effects of chemical complexing, taking into account both the species at the solid/solution interface and those in the solution in bulk. Our purpose has been to clarify the changes in colloid stability as they are influenced by the composition of the solutes in the dispersion medium and/or in the interfacial region. Some of these effects are rather dramatic. For example, a shift in pH of only 0.02 unit suffices to change an unstable silver bromide sol into an extremely stable sol in the presence of a very small amount (e.g. $10^{-4}M$) of an aluminum salt[21]. At present, such exceedingly sensitive effects cannot be predicted by any theory, yet they can be readily explained from the changes in chemical composition of the electrolyte medium caused by these minute variations of pH. As another example, Ni^{2+} ion coagulates a negatively charged silver bromide sol at a concentration of $\sim 1 \times 10^{-3}M$, whereas in the presence of only $3 \times 10^{-6}M$ 2,2'-dipyridyl (dipy) the same sol is coagulated by $3 \times 10^{-7}M$ nickel salt, which represents an enhancement of the coagulation ability of nickel ion of nearly four orders of magnitude. In addition, in the absence of dipy, nickel does not reverse the charge, yet with dipy $(3 \times 10^{-6}M)$ the sol is restabilized due to charge reversal at a concentration as low as $4 \times 10^{-5}M$ $Ni(NO_3)_2$[22]. Again, these rather dramatic phenomena can be explained if the complex chemistry of the system is understood.

Using a variety of experimental techniques we have been able to elucidate many cases and to arrive at some general conclusions relating the formation and stability of colloidal particles to the chemical composition of the electrolyte environment. This experience has enabled us to interpret many rather complicated systems of practical significance and to explain some apparently unusual phenomena in colloid dispersions. It is important to recognize that it should be possible to apply the theory of colloid stability to "real" systems once the chemical picture is clarified. The conclusions we have reached are also of interest in areas other than sol properties; they contribute to better understanding of processes involving solid/solution interfaces, such as corrosion, heterogeneous catalysis, flotation, paper sizing, and adhesion, to mention just a few. Finally, in some fortuitous cases, it was possible to utilize the observed stability effects in order to formulate the composition of counterion complexes in solution, or even to determine their formation constants.

To illustrate the chemical complexing effects with respect to colloid dispersions in general, I shall describe some of our results dealing with the following topics:

    A. Colloid formation and complex chemistry
    B. Sol stability and solute complexing
        1. Weak interactions of solutes with solids
        2. Strong interactions of solutes with solids
    C. Stability domains

Colloid Formation and Complex Chemistry

Whenever colloidal particles are produced by condensation of solutes the embryonic stage may be considered as chemical complexing. In most cases, this consists of ionic or molecular association[23]. However, in some instances well-defined coordination complexes are precursors to the formation of nuclei and particles. Although this has been indicated[24], it seems that no convenient systems have been found which would permit a correlation between solute coordination and particle formation.

Recently, we have been able to generate hydrous metal oxide sols consisting of spherical particles of narrow size distributions[25-27]. The sols are sufficiently monodispersed to show higher order Tyndall spectra (HOTS) and, therefore, their size distribution and number concentration can be determined *in situ* by light scattering[28]. Such sols of uniform size are produced by aging metal salt solutions at elevated temperatures, but only if certain anions are present, indicating that complexing is an essential step in this process. Specific effects of anions on homogeneous precipitation have been observed before. Particularly, "denser" precipitates were obtained when metal hydroxides were precipitated by hydrolyzing urea in the presence of "suitable anions"[29]. No attempts were made to investigate the causes of these effects.

Monodispersed metal hydroxides, as we have prepared them, lend themselves well to the study of the relationship between the complex chemistry of the solute "precursors" and the concentration and size of the sol particles produced by aging. This is of interest because it permits the elucidation of the processes leading to the formation of metal hydroxides which are "very much more complicated than the precipitation of ionic crystals"[24]. In addition to numerous applications of hydrous metal oxides, this work should contribute to the understanding of corrosion processes resulting in colloidal products.

Figure 1a gives an electron micrograph of a chromium hydroxide sol prepared by aging for 18 hours at 75°C of a $4 \times 10^{-4}M$ solution of chrom alum, $KCr(SO_4)_2$. The histogram in Figure 2 (lower part) shows the size distribution of this sol from electron microscopy. Solid and dashed lines are obtained by light scattering using the polarization ratio[28] and specific intensity methods, respectively[30]. The agreement between the data is quite satisfactory. "Monodispersed" hydroxide sols could be obtained by aging solutions of chrom alum but not those of some other salts, such as chromium(III) nitrate or perchlorate. However, if potassium sulfate or potassium phosphate was added to the latter, the sols again formed. These anions are known to act as "penetrating ions"[31]; i.e., they form basic metal ion sulfates (or phosphates) by being incorporated in the coordination sphere of the cation.

The equilibrium modal particle diameter, D_M, depends on the chromium to sulfate (or phosphate) ratio in the aging solution (Table I). Figure 3 shows in the upper part the changes in D_M and the number concentration, N, of sols obtained by keeping a $4 \times 10^{-4}M$ solution of $KCr(SO_4)_2$ at 75° for the lengths of time indicated on the abscissa. The lower part of Figure 3 gives the accompanying change of pH and of chromium concentration in the dispersing medium. Additional analyses[27] of the filtrates showed that during the heating a certain amount of sulfate ion was bound in solute complexes, whereas polarographic data indicated

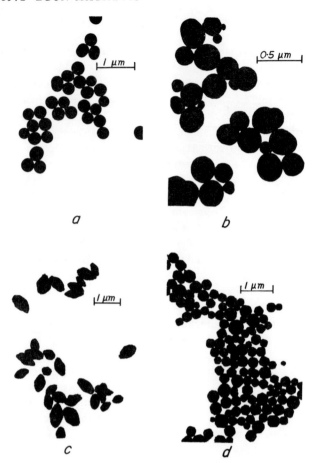

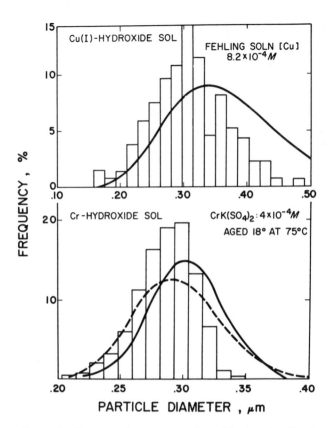

Figure 1. Electron micrographs of several metal hydroxide sol particles.

(a) *Chromium(III) hydroxide sol obtained by aging of a $8 \times 10^{-4} M$ solution of $CrK(SO_4)_2$ at $75°C$ for 18 hours.*

(b) *Zirconium(IV) hydroxide sol obtained by aging of a $4 \times 10^{-3} M$ solution of $ZrOSO_4 \cdot H_2SO_4 \cdot 3H_2O$ solution at $75°C$ for 22 hours.*

(c) *Copper(II) hydroxide sol obtained by aging of a solution containing $1 \times 10^{-3} M$ $CuSO_4$ and $2 \times 10^{-3} M$ NaOH at $75°C$ for 18 hours.*

(d) *Copper(I) oxide sol obtained by reduction of a Fehling's solution containing $8.2 \times 10^{-4} M$ copper with $1.33 \times 10^{-3} M$ glucose heated to $95°C$.*

the existence of two complexes in equilibrium with each other; these might have been[32,33]:

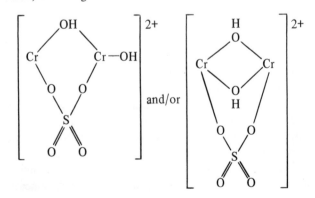

Figure 2. Lower: Comparison of particle size distributions obtained by electron microscopy (histogram) and by light scattering using the polarization ratio method (solid line) and the specific intensity method (dashed line) for the chromium hydroxide sol shown in Figure 1(a).

Upper: Comparison of particle size distributions obtained by electron microscopy (histogram) and by light scattering using the polarization ratio method for the copper(I) hydroxide sol shown in Figure 1(d).

Table I

Modal Particle Diameters, D_M, of Chromium Hydroxide as a Function of $[Cr^{3+}]$ to [Anion] Ratio

$[Cr^{3+}] : [SO_4{}^{2-}]$	$2:3^a$	$1:2$	$1:3$
D_M (nm)	204	286 297*	423

$[Cr^{3+}] : [H_2PO_4{}^-]$		$1:3$	$1:5$
D_M (nm)		220	327

[a]All sols were prepared by heating solutions of $Cr(NO_3)_3$ ($4 \times 10^{-4} M$) and K_2SO_4 (or KH_2PO_3) in molar concentration ratios as indicated except the sol indicated by asterisk which was obtained from $4 \times 10^{-4} M$ solution of $KCr(SO_4)_2$.

Another determination showed that the percentage of sulfate in the particles themselves is considerably lower than would be expected if the solids consisted entirely of the basic chromic sulfate with the $Cr^{3+}:SO_4^{2-}$ ratio of 2:1.

All of this would indicate that the role of sulfate ion seems to be restricted to the nucleation stage or to the cross-linking of the polymeric chromium hydroxide chains. Thus

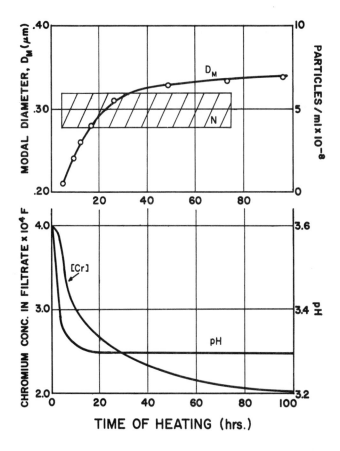

Figure 3. Upper: Change in the modal particle diameter (D_M) and the number concentration (N) of a chromium hydrous oxide sol as a function of time of heating at 75°C of a 4×10⁻⁴ M solution of chrom alum.

Lower: Change of pH and of chromium content in the electrolyte medium in the system described above.

the initial step in particle formation may consist of the formation of embryos and nuclei from basic chromium sulfate species that are weakly charged or electrically neutral; onto these nuclei the polymeric hydrolyzed complexes, produced continuously during the heating, aggregate with further coordination of sulfate ions. The fact that the maximum modal diameter of particles depends on the $Cr^{3+}:SO_4^{2-}$ ratio further supports the suggested role of the anion in the embryonic step.

The chemistry of basic chromium phosphates is poorly understood, but it is likely that phosphate ion would behave similarly to sulfate ion, for these two ions should coordinate in a similar way with the basic chromium species. Indeed the "monodispersed" sols can be generated by aging chromium salt solutions in the presence of phosphate [26].

Should the entire concept be correct, one would further expect other metal ions that form strong basic sulfate complexes to give sols similar to those of chromium. In this regard zirconium seemed an obvious choice as it hydrolyzes readily and also coordinates sulfate[34,35]. Preliminary experiments showed that sols consisting of spherical particles are indeed generated by aging zirconium sulfate solutions (Figure 1b).

Copper(II) sulfate solutions also produced sols with reasonably uniform, but non-spherical, particles (Figure 1c). However, when Fehling's solution, containing copper(II), sulfate, and tartrate ions was heated in the presence of glucose to reduce copper(II) to copper(I), particles were formed that were sufficiently compact and uniform for the sols to show HOTS (Figure 1d). In this case, however, it is most likely that the basic copper tartrate complex acts as the embryonic precursor. An attempt was made to analyze this sol for distribution using the Mie light scattering functions and the polarization ratio method[29]. Figure 2 (upper) gives a comparison of the histogram obtained by electron microscopy with the size distribution curve from light scattering measured at 546 nm and taking the refractive index relative to water to be 2.03-0.52i. Recognizing that the particles are neither perfectly spherical nor isomorphous the agreement is quite reasonable.

In continuing these studies we hope to be able to correlate even more precisely the complex solute composition with homogeneous precipitation of various metal hydroxides.

Sol Stability and Solute Complexing

The effects of solute complexing on sol stability arise from two main factors: the change in counterion charge resulting from the coordination, and the difference between the types and degrees of the interactions of the complexed and the uncomplexed species with the particles. Thus a logical classification seems to be into cases in which the interactions between counterion complexes and the particles are weak and cases in which these are strong.

1. Weak interactions of solutes with solids

The systems discussed in this section consist of colloidal particles dispersed in a medium containing complex solutes (either added or formed *in situ*) which show only negligible interactions (such as absence of ion exchange, adsorption, or coordination) with the particulate matter. The interactions may involve counterion adsorption in the Stern layer. The case of weak interactions could either apply to a broad concentration range of the solute (mainly counterion) complexes, or be limited to a narrow (and usually low) region of concentrations.

The most striking aspect of colloid stability is the effect of the concentration of counterion on the rate of coagulation. The minimum concentration of counterions necessary to initiate fast coagulation is known as the critical coagulation concentration (c.c.c.). When the interactions between the solutes and the particles are weak, the c.c.c. for lyophobic sols depends dramatically on the counterion charge (the Schulze-Hardy rule). This effect is so pronounced that it is observed regardless of how the c.c.c. is determined or what criteria for colloid stability are employed. The DLVO theory predicts, at least for counterions of lower charge, an inverse sixth power law; i.e., the ratios of coagulation concentrations for 1:1, 2:2, and 3:3 electrolytes should be $c_1:c_2:c_3 = 1/1^6 : 1/2^6 : 1/3^6$. Although it is gratifying to see the Schulze-Hardy rule follow quantitatively from the theory, the simplifications that have to be made to arrive at the above formulation are such to make the $1/z^6$ law inapplicable in most cases[36]. On the other hand there is ample experimental evidence that a linear relationship is observed if the logarithm of the molar c.c.c. is plotted against the counterion charge. Figure 4 gives several examples of this relationship for a number of sols which vary in chemical composition, sign of the charge, and solvent. Many other sols could have been shown to give the same relationship. As a curiosity, coagulation data for *Escherichia Coli* obtained by Rubin *et al.*[37] are also included, although we do not suggest that the bacteria are lyophobic colloids.

It is obvious that any complexing which changes the charge of the counterion will have a profound effect upon its coagulation ability, even when no specific interactions between the complex species and the particles take place. Decreasing the charge of the counterien will considerably increase the concentration of the electrolyte required to coagulate a sol and vice versa.

Several examples will be given here in which quantitative relationships between counterion complexing and the c.c.c.

have been established. The systems to be shown will include negative sols with cations as counterions. The effect of counterion complexing on sol stability will be illustrated for cases where the ligands are either inorganic anions (with special emphasis on hydroxylation) or organic species of various compositions.

a. Complexing with inorganic ligands except hydroxyl

The first case to be discussed deals with aluminum ions. Few metal salts of polyvalent cations find as many applications as those of aluminum. Furthermore, the solution chemistry of aluminum, complex though it may appear, is still less involved than that of other cations having higher charges.

In Figure 5, the critical coagulation concentrations of aluminum nitrate and sulfate (expressed as the total molar concentration of aluminum) are plotted against pH for a negatively charged silver iodide sol ($1 \times 10^{-4} M$). When acidification was necessary, nitric acid was used in the presence of $Al(NO_3)_3$ and sulfuric acid in the presence of $Al_2(SO_4)_3$. For each salt the c.c.c. decreases at pH values above about 4, for a reason that will be discussed later. Below this pH, the coagulation concentration for aluminum nitrate remains constant and its value is characteristic for counterions of 3+ charge for this sol, indicating that the aluminum ion is present in its unhydrolyzed form[39]. When aluminum sulfate is used as coagulant in solutions acidified with H_2SO_4, the c.c.c. increases as the pH is decreased below 3.3 (dashed line). The only difference between these two systems is in the content of the sulfate ion. It was suggested that in acidic solutions the complex $AlSO_4^+$ is formed. Since no enhancement in adsorption of the counterien at low pH could be detected, the increase of the c.c.c. must be due to removal

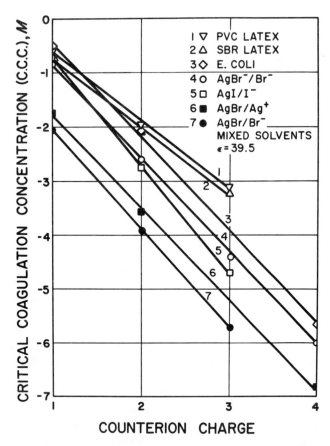

Figure 4. Critical coagulation concentration (c.c.c. in mole/liter) as function of counterion charge for a polyvinyl chloride (PVC) latex, ▽ ; styrene-butadiene (SBR) latex, △ ; Escherichia Coli, ◇ ; negatively charged silver bromide sol, ○ ; negatively charged silver iodide sol, □ ; positively charged silver bromide sol, ■ ; and a negatively charged silver bromide sol at dielectric constant ε = 39.5, ● .

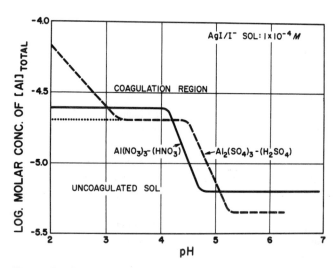

Figure 5. Critical coagulation concentrations of aluminum nitrate (acidified with HNO_3 when necessary) (solid line), and of aluminum sulfate (acidified with H_2SO_4 when necessary) (dashed line) as a function of pH for a negatively charged silver iodide sol (AgI: $1 \times 10^{-4}M$, KI: $4 \times 10^{-4}M$). Dotted line gives the concentrations of uncomplexed aluminum ions in aluminum sulfate solutions.

of Al^{3+} to form the singly charged complex. Whatever the fraction of the total aluminum ion bound as this complex may be, the concentration of $AlSO_4^+$ is much too low for it to have any effect on the rate of coagulation (Figure 4). Thus, it was suggested that in the presence of $Al_2(SO_4)_3$ at all pH values below about 4, it is Al^{3+} that acts as the coagulant, as it does with $Al(NO_3)_3$. To prove this, it would have been desirable to calculate the concentration of free Al^{3+} ions from the known formation constant for $AlSO_4^+$. Unfortunately the reported values differed considerably and, therefore, the formation constant was evaluated from the stability data for the silver iodide sol in the presence of $Al_2(SO_4)_3$ assuming an unchanging coagulation concentration for Al^{3+} over the pH range < 4. This constant concentration is represented by the dotted line in Figure 5. The calculation yielded a value of $K_{AlSO_4^+} = 370$ [39].

The effect of Al^{3+}-SO_4^{2-} complexing described here explained one of the old but poorly understood coagulation phenomena, namely the "antagonistic effect". This refers to the observation that the coagulation concentration of one electrolyte may be substantially increased by the addition of a second electrolyte. Various explanations have been offered for this effect. Wo. Ostwald claimed that ionic antagonism follows from his activity coefficient theory of coagulation[40]. To prove it, he cited coagulation data obtained with mixtures of $Al(NO_3)_3$ and K_2SO_4 and treated this mixture as a $3-1_3 + 1_2-2$ electrolyte in calculating the activity coefficients. More recently, Levine and Bell[41] interpreted ionic antagonism as a consequence of the discreteness-of-charge effect in the ionic double layer.

Our results indicate that the "antagonistic effect" is essentially due to counterion complexing. To prove this, the open symbols in Figure 6 show the effect of K_2SO_4 concentration on the c.c.c. of $Al(NO_3)_3$ for negatively charged silver iodide and silver bromide sols. Full symbols give the concentrations of free Al^{3+} calculated by using the previously derived value, $K = 370$, of the complex constant for $AlSO_4^+$. In both cases the concentration of free Al^{3+} remains constant and equal to the coagulation concentration of this ion in the absence of K_2SO_4. Thus, the higher concentrations of aluminum salt necessary to coagulate the silver halide sols in the presence of SO_4^{2-} provide the system with just the concentration of the free Al^{3+} that suffices to cause fast coagulation. The rest of the aluminum ion is converted into the singly charged $AlSO_4^+$ species, the content of which is too low to affect the rate of coagulation. To illustrate the generality of this complexing effect, Figure 7 shows the dependence on pH of the c.c.c. (expressed as molarity of aluminum ions) of $Al(NO_3)_3$ (dashed line) and of $Al_2(SO_4)_3$ (full line) for a Ludox HS silica sol. Over the entire pH range the total concentration of aluminum needed to coagulate the sol is higher in the sulfate solutions. Circles give the corresponding concentrations of the Al^{3+} ion as calculated for the sols coagulated by $Al_2(SO_4)_3$, again assuming that the remainder of the aluminum is bound as the $AlSO_4^+$ complex. These agree very well with the coagulation curve for the noncomplexed aluminum ions in the systems containing

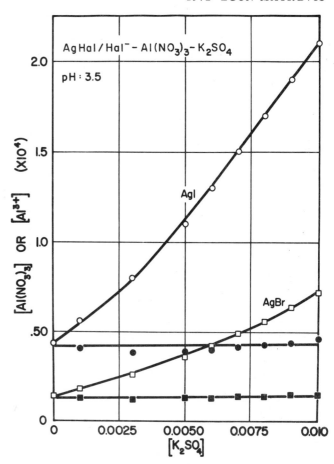

Figure 6. Critical coagulation concentrations of aluminum nitrate as a function of added potassium sulfate (in moles/liter) for negatively charged silver iodide (○) and silver bromide (□) sols, respectively. Full symbols give the concentrations of uncomplexed aluminum ions in the corresponding sols.

$Al(NO_3)_3$.

One would expect the antagonistic effect to become more pronounced as the complexation of highly charged counterions by anions becomes stronger. Indeed, this can be demonstrated with aluminum salts in the presence of fluoride ions. It has been established that a series of complexes of the type $AlF_n^{(3-n)+}$ with $n = 1$-6 exist in acidic solutions, and the formation constants are known for all of these species. Figure 8 shows the effects of the molar concentration of KF on the c.c.c. of $Al(ClO_4)_3$ for a negatively charged silver bromide sol at two different pH values[42]. Since all of the fluoride complexes are either of positive charge lower than the free Al^{3+}, neutral, or negatively charged, it is expected that only the triply positively charged uncomplexed aluminum ion will affect the rate of coagulation. The calculated concentration of free Al^{3+} is given by the dashed line. The two diamonds indicate the range of calculated values. This is in good agreement with the c.c.c. of Al^{3+} in the absence of KF for the same sol as shown by the arrow. The magnitude of the complexing effect is quite impressive: in the presence of $10^{-2} M$ KF the concentration

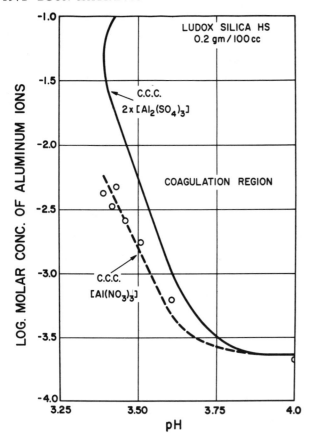

Figure 7. *Critical coagulation concentrations of aluminum sulfate (solid line) and of aluminum nitrate (dashed line) for a Ludox HS silica sol as a function of pH. Circles give the concentration of free aluminum ions in the silica-aluminum sulfate systems at corresponding pH values.*

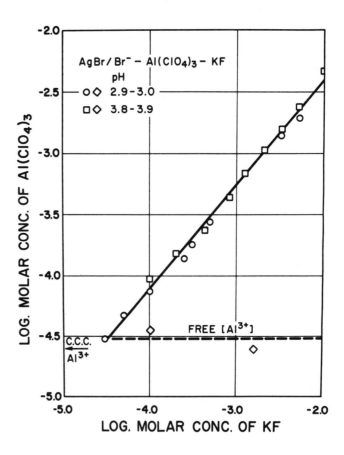

Figure 8. *Critical coagulation concentrations of aluminum perchlorate as a function of added potassium fluoride for a negatively charged silver bromide sol at two different pH values. Dashed line and diamonds give the concentrations of uncomplexed aluminum ions in the corresponding sols. Arrow indicates the c.c.c. of Al^{3+}_{aq} for the same sol.*

of the free Al^{3+} ions is only about 2% of the total concentration of aluminum salt necessary to coagulate the sol. This means that 98% of the aluminum counterion is "inactivated" due to fluoride complexing.

It is worth noting that all mixed salt systems which exhibit pronounced antagonistic effect contain cations and anions known to form strong complexes, such as $Th(NO_3)_4$ – K_2SO_4, $La(NO_3)_3$–K_2SO_4, etc.[43]. Our results not only explain the antagonistic effect in terms of counterion complexing; they also point the way to inhibiting the coagulation by polyvalent counterions when it is desired to prevent sol destabilization.

b. Hydroxylated complexes

Probably the most important ligand to be considered in complexing effects of counterions is the hydroxyl ion. Hydrolyzable ions are present in most natural sols and are used in inumerable applications, which makes the understanding of their effects on colloid stability essential.

It has been established beyond any doubt that replacing at least one coordinated water molecule of a hydrated metal ion by a hydroxyl ion dramatically enhances the adsorption of this ion on lyophobic surfaces of various chemical compositions. Despite this general phenomenon, some hydro-

lyzed complexes may still observe the Schulze-Hardy rule. This will be the case when the adsorption of hydrolyzed counterions is negligible even from solutions containing these species at concentrations sufficient to coagulate the sol[44]. Such conditions may be met when highly charged polynuclear hydrolysis products are formed. According to the Schulze-Hardy rule, the c.c.c. of a highly charged counterion is extremely low; at such low concentrations the adsorption of a hydrolyzed counterion may be so slight that it hardly affects the particle charge. As a consequence, these ions cause coagulation by the same mechanism as any other nonadsorbed ions of the same charge. At higher concentrations of the same hydrolyzed ions the adsorption increases, and as a result the stability of sols is affected as will be described in a later section. Here the effects of pH on critical coagulation concentrations of lyophobic sols for several hydrolyzable cations will be explained in terms of counterion complexing in the electrolyte environment.

As was shown in Figure 5, the c.c.c. of aluminum nitrate for a negatively charged silver iodide sol decreases at pH ∿4 and then remains constant at a lower value for pH >∿4.7. At this lower c.c.c., the mobility of the sol particles remains essentially the same as for uncoagulated sols, indicating that no detectable adsorption of counterions occurs (see Figure 19). Thus, the enhanced coagulation ability of the hydrolyzed aluminum species must be explained in terms of increased charge of the counterions. In general, it is agreed that in the hydrolysis products of aluminum ion the ratio of Al^{3+} to OH^- is 1:2.5. Assuming the charge of the complex to be 4+, the coagulating hydrolyzed species between pH $\sim$4.7 and 7 should be consistent with the formulation $Al_8(OH)_{20}^{4+}$ [38]. Indeed, if the coagulation concentration of $Al(NO_3)_3$ at pH > 4.7 (Figure 5) is divided by 8 to account for the polymerization of the complex counterion, a value or $8 \times 10^{-7} M$ is obtained which is in very good agreement with the expected c.c.c. of 4+ ions for negatively charged silver halide sols (Figure 4). A tetrapositively charged polymeric hydrolysis product of aluminum, $Al_7(OH)_{17}^{4+}$, was later on proposed by Biedermann[45] from potentiometric titration data. The similarity of the formulations for the dominant aluminum hydrolyzed species, established from stability and from electrochemical data, is striking.

It is rather interesting that the c.c.c. for aluminum sulfate shows about the same trend as that for aluminum nitrate at pH >$\sim$4.5. This suggests that the hydrolyzed species in sulfate solutions are also of 4+ charge. However, in the presence of sulfate ions the ratio of Al^{3+} to OH^- was found to be 1:1.25 which must mean that basic aluminum sulfate ions act as coagulants at pH >$\sim$4.5. The effect of pH on the c.c.c. and the ligand ratio are consistent with the formulation $[Al_8(OH)_{10}(SO_4)_5]^{4+}$ [46].

Another case to be considered deals with the stability data caused by hydrolysis of beryllium ions. The change of the coagulation concentration of beryllium nitrate with pH for a negatively charged silver bromide sol is shown by the open circles in Figure 9. As the pH increases the c.c.c. decreases, which is in agreement with the generally accepted formulation of the hydrolyzed species as $Be_3(OH)_3^{3+}$ [47]. The full circles give the ratio of the concentrations of hydrolyzed to total beryllium ion as calculated using Sillén's value for the formation constant $\log K_{33} = -8.66$ [48]. The agreement with the coagulation data is rather good. Up to pH $\sim$4 the species responsible for coagulation is the divalent unhydrolyzed Be_{aq}^{2+} ion, whereas at pH >$\sim$6 the trimer of 3+ charge acts as coagulant. It is interesting that Sillén also suggested the existence of a neutral $Be(OH)_2$ hydrolysis product, for which he assigned $\log K_{21} = -10.87$. If this were correct, at pH >6 almost all of the beryllium would be in the form of this neutral species, which is entirely inconsistent with the coagulation data. More recently[49] the most basic hydrolysis product proposed by Sillén, $Be(OH)_2$, was found to be incorrect and only the assumption of hydrolyzed beryllium species of charge 3+ sufficed to interpret the potentiometric measurements, confirming the conclusions based upon stability results.

The previous two cases dealt with ions having one predominant polynuclear hydrolysis product, the charge of which was higher than that of the corresponding nonhydrolyzed metal ions. Only in such cases could one hope to use stability effects to identify counterion species. However,

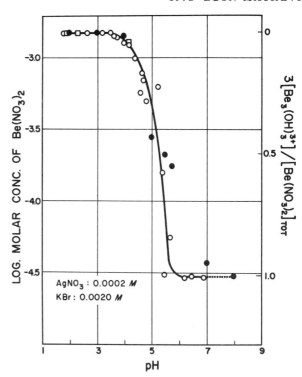

Figure 9. Critical coagulation concentrations of beryllium nitrate (○) and beryllium perchlorate (□) as a function of pH for a negatively charged silver bromide sol. Full circles correspond to the $3[Be_3(OH)_3^{3+}]/[Be(NO_3)_2]_{tot}$ vs. pH plot using $\log K_{33} = -8.66$ [52].

coagulation results with counterions that give a series of hydrolyzed ions may be explained if reliable complex constants are available. For illustration, data obtained with scandium ions will be shown here. The effect of pH on the c.c.c. of scandium nitrate for a negatively charged silver bromide sol is shown by the open circles in Figure 10[50]. Aveston[51] identified the following complex hydrolyzed scandium ions: $ScOH^{2+}$, $Sc_2(OH)_2^{4+}$, $Sc_3(OH)_4^{5+}$, and $Sc_3(OH)_5^{4+}$ and evaluated their formation constants. Using these, the concentrations of all the scandium species were calculated for the conditions along the c.c.c.-pH curve in Figure 10. It was found that the concentrations of $ScOH^{2+}$ and of $Sc_3(OH)_4^{5+}$ are negligible, and, therefore, these species cannot affect the rate of coagulation. The concentration of the unhydrolyzed ions Sc_{aq}^{3+} and the sum of concentrations of the two tetrapositively charged species $Sc_2(OH)_2^{4+}$ and $Sc_3(OH)_5^{4+}$ are given by triangles and squares, respectively. Up to pH $\sim$4, the coagulation is obviously due to the presence of the unhydrolyzed scandium ions. Above pH $\sim$4.5 the total concentration of the tetrapositive ions is fairly constant and characteristic of the c.c.c. of counterions of 4+ charge (see arrow), indicating that these species additively cause coagulation. Over the same pH range, the concentration of the Sc^{3+} ions drops off rapidly and is well below the c.c.c. for 3+ counterions. Thus, we have a rather remarkable case in which one can identify the coagulating species despite the complexity of the electrolyte environment.

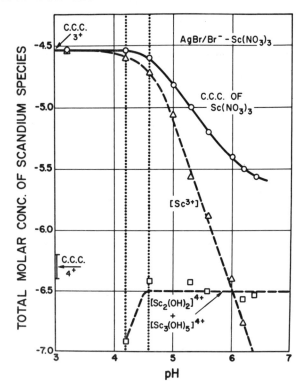

Figure 10. Critical coagulation concentrations of scandium nitrate (circles) as a function of pH for a negatively charged silver bromide sol. Triangles give the concentrations of the unhydrolyzed scandium ion, Sc_{aq}^{3+}, and squares represent the sum of the molar concentrations of $Sc_2(OH)_2^{4+}$ and $Sc_3(OH)_5^{4+}$ in the corresponding sols as calculated using the hydrolysis constants by Aveston[51]. Arrows indicate the c.c.c. of 3+ and 4+ counterions for the same sol.

The final example will refer to the interesting situation of complexing when polymerization takes place without affecting the charge or the adsorptivity of the counterions. The net result should be a moderate decrease in the coagulation ability because, for a given total concentration, the number of counterions will decrease with complexing. The hydrolysis of uranyl ion provides us with such a system. In discussing this hydrolysis, two processes need to be considered[53]:

$$UO_2^{2+} + H_2O \rightleftharpoons UO_2OH^+ + H^+$$

and $$2UO_2^{2+} + H_2O \rightleftharpoons (UO_2)_2O^{2+} + 2H^+$$

Table II shows that the c.c.c. of $UO_2(NO_3)_2$ for a negatively charged silver iodide sol, increases with increasing pH[53]. The corresponding concentrations of all species in solution are also given in the table and it can be seen that the concentration of singly charged complex is much too low to have an effect on the coagulation. Thus, only the two 2+ species have to be considered. As the pH rises the concentration of the monomeric ion decreases, whereas that of the dimeric ion increases. The sum of the two remains reasonably constant over the entire pH range (column 5). It is then obvious that the rate of coagulation is determined by the combined effect of the two doubly charged species, UO_2^{2+} and

Table II

Critical Coagulation Concentrations of $UO_2(NO_3)_2$ and the Concentrations of Various Counterion Species as a Function of pH for a Silver Iodide Sol in Statu Nascendi
($AgNO_3$ $1\times10^{-4}M$, KI $2\times10^{-3}M$)

1 pH	2 Exptl. c.c.c. $\times 10^3$ $[UO_2(NO_3)_2]_t$ (M/1)	3 $[UO_2^{2+}]\times 10^3$ (M/1)	4 $[U_2O_2^{2+}]\times 10^3$ (M/1)	5 (3+4) (M/1)$\times 10^3$	6 log. c.c.c. (3+4)	7 $[UO_2OH^+]$ (M/1)
3.50	1.35	1.30	0.017	1.32	-2.88	1.64×10^{-5}
3.85	1.29	1.13	0.063	1.19	-2.92	3.19×10^{-5}
4.00	1.48	1.16	0.135	1.30	-2.89	4.65×10^{-5}
4.25	1.66	0.98	0.304	1.28	-2.89	6.98×10^{-5}
4.38	1.95	0.906	0.472	1.38	-2.86	8.69×10^{-5}
4.70	2.63	0.615	0.946	1.56	-2.81	1.26×10^{-4}
5.00	3.09	0.358	1.282	1.54	-2.81	1.43×10^{-4}
5.50	3.16	0.120	1.457	1.58	-2.80	1.52×10^{-4}

$(UO_2)_2O^{2+}$, and that the increase in the c.c.c. is directly related to the dimerization of the uranyl ion. It is noteworthy that the sol stability is sufficiently sensitive to reflect this type of counterion complexing in solution.

Coagulation effects with positive sols and anions as counterions give analogous results. For example, the question whether 6-molybdocobaltic acid is in solution in the monomeric form, $H_3CoMo_6O_{21}$, or in the dimeric form, $H_6Co_2Mo_{12}O_{42}$, was resolved by determining the c.c.c. of this acid for a positively charged silver bromide sol. The results definitely indicated a counterion charge of minus three, which favors the monomeric formulation[54].

c. Complexing with organic ligands

In view of the rather pronounced phenomena that result from complexing of counterions with inorganic ligands, it stands to reason to expect even larger effects if coagulating ions are coordinated with organic ligands. The latter vary so much in charge, size, shape, degree of hydration, etc., that, upon coordination, many different reactions can be anticipated between the complex counterions and the particle surface constituents. Systematic studies of interactions of a number of metal chelates with colloidal sols have indeed shown significant enhancements in coagulation ability and even reversal of charge by metal counterions as a consequence of chelation. Examples of these effects will be given later. Here, only the coagulation of silver halide sols by ethylenediamine, en, and by its nickel chelate will be discussed as it could be shown that the adsorptivity of these species over the c.c.c. range is negligible, yet the sol destabilization is strongly dependent on the chemical composition of the electrolyte medium.

The upper part of Figure 11 shows the effect of pH on the c.c.c.s of Ni^{2+}, bis-ethylenediamine Ni(II)-ion, $Ni(en)_2^{2+}$, and tris-ethylenediamine Ni(II)-ion, $Ni(en)_3^{2+}$, for a negatively charged silver bromide sol[55]. At first, it would seem that the chelation of Ni^{2+} with en leads to a considerable lowering in the c.c.c. However, it could be shown that the latter is actually due to the presence of the protonated en. The nickel chelates with en are rather unstable at low pH and decompose according to this schematic sequence of reactions:

$$Ni(en)_3^{2+} \xrightarrow{H^+} Ni^{2+} + enH^+$$

$$enH^+ \xrightarrow{H^+} enH_2^{2+}$$

294

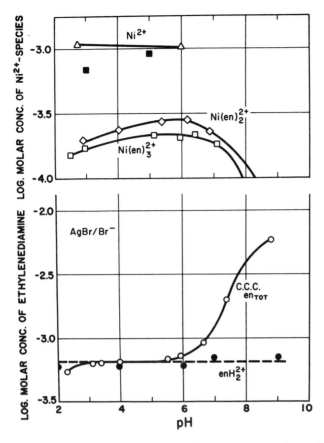

Figure 11. Upper: Critical coagulation concentrations of Ni_{aq}^{2+} (△), $Ni(en)_2^{2+}$ (◇), *and* $Ni(en)_3^{2+}$ (□), *respectively, as a function of pH for a negatively charged silver bromide sol. Full squares give the sum of the concentrations of all 2+ charged species in the silver bromide-*$Ni(en)_3^{2+}$ *system.*

Lower: Critical coagulation concentration of ethylenediamine (en) (○) *as a function of pH for a negatively charged silver bromide sol. Full circles give the concentration of* enH_2^{2+} *ions in the corresponding sols.*

In the lower part of Figure 11 the c.c.c. of ethylenediamine is plotted against pH for the same silver bromide sol. Up to pH 6 the c.c.c. is fairly constant, but it increases rather sharply above this pH. The calculated concentration of enH_2^{2+} along the coagulation line is given by full circles. This remains unchanged over the entire pH range and one must conclude that the coagulation of the sol is caused by diprotonated species. The c.c.c. of enH_2^{2+} is characteristic of counterions of 2+ charge of somewhat larger size. This is substantiated by the fact that the isoelectric point (IEP) is far removed from the c.c.c. for the silver bromide sol in the presence of en (Table V). The increase at pH>6 is due to deprotonation of the coagulating ion.

These results also explain the behavior of the $Ni(en)_3^{2+}$ chelate. At low pH the solution actually contains Ni^{2+} and enH_2^{2+} ions rather than the complex. The concentrations of all ions in solution, including the lower charged enH^+ and en species, were calculated for the c.c.c. of $Ni(en)_3^{2+}$ at different pH values[22]. The sum of $[Ni^{2+}]$ and $[enH_2^{2+}]$ is given by

full squares; this is rather close to the c.c.c. of $Ni(NO_3)_2$. It is quite apparent that the coagulation of the sol is caused by these two ions in an additive manner. The somewhat lower total coagulation value is due to more efficient coagulation effect of enH_2^{2+} as shown in the lower portion of Figure 11. Thus, a rather complicated situation could be well explained by a careful analysis of the composition of the electrolyte medium.

2. Strong interactions of solutes with solids

When counterions interact with particles, two major effects may be expected: sol destabilization due to surface charge reduction and, in some cases, charge reversal. If the interactions are strong, the sols may be destabilized in the presence of counterions at concentrations significantly lower than would be needed if the coagulation were caused by double layer compression (i.e., if the Schulze-Hardy rule were applicable). When charge reversal takes place, the stability of the sol depends on the surface charge density of the recharged sol and on the characteristics of the accompanying ions, i.e. ions which have charges of the same sign as the original sol. Once the charge of the sol is reversed these accompanying ions assume the role of counterions. In general, the Schulze-Hardy rule is expected to apply if the sols of the reversed charge remain hydrophobic[56].

The nature of the interactions between the solutes and sol particles will vary considerably, depending on the properties of the potential-determining complexes and the constituent ions of the solid phase and the types of species in the solution phase. For example, processes such as ion exchange, condensation, coordination, and polymerization have been found to take place at solid/solution interface. In many cases the surface reactions are still not properly identified. In this discussion the effects of ion exchange, surface coordination, and adsorption will be demonstrated. This subdivision is somewhat arbitrary, particularly as cases will be presented under the heading of "adsorption" in which the natures of the interfacial reactions have not been fully identified.

a. Ion exchange

Figure 12 illustrates how differently two sols may behave in the presence of the same electrolytes. The diagram on the left-hand side shows the effects of pH and excess bromide concentration on the c.c.c. of Na^+, Ca^{2+}, and La^{3+} for a negatively charged silver bromide sol. In each case the c.c.c. depends only on the counterion charge (the sols are coagulated at concentrations above the lines as indicated by shading) in agreement with the Schulze-Hardy rule. The right-hand side gives the coagulation concentrations of the same counterions for a Ludox silica sol. One observes the tremendous effect of pH and it is quite apparent that the charge rule does not apply in this case. Particularly remarkable is the fact that the c.c.c. of mono- and divalent counterions decrease rapidly with increasing pH, which means that the silica sols are less stable when they are more strongly charged. Obviously the two sols are destabilized by different

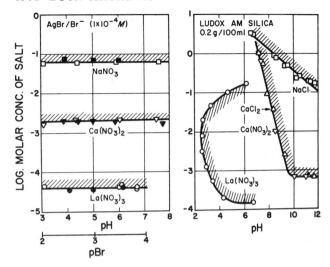

Figure 12. *Left: Critical coagulation concentrations of* NaNO$_3$ ($\square$, $\blacksquare$), Ca(NO$_3$)$_2$ (∇, $\blacktriangledown$), *and* La(NO$_3$)$_3$ ($\bigcirc$, $\bullet$), *respectively, as a function of pH (open symbols) and of pBr (full symbols) for a negatively charged silver bromide sol.*

Right: Critical coagulation concentrations of NaCl ($\square$), CaCl$_2$ ($\triangle$), Ca(NO$_3$)$_2$ (∇), *and* La(NO$_3$)$_3$ ($\bigcirc$), *respectively, as a function of pH for a Ludox AM silica sol.*

mechanisms. Silica particles have two distinct characteristics: they are highly hydrated, especially at low pH values, and their surfaces contain silanolic groups, $\equiv$SiOH, whose protons are exchangeable for counterions in solution. It could be shown that these properties of the particle surface are responsible for the stability behavior of silica sols.

Firstly, the exchange of the silanolic protons was studied in detail and explained quantitatively. There is considerable evidence that two distinct types of silanol groups (to be designated as I and II) are available on silica[57]. Assuming that the free energies of exchange differ so much that type I sites first fill completely, the relevant equilibria can be written

$$K_I = \frac{[\equiv SiO_I Na]\,[H^+]}{[\equiv SiO_I H]\,[Na^+]}$$

$$K_{II} = \frac{[\equiv SiO_{II} Na]\,[H^+]}{[\equiv SiO_{II} H]\,[Na^+]}$$

It is convenient to suppose that the difference between the free energies of a hydrogen and a sodium ion at the surface depends on the total number of exchanged sites:

$$RT \ln K_i = RT \ln K_{i,o} - A_i f_i$$

where $K_{i,o}$ is for that kind of site when it is fully occupied by protons,

$$f_i = [\equiv SiO_i Na]/[\equiv SiO_i H]_o$$

is the fraction of sites on which protons have been replaced by sodium ions, and A is a proportionality constant.

Letting ϕ represent the fraction of type I sites, one can write

$$\log f_I/(1-f_I) + A_I f_I/2.303\,RT = pH + \log K_{I,o} + \log[Na^+]$$

$$\log f_{II}/(1-f_{II}) + A_{II} f_{II}/2.303\,RT = pH + \log K_{II,o} + \log[Na^+]$$

and

$$f = f_I \phi + f_{II}(1-\phi).$$

Assuming $A_I = A_{II}$ one obtains the relationship given by the solid line in Figure 13, which is an exceedingly good fit to the experimental points (circles)[58].

Since water is bound to silica by hydrogen bonding to silanolic protons[59], ion exchange should have a strong effect on the degree of hydration, and consequently on the sol stability. Indeed, it was possible to show that destabilization of silica sols at a given pH occurs when a critical fraction of silanolic protons is exchanged. Figure 14 gives the measured exchange of silanolic protons for a number of different counterions at their respective critical coagulation concentrations as a function of pH. All counterions fall on the same "master curve", although for a given pH the c.c.c.

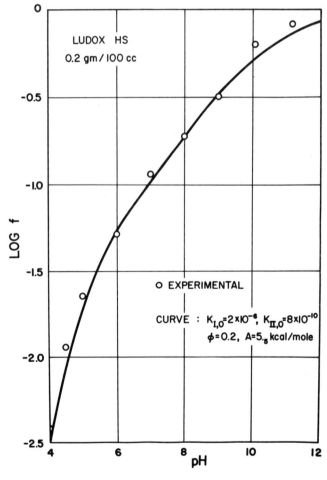

Figure 13. *Plot of the logarithm of the fraction of the total number of exchange sites on Ludox HS silica occupied by* Na$^+$ *against pH. Circles denote experimental data and the curve is computed with numerical parameters shown on the figure.*

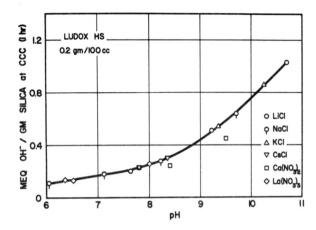

Figure 14. Critical exchange curve for Ludox HS silica. The plot shows the number of milliequivalents of hydroxide ions consumed by 1 gm of silica as a function of pH at the 1 hour critical coagulation concentrations of LiCl (○), NaCl (♀), KCl (△), CsCl (▽), Ca(NO₃)₂ (□), and La(NO₃)₃ La(NO₃)₃ (◇).

may differ rather markedly (Figure 12). Thus, the coagulation concentration of a salt is that amount of a counterion which will produce the same critical exchange. With increasing pH this corresponds to a decreased degree of hydration which results in reduced stability and consequently lower c.c.c.[60].

The critical extent of ion exchange explains the stability of silica sols in situations which would be otherwise difficult to understand. For example, the c.c.c. of K^+ ions for the Ludox AM sol shows a pronounced minimum when plotted against pH, yet all the values fall on the same exchange "master curve"[61]. The same principle applies to coagulation by mixtures of electrolytes. Table III gives the exchanged quantities at the c.c.c. of KCl, NaCl, and KCl + NaCl for the Ludox AM sol at pH 11.30. In all three cases, the exchanged amount remains constant (within ± 1%), showing excellent additivity in the case of mixed salts. The ion-exchange behavior of Ludox sols seems to be characteristic of silicas prepared in many different ways[62].

Table III

Amounts of OH⁻ Consumed per Gram of Ludox AM at pH 11.3 at c.c.c. of Monovalent Counterions

Coagulating Electrolyte	c.c.c.	Meq. OH⁻/gm silica
KCl	0.63 M	1.06
NaCl	0.21 M	1.07
KCl + NaCl	0.20 + 0.44 M	1.05

b. Surface coordination

The surface characteristics, and consequently the stability of a sol may be influenced by reactions other than those between counterions and potential-determining ions, although these are the reactions whose effects are the most commonly observed. In principle, any ion contained in the particle could interact with a solute species with a resulting change in the sol stability. A rather unique example of the effects of surface coordination on sol stability is found when certain organic ligands such as 2,2'-dipyridyl (dipy) or 1,10-phenanthroline (phen) are added to silver halide sols.

Figure 15 gives the c.c.c. boundary (circles) and the critical stabilization boundary (c.s.c.) due to charge reversal (squares) for a negatively charged silver bromide sol in the presence of dipy over the pH range 2–8[22]. Two effects are noteworthy: the c.c.c. and the c.s.c. decrease with increasing pH and the restabilization due to charge reversal does not take place in acidic solutions (at pH < 3). This behavior of dipy is distinctly different from that of ethylenediamine (en) for the same sol (Figure 11). The latter was a better coagulant at low pH and no restabilization with en was accomplished. Like en, 2,2'-dipyridyl is an organic base, i.e., it undergoes protonation in acidic aqueous solutions according to

$$dipy \xrightarrow{\text{H}^+} dipyH^+ \xrightarrow{\text{H}^+} dipyH_2^{2+}.$$

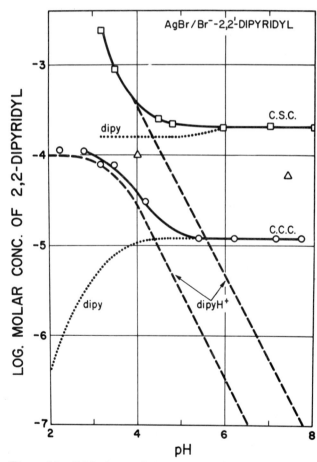

Figure 15. Critical coagulation concentrations (circles) and critical stabilization concentrations due to charge reversal (c.s.c., squares) of 2,2'-dipyridyl (dipy) as a function of pH for a negatively charged silver bromide sol. Dotted and dashed lines give the molar concentrations of dipy and dipyH⁺, respectively, as calculated for the corresponding c.c.c.s and c.s.c.s. Triangles designate the points of zero charge.

From the known dissociation constants the concentrations of dipy and its protonated forms were calculated for the conditions along the c.c.c. and c.s.c. curves. In both cases the concentrations of dipyH$_2^{2+}$ were negligible. The rather unexpected result is that for the negatively charged sol the *neutral* dipy species is a more efficient coagulant and charge reversal agent than the protonated species dipyH$^+$. This can be explained by the fact that dipy coordinates with silver ions of the silver bromide crystal lattice by replacing the bromide ions, resulting in charge reduction and destabilization. When all excess bromide ions are exchanged for dipy, the zero point of charge is reached (triangles in Figure 15). Further substitution of neutral ligands for the bromide ions leaves silver ions in excess and charge reversal takes place, finally leading to sol restabilization. The protonated ligands cannot coordinate with positively charged silver ions; thus, at lower pH the coagulation ability of 2,2'-dipyridyl decreases and restabilization does not occur. Very similar results are obtained with 1,10-phenanthroline and silver halide sols. In neither case do the counterion charges play a significant role in sol destabilization.

The formation of surface coordinated silver complexes with dipy and phen was confirmed by infrared spectra. Table IV shows a great similarity of characteristic frequencies for phen and dipy adsorbed on silver bromide and for Ag(phen)$_2^+$ and Ag(dipy)$_2^+$ ions in a KBr matrix[63].

Table IV

Characteristic Infrared Frequencies between 700 and 900 cm^{-1} of 1,10-phenanthroline, 2,2'-dipyridyl on a AgBr Sol and of [Ag(phen)$_2$]$^+$ and [Ag(dipy)$_2$]$^+$

1,10-phenanthroline in KBr matrix	724 sh	739 s	778 m		
adsorbed on AgBr	732 vs		776 vs	847 vs	862 s
[Ag(phen)$_2$]$^+$ in KBr matrix	727 vs		764 s	840 vs	860 s

2,2'-dipyridyl in KBr matrix	741 sh	758 vs
adsorbed on AgBr	737 m	755 vs
[Ag(dipy)$_2$]$^+$ in KBr matrix	736 ms	753 vs

Abbreviations: sh = shoulder, m = medium, s = strong, vs = very strong

c. Counterion adsorption

As pointed out earlier the adsorption of counterions on colloidal particles may exercise two effects; it can lower the charge density of the potential-determining ions and thus decrease the stability of the sol, or the counterions may be adsorbed in sufficient amounts to take up the role of potential-determining ions, i.e. cause charge reversal. The adsorptivities of counterions depend on their ligands and on the type of the potential-determining or constituent ions of the adsorbent. The exact nature of the adsorption process is in most cases not understood. However, it has been shown that—particularly in the case of hydrosols—water

molecules, either as counterion ligands or in the form of solvation layer, play a most significant role.

In this work, examples of stability effects resulting from strong adsorption of counterions will be discussed using some metal chelates and hydrolyzed metal ions as examples.

Figure 16 shows the effects of pH on the critical coagulation concentrations and critical stabilization concentrations for 2,2'-dipyridyl and 1,10-phenanthroline chelates of Co^{3+} and Ni^{2+} for a negatively charged silver bromide sol. Several observations are significant. The c.c.c. of cobalt(III)-chelates is nearly three orders of magnitude lower than that for the uncomplexed Co^{3+} ion, which is indicated by the arrow. The coagulation ability of nickel is even more strongly enhanced by chelation as compared to the simple Ni^{2+} ion (Figure 11). Furthermore, the effect of counterion charge is diminished in the case of dipy complexes and it completely disappears with phen chelates; Ni(phen)$_3^{2+}$ and Co(phen)$_3^{3+}$

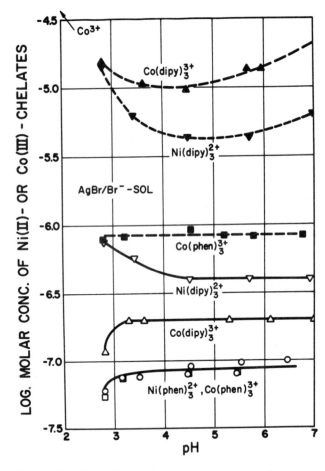

Figure 16. Critical coagulation concentrations (open symbols, full lines) and critical stabilization concentrations due to charge reversal (full symbols, dashed lines) of several metal chelates as a function of pH for a negatively charged silver bromide sol. Chelates: tris(2,2'-dipyridyl) Ni^{2+} (∇ ▼); tris (1,10-phenanthroline) Ni^{2+} (○); tris (2,2'-dipyridyl) Co^{3+} (△ ▲); tris (1,10-phenanthroline) Co^{3+} (□ ■). Arrow indicates the c.c.c. of Co$_{aq}^{3+}$ for the same sol; for the c.c.c. of Ni$_{aq}^{2+}$ see Figure 11.

show the same c.c.c. Finally, although Co(dipy)$_3^{3+}$ is a somewhat better coagulating agent, it restabilizes the sol less efficiently than the corresponding Ni-complex. All of this clearly indicates that the coagulation is not caused by a double-layer compression mechanism, but must be related to strong interactions of these chelates with silver halide particles leading to charge neutralization. Restabilization is due to charge reversal. To substantiate this, Table V lists the critical coagulation concentrations and the electrophoretic isoelectric points (IEP) of several chelates and chelating agents for the same silver bromide sol. The c.c.c. of every solute except ethylenediamine (en) is just a little lower than the IEP, clearly indicating that the onset of coagulation is due to particle-charge neutralization. The case of en was discussed earlier.

Both infrared spectra and tracer adsorption measurements with cobalt chelates containing Co^{60} have confirmed the adsorption of these chelates on silver bromide particles[63]. One possible mechanism for this adsorption involves association of the potential-determining halide ion with the highly charged chelate counterion, subsequent coordination to the central Co(III) atom, and the formation of an intermediate or a transition state complex[64]. This could then be followed by a partial dissociation of the chelating ligand(s).

The suggestion that the exceptionally strong effects of metal chelates (and chelating agents) on silver halide sols are due to coordination reactions between these solutes and the crystal lattice ions is indirectly substantiated by the fact that the stabilities of some other lyophobic colloids are much less strongly affected by the same counterions.

Figure 17 is a plot of the c.c.c. of Co(dipy)$_3^{3+}$ and

Table V

Critical Coagulation Concentrations (c.c.c.) and Isoelectric Points (IEP) of a Silver Bromide Sol (AgBr: $1.0 \times 10^{-4} M$, pBr$^-$: 3.72) in the Presence of Various Metal Chelates and Chelating Agents

Solute Species	pH	log c.c.c. (M/1)	IEP
Ni(dipy)$_3^{2+}$	3.6	-6.2	-6.1
	6.4	-6.4	-6.2
Co(dipy)$_3^{3+}$	3.6	-6.7	-6.5
Co(en)$_3^{3+}$	3.7	-6.4	-6.0
	6.4	-6.4	-6.0
Cr(en)$_3^{3+}$	3.6	-6.2	-6.1
	6.4	-6.0	-5.9
en	3.4	-3.2	-1.7
	5.2	-3.2	-1.9
dipy	4.4	-4.6	-4.3
	7.4	-4.9	-4.1
phen	3.5	-6.0	-5.7
	7.8-8.6	-6.0	-5.8

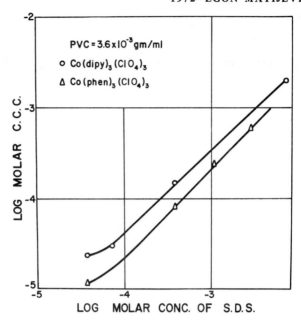

Figure 17. Critical coagulation concentrations of Co(dipy)$_3^{3+}$ *(○) and of* Co(phen)$_3^{3+}$ *(Δ) as a function of the total molar concentration of sodium dodecyl sulfate (SDS) for a polyvinyl chloride (PVC) latex.*

Co(phen)$_3^{3+}$ for a polyvinyl chloride (PVC) latex against the molar concentration of the stabilizer sodium dodecyl sulfate[65]. One notices that the coagulation ability of these complex counterions is much less for PVC than for the silver bromide sol. Also, the c.c.c. increases considerably as the surfactant concentration becomes higher.

It has been shown that electrolytes enhance the adsorptivity of dodecyl sulfate ions on latex[66], and that chelates associate with this surfactant. Figure 18 gives two adsorption isotherms obtained using cobalt chelates tagged with Co^{60} along with the areas per molecule of the two chelates on the "monodispersed" PVC calculated from the known

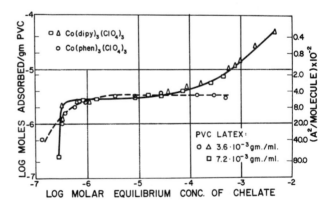

Figure 18. The adsorption isotherm of Co(dipy)$_3^{3+}$ *and* Co(phen)$_3^{3+}$ *on the polyvinyl chloride (PVC) latex. The log of the moles of adsorbed chelate ions and the area per chelate ion are plotted vs. the equilibrium molar concentration of the chelates. PVC latex concentrations:* 3.6×10^{-3} *gm/ml (Δ, ○);* 7.2×10^{-3} *gm/ml (□).*

size of the spherical latex particles. At saturation, the area is 450 Å²/molecule as compared to ∼120 Å² for the cross-sectional area of such complexes[67]. Thus, two compensating phenomena define the stability of the PVC latex in the presence of chelate counterions: the adsorption of chelates decreases the particle charge density, whereas the enhanced adsorption of dodecyl sulfate, caused by the presence of these complexes, increases the charge. The measured c.c.c. reflects the net effect of these interfacial interactions. Electrophoretic mobility measurements are in line with these observations[67].

The changes in stability due to adsorption of hydroxylated metal ions will be illustrated using aluminum ions as an example. The lower part of Figure 19 shows turbidities of a silver iodide sol in the presence of aluminum nitrate in decreasing concentration for two different pH values. High and low turbidities indicate coagulated and stable sols, respectively. The pH effect is rather dramatic: at pH 3 there is only one coagulation boundary (A) which is characteristic for counterions of 3+ charge. At all concentrations higher than the c.c.c. the sols are coagulated, and the particles remain negatively charged over the entire range of salt concentrations. At pH 5 the coagulation boundary A is shifted although the sol is still negatively charged at the c.c.c. This shift was discussed earlier (Figure 5). At somewhat higher concentrations the charge of the sols is first neutralized and then reversed resulting in a restabilization boundary B above which stable positively charged sols exist until at C the concentration of anions is sufficiently high to produce the second coagulation maximum.

If hydrolysis of aluminum ions is responsible for the mobility change and the resulting restabilization, it should be possible to observe corresponding effects if the concentration of the aluminum ion is kept constant while the pH is increased systematically. The dotted line in the upper part of Figure 19 shows the effect of pH on the mobility of a silver iodide sol, which remains reasonably constant as expected. In the presence of $Al(NO_3)_3$ $(9 \times 10^{-4} M)$, the mobility shows strong charge reversal (dashed line) accompanied by sol restabilization (solid curve, right). Aging aluminum salt solutions at elevated temperatures enhances the hydrolysis of aluminum ion and should, in turn, result in more efficient charge reversal. The left coagulation curve and the corresponding mobility curve in Figure 19 (upper) were obtained with systems identical with those described except that the $Al(NO_3)_3$ solution was aged at 90°C for 24 hours, cooled, and then used in coagulation experiments. The effect of this aging of the salt solution upon mobility and stability of the sol is rather striking[21].

Since the restabilization due to charge reversal by metal counterions is caused by adsorption of hydrolyzed species, the c.s.c. of a counterion for a given sol must change with pH. This is necessary to produce a constant concentration of hydrolyzed species that just suffices to establish a repulsion barrier on the particles which insures sol stability. For the case of aluminum ions, for which the hydrolysis product was shown earlier to be consistent with the composition

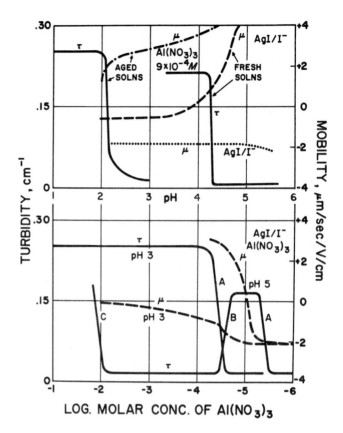

Figure 19. Lower: Turbidities (τ, solid lines) and mobilities (μ, dashed lines) of a silver iodide sol in the presence of various molar concentrations of aluminum nitrate at pH 3 and 5, respectively. A, coagulation boundaries of negatively charged sols, B, restabilization boundary due to charge reversal, C, coagulation boundary of recharged sol.

Upper: Turbidities (τ, solid lines) of a silver iodide sol as a function of pH in the presence of a constant amount of $Al(NO_3)_3$ $(9 \times 10^{-4} M)$ added from a freshly prepared solution (right curve) and from a solution aged at 90°C for 24 hours. Mobilities of the same silver iodide sol as a function of pH in the absence of $Al(NO_3)_3$ (dotted line), and in the presence of $9 \times 10^{-4} M$ $Al(NO_3)_3$ added from a freshly prepared solution (dashed line) and a solution of aluminum nitrate aged at 90°C for 24 hours (dashed-dotted line).

$Al_8(OH)_{20}^{4+}$, one may consider the following equilibrium

$$8Al^{3+} + 20H_2O \rightleftharpoons Al_8(OH)_{20}^{4+} + 20H^+$$

$$K = \frac{[Al_8(OH)_{20}^{4+}] \ [H^+]^{20}}{[Al^{3+}]_8}.$$

For any combination of aluminum salt concentration-pH conditions restabilization due to charge reversal will take place when $[Al_8(OH)_{20}^{4+}]$ = constant. If a small fraction of total aluminum ions is hydrolyzed

$$[Al^{3+}]_{equil.} = [Al^{3+}]_{total}$$

and as a result

$$\log[Al^{3+}] = -2.5\,pH + K'$$

where $\qquad K' = 1/8 \log\left\{[Al_8(OH)_{20}^{4+}]/K\right\}$

Thus a plot of log $[Al^{3+}]$ vs. pH for the c.s.c. of a given sol should give a straight line with a slope of -2.5 [21]. Figure 20 shows such plots for five different sols. In each case within experimental error the slope has—as expected—a value of -2.5

The reasons for the enhanced adsorptivity of hydrolyzed metal ions on lyophobic surfaces are not understood. The effect is observed with particles of varied chemical composition and must be related to the inclusion of one or more hydroxyl ions into the coordination sphere of the counterion. Just why the hydroxyl ion promotes counterion adsorption is an unresolved question. It is certain that the electrostatic contribution to counterion adsorptivity is negligible. Hafnium ions hydrolyze under certain conditions to neutral, soluble species, Hf(OH)$_4$, yet radioactive tracer experiments using Hf[181] showed that these molecular species are efficiently adsorbed onto various lyophobic sur-

faces[68,69]. The fact that unhydrolyzed ions, regardless of their charge, are not adsorbed onto such surfaces (unless, for example, they form a sparingly soluble salt with the potential-determining ion) is explained by ion hydration. All metal ions, and particularly polyvalent ones, are strongly hydrated in aqueous solutions. Thus, hydrophobic colloidal particles will not adsorb unhydrolyzed ions coordinated with water for the same reason they do not adsorb single water molecules; especially since the charge of the counterion was shown not to be a sufficient cause for adsorption.

Hydrolyzed metal ions are not adsorbed onto hydrophilic colloids unless some specific interfacial reaction takes place. This could be ion exchange or a condensation reaction with some surface groups, such as in the case of silica.

$$\equiv SiOH + OH-M(H_2O)_n^{z+} \rightleftharpoons \equiv SiO-M(H_2O)_n^{z+} + H_2O$$

where $M^{(z+1)+}$ is the hydrolyzable metal ion.

Obviously, water molecules, either coordinated with an unhydrolyzed counterion or present as a solvation layer around the particles, represent the most efficient barrier to interactions between the ions and the solid surfaces.

An interesting aspect of the processes between hydrolyzed ions and lyophobic surfaces is the desorbability of the adsorbed species. This problem might shed some light on the mechanism of the adsorption phenomena due to counterion hydrolysis. Only recently have desorption studies been undertaken. Figure 21 shows how the pH affects the mobilities of a styrene-butadiene (SBR) latex and of a silver iodide sol in solutions containing Al(NO$_3$)$_3$ (open symbols). In each case rather strong charge reversal is observed at pH $> \sim 4.3$. Full symbols represent the mobilities of sols that were first recharged with the same concentrations of aluminum ions at pH ~ 6 and then acidified with nitric acid. Dashed curves are for different times after acidification. The charge in each case becomes less positive with time; if the sols are made sufficiently acidic the charge of the particles is eventually reversed back to negative, indicating the desorption of the dehydrolyzed aluminum ions. It is noteworthy that the desorption process is rather slow, particularly if one recognizes that the adsorption of hydrolyzed species is almost instantaneous. This would indicate a slow surface decomplexation of the adsorbed counterions upon acidification.

In view of the fact that mobilities change on lowering the pH values of recharged sols, the stabilities should also be affected. Figure 22 shows this to be the case. One hour after acidification, SBR latex sols are stable, which is to be expected since at that time the mobility is still high and essentially independent of pH. However, after several hours the sols become less strongly charged at lower pH values and consequently unstable, giving rise to a coagulation boundary. If sufficiently long time is allowed, the c.s.c.s obtained by acidification of the recharged sols and by increasing the pH of a freshly prepared sol in the presence of the same concentration of Al(NO$_3$)$_3$ are in excellent agreement, indicating complete reversibility of the solid/solute interfacial process[70].

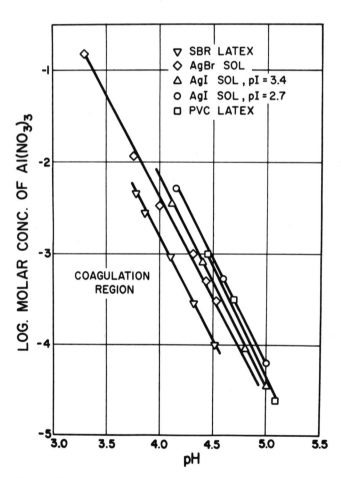

Figure 20. Boundaries between the coagulation region and the restabilization region due to charge reversal by aluminum hydrolyzed species for several sols plotted as a function of pH and the log molar concentration of Al(NO$_3$)$_3$. Sols: styrene-butadiene rubber (SBR) latex ($\triangledown$); silver bromide ($\diamond$); silver iodide (pI 3.4, $\triangle$; pI 2.7, $\bigcirc$); polyvinyl chloride (PVC) latex ($\square$).

301

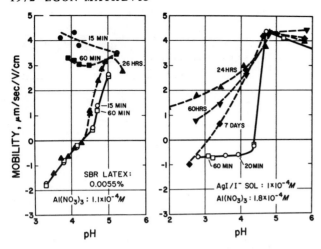

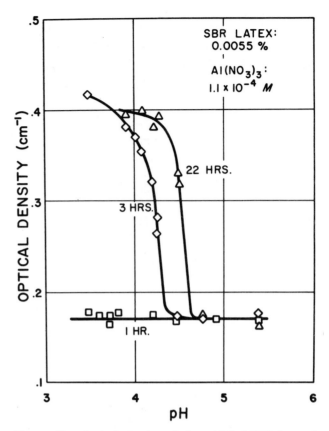

Figure 21. Left: Electrophoretic mobilities of a styrene-butadiene rubber (SBR) latex (0.0055% solids by wt.) in the presence of $1.1 \times 10^{-4} M$ Al(NO₃)₃ as a function of pH. Solid line gives the mobilities 15 min. (○) and 60 min. (□) after adding the salt and adjusting the pH of the latex. The other curves are for latex the charge of which was first reversed by Al(NO₃)₃ at pH 5.6 and subsequently acidified to various lower pH values. The measurements were carried out 15 min. (●), 60 min. (■), and 26 hours (▲) after acidification.

Right: Electrophoretic mobilities of a silver iodide sol ($1 \times 10^{-4} M$ AgI; $1.9 \times 10^{-3} M$ excess I⁻) in the presence of $1.8 \times 10^{-4} M$ Al(NO₃)₃ as a function of pH. Solid line gives mobilities 20 min. (○) and 60 min. (□) after adding salt and adjusting the pH. Dashed lines are for sols recharged at pH ∿ 6, 24 hours (▲), 60 hours (▼), and 7 days (♦) after acidification with HNO₃.

Figure 22. Optical density of the acidified SBR latex the charge of which was reversed by $1.1 \times 10^{-4} M$ Al(NO₃)₃ (the same systems as in Figure 21 left) 1 hour (○), 3 hours (◇) and 22 hours (△) after lowering the pH with HNO₃.

Stability Domains

The foregoing examples have shown that the condition of a sol in the presence of complexing ions depends most strongly on the concentrations of the counterion and the complexing agent. The so-called "stability domains" represent in a convenient way various stability regions for a given sol as functions of the concentrations of the coagulating and the coordinating ions. In the case of hydrolyzable metal ions, the content of the coordinating OH⁻ group is best controlled by pH; thus, the stability domains are given as log[salt]–pH plots.

Figure 23 represents such stability domains for three different sols in the presence of aluminum salts. For silver iodide sol, the c.c.c. curve was described in detail in Figure 5 and the c.s.c. curve in Figure 20. In this domain sols remain stable below the c.c.c. boundary and the conditions producing coagulated sols are indicated by hatching. To the right of the c.s.c. boundary, the sols are restabilized due to charge reversal. The dotted line shows the conditions at which the first trace of precipitated aluminum hydroxide is detected. Thus, all of the stability effects described are caused by soluble aluminum species. The similarity of the domains for three so chemically different lyophobic sols

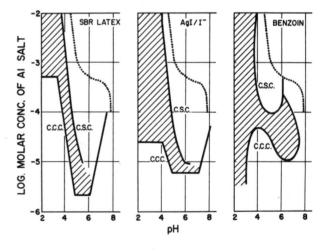

Figure 23. The entire log molar concentration Al(NO₃)₃ (or AlCl₃)–pH stability domain for styrene-butadiene rubber (SBR) latex (left), silver iodide sol (middle), and benzoin sol (right). Shaded area designates the coagulation region; below the c.c.c. line the sols remain stable, above the c.s.c. line the sols are restabilized due to charge reversal. The dotted line indicates the formation of aluminum hydroxide precipitate in the absence of sol particles.

clearly indicates that it is the state of aluminum ions in solution which is responsible for the stability conditions of these colloidal systems. The case of colloidal "benzoin", which is a sol prepared from powdered Sumatra gum, is particularly interesting as the domain is redrawn from work published nearly half a century ago[71]. Except for the lowest pH value, at which the sol is unstable, the plot shows a striking resemblance to the domains established recently using considerably more sophisticated instrumentation.

Since each metal ion shows different hydrolysis characteristics, the stability domains vary significantly from case to case. As an example, Figure 24 represents the effects of pH on the stability regions of a silver bromide sol in the presence of $Cr(NO_3)_3$[72]. The circles show the c.c.c. and c.s.c. boundaries as obtained with freshly prepared solutions of chromium(III) nitrate. Hatching again designates coagulated sols. If the experiments are repeated using chromium salt solutions which were aged by heating at 90° for several days, then cooled and used in coagulation experiments, the region delineated by triangles and squares is obtained. The tremendous effect of aging, which is known to produce soluble polynuclear hydrolysis products of chromium, upon the sol stability is another good example of the great sensitivity of the sols to counterion complexing.

It is also to be expected that a change in the nature of the colloidal sols, particularly in regard to surface characteristics (such as the degree of solvation), and consequently the change in the interfacial reactions with counterions will reflect itself in the shape of the stability domains.

Figure 25 illustrates the effect of pH on the behavior of a Ludox AM silica sol in the presence of $Al(NO_3)_3$[73]. In this case, the domain is extended so that it also includes conditions which lead to precipitation of aluminum hydroxide (the range within the dotted lines). Six different stability regions can be distinguished: I, stable sols; II, sols coagulated by solute aluminum species; III, large flocs formed by mutual coagulation of silica with aluminum hydroxide; IV, V, stable silica sols in the presence of aluminum hydroxide; VI, recharged (positive) stable silica sols. The dashed line gives the electrophoretic points of zero charges. The difference between the domains of lyophobic sols (Figure

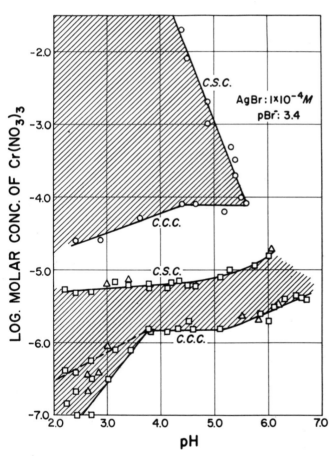

Figure 24. The entire log molar concentration of $Cr(NO_3)_3$ -pH stability domain for silver bromide sol in statu nascendi using fresh solutions of chromium(III) nitrate (○) and solutions of chromium(III) nitrate aged by heating at 90°C for 4 days (□) and 5 days (△). Shaded areas indicate coagulation regions; below the c.c.c. lines the sols remain stable, above the c.s.c. lines the sols are restabilized due to charge reversal.

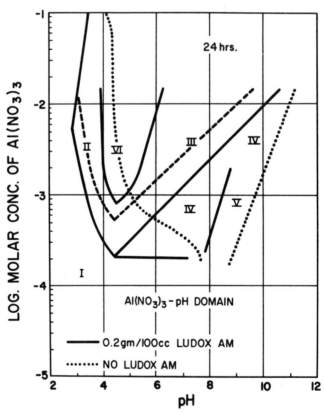

Figure 25. The entire log molar concentration $Al(NO_3)_3$ -pH stability domain for the Ludox AM silica sol. Regions: I, stable sols; II, sols coagulated by solute aluminum species; III, large flocs formed by mutual coagulation of silica with aluminum hydroxide; IV, V, stable silica sols in the presence of aluminum hydroxide; VI, recharged (positive) stable silica sols. Dashed line gives the electrophoretic points of zero charges. Dotted line indicates range of aluminum hydroxide precipitation.

23) and those of the hydrophilic silica sol (Figure 25) is rather remarkable. The increasing complexity of the stability domains when more than one colloidal system is present is also noteworthy.

If one is to fully understand the behavior of a sol in the presence of complexing electrolytes, it is essential to establish the stability domains. For "typical" lyophobic dispersions with electrolytes whose chemical behaviors in solution are known, general stability effects can now be predicted, although specific stability boundaries will still have to be delineated experimentally in each case.

It is fair to state that a great many interfacial reactions responsible for stabilization/destabilization effects of colloids have now been explained, particularly for lyophobic sols. The great majority of the studies involved sols containing only one colloidal system. Many naturally occuring dispersions, or sols produced in various practical applications, consist of a mixture of colloidal particles of different chemical compositions and surface characteristics. The behavior of such mixed systems in a given electrolyte environment and, particularly, the problems of mutual particle interactions deserve careful attention. The progress made in the understanding of simpler systems, for which some examples have been given in this paper, makes it now possible to extend the studies to the much more complicated mixed colloidal sols with the hope of arriving at rational explanations of their stability properties.

Acknowledgement

This address is based on work of many students and associates of mine to whom I am greatly indebted for their patience, perseverance, and friendship. Here, the names are mentioned of those whose results are included in the foregoing paper: Dr. L. H. Allen, Dr. S. Lee Allen, Mr. A. Bell, Mr. A. A. Bibeau, Mr. A. Bleier, Mr. D. L. Catone, Dr. R. Demchak, Mr. L. Eriksson, Dr. C. G. Force, Dr. G. E. Janauer, Mr. E. P. Katsanis, Dr. N. Kolak, Dr. S. Kratohvil, Dr. R. V. Lauzon, Mr. A. B. Levit, Dr. A. D. Lindsay, Dr. F. J. Mangravite, Jr., Dr. K. G. Mathai, Dr. P. McFadyen, Dr. R. Nemeth, Dr. L. J. Stryker, and Mr. D. White.

Three persons deserve a special mention: Professor Božo Težak, Univ. of Zagreb, who introduced me to colloid science; Dean Milton Kerker, Clarkson College, who gave me the opportunity to continue with my work in the U. S. and with whom I have had many years of most fruitful collaboration; and Professor Josip Kratohvil, Clarkson College, who, with his unique ability to absorb and store facts, has been an invaluable help in discussions and as a source of information.

References

1. Faraday, M., *Phil. Trans.,* **147**, 145 (1857)
2. Freundlich, H., *"Kapillarchemie",* 3rd Ed., Akademische Verlagsgesellschaft, Leipzig 1923, 580 ff.; English translation *"Colloid and Capillary Chemistry",* Dutton, New York, 1926.
3. Freundlich, H., Joachimsohn, K. and Ettisch, G., *Z. physik. Chem.,* **A141**, 249 (1929)
4. Powis, F., *Z. physik. Chem.,* **89**, 186 (1915)
5. Ostwald, Wo., *Kolloid-Z.,* **73**, 301 (1935); **94**, 169 (1941); *J. Phys. Chem.,* **42**, 981 (1938)
6. Težak, B., *Z. physik. Chem.,* **A191**, 270 (1942); *Arhiv kem.,* **22**, 26 (1950)
7. Schulze, H., *J. prakt. Chem.,* **25**, 431 (1882); **27**, 320 (1883)
8. Hardy, W. B., *Proc. Roy. Soc.,* London, **66**, 110 (1900); *Z. physik. Chem.,* **33**, 385 (1900)
9. Derjaguin, B. V. and Landau, L., *Acta Physicochim. U.S.S.R.,* **14**, 663 (1941)
10. Verwey, E. J. and Overbeek, J. Th. G., *"Theory of the Stability of Lyophobic Colloids",* Elsevier, Amsterdam, 1948
11. Vouk, V. B., *Nature,* **170**, 762 (1952)
12. Vincent, B., Bijsterbosch, B. H., and Lyklema, J., *J. Colloid Interface Sci.,* **37**, 171 (1971)
13. Levine, S. and Bell, G. M., *J. Colloid Interface Sci.,* **17**, 838 (1962)
14. Shaw, D. J., *"Introduction to Colloid and Surface Chemistry",* 2nd Ed., Butterworth, London 1970, 168.
15. Sheludko, A., *"Colloid Chemistry",* Elsevier, Amsterdam, 1966, 211-213.
16. Jirgensons, B. and Straumanis, M. E., *"A Short Textbook of Colloid Chemistry",* 2nd Ed., Pergamon Press, Oxford 1962, 329, 330.
17. Stauff, J., *"Kolloidchemie",* Springer Verlag, Berlin, 1960, 475
18. Mysels, K. J., *"Introduction to Colloid Chemistry",* Interscience, New York, 1959, 372, 373
19. Overbeek, J. Th. G., in *"Colloid Science",* editor H. R. Kruyt, Elsevier, Amsterdam, 1952, 82.
20. Wenning, H., *Kolloid-Z.,* **154**, 154 (1957)
21. Matijević, E., Janauer, G. E. and Kerker, M., *J. Colloid Sci.,* **19**, 333, (1964)
22. Matijević, E., Kolak, N. and Catone, D. L., *J. Phys. Chem.,* **73**, 3556 (1969)
23. LaMer, V. K., *Ind. Eng. Chem.,* **44**, 1270 (1952)
24. Lieser, K. H., *Angew. Chem.* (Intern Ed.), **8**, 188 (1969)
25. Demchak, R. and Matijević, E., *J. Colloid Interface Sci.,* **31**, 257 (1969)
26. Matijević, E., Lindsay, A. D., Kratohvil, S., Jones, M. E., Larson, R. I. and Cayey, N. W., *J. Colloid Interface Sci.,* **36**, 273 (1971)
27. Matijević, E. and Bell, A., *Proc. Soc. Chem. Ind.* (London) (in press)
28. Kerker, M., Matijević, E., Espenscheid, W. F., Farone, W. A., and Kitani, S., *J. Colloid Sci.,* **19**, 213 (1964)
29. Gordon, L., Salutsky, M. L. and Willard, H. H., *"Precipitation from Homogeneous Solution",* Wiley and Sons, New York (1959), p. 6 ff.
30. Larson, R. I., Fullam, E. F., Lindsay, A. D. and Matijević, E., *Preprint 19b,* 71st Natl. Meeting, AIChE, Dallas, Texas, February, 1972
31. Thomas, A. W. and Vartanian, R. D., *J. Am. Chem. Soc.,* **57**, 4 (1935)
32. Shuttleworth, S. G., *J. Soc. Leather Trades' Chem.,* **38**, 58 (1954)
33. Erdmann, H., *Angew. Chem.,* **64**, 500 (1952)
34. Matijević, E., Watanabe, A. and Kerker, M., *Kolloid Z. Z. Polym.,* **235**, 1200 (1969)
35. Blumenthal, W. B., *"The Chemical Behavior of Zirconium",* Van Nostrand Co., Princeton, N. J. 1958
36. Matijević, E., *Discussions Faraday Soc.,* **42**, 106 (1966)
37. Rubin, A. J., Hayden, P. L. and Hanna, G. P., Jr., *Water Res.,* **3**, 843 (1969)
38. Matijević, E., Mathai, K. G., Ottewill, R. H. and Kerker, M., *J. Phys. Chem.,* **65**, 826 (1961)
39. Stryker, L. J. and Matijević, E., *J. Phys. Chem.,* **73**, 1484 (1969)
40. Ostwald, Wo. and Hoffmann, K., *Kolloid-Z.,* **80**, 186 (1937)
41. Levine, S. and Bell, G., *J. Colloid Sci.,* **20**, 695 (1965)

42. Matijević, E., Kratohvil, S. and Stickels, J., *J. Phys. Chem.*, **73**, 564 (1969)

43. Težak, B., Matijević, E., Schulz, K. F., Hallassy, R. and Kostinčer, I., *Proc. 2nd Int. Congress Surface Activity*, **3**, 607 (1957)

44. Matijević, E. and Allen, L. H., *Environ. Sci. Technol.*, **3**, 264 (1969)

45. Biedermann, G., *Svensk. Kem. Tidskr.*, **76**, 362 (1964)

46. Matijević, E. and Stryker, L. J., *J. Colloid Interface Sci.*, **22**, 68 (1966)

47. Matijević, E., *J. Colloid Interface Sci.*, **20**, 322 (1965)

48. Hietanen, S. and Sillén, L. G., *Acta Chem. Scand.*, **18**, 843 (1964)

49. Mesmer, R. E. and Baes, C. F., Jr., *Inorg. Chem.*, **6**, 1951 (1967)

50. Matijević, E., Levit, A. B. and Janauer, G. E., *J. Colloid Interface Sci.*, **28**, 10 (1968)

51. Aveston, J., *J. Chem. Soc.*, **A1966**, 1599

52. Sillén, L. G. and Martell, A. E., *"Stability Constants of Metal Ion Complexes"*, The Chemical Soc., London, 1964

53. Bleier, A., Matijević, E., McGuire, J. and Wear, J. O., *J. Colloid Interface Sci.*, **38**, 647 (1972)

54. Matijević, E. and Kerker, M., *J. Amer. Chem. Soc.*, **81**, 5560 (1959)

55. Matijević, E. and Kolak, N., *J. Colloid Interface Sci.*, **24**, 441 (1967)

56. Matijević, E., Broadhurst, D. and Kerker, M., *J. Phys. Chem.*, **63**, 1552 (1959)

57. Olson, F. A., Reese, D. A., and Wadsworth, M. F., *The Inst. of Metals and Explosives Research Rept. No. 5*, 1963 (Office of Technical Services, Dept. of Commerce, Washington, D. C.)

58. Allen, L. H., Matijević, E. and Meites, L., *J. Inorg. Nucl. Chem.*, **33**, 1293 (1971)

59. Kiselev, A. V., *Proc. Symp. Colston Res. Soc.*, **10**, 195 (1958); *"The Structure and Properties of Porous Materials"*, Butterworths, London, 1958, p. 195

60. Allen, L. H. and Matijević, E., *J. Colloid Interface Sci.*, **33**, 420 (1970)

61. Allen, L. H. and Matijević, E., *J. Colloid Interface Sci.*, **31**, 287 (1968)

62. Hockey, J. A., private communication

63. Matijević, E., *Kolloid Z. Z. Polym.*, **250**, 646 (1972)

64. Vaska, L., private communication

65. Lauzon, R. V. and Matijević, E., *J. Colloid Interface Sci.*, **37**, 296 (1971)

66. Bibeau, A. A. and Matijević, E., *J. Colloid Interface Sci.* (to be published)

67. Lauzon, R. V. and Matijević, E., *J. Colloid Interface Sci.*, **38**, 440 (1972)

68. Stryker, L. J. and Matijević, E., *Adv. Chem. Ser.*, **79**, 44 (1968)

69. Stryker, L. J. and Matijević, E., *J. Colloid Interface Sci.*, **31**, 39 (1969)

70. Eriksson, L., Matijević, E. and Friberg, S., *J. Colloid Interface Sci.* (submitted)

71. Kermack, W. O. and Voge, C. I. B., *Proc. Roy. Soc.*, Edinburgh, **45**, 90 (1924/25)

72. Kratohvil, S. and Matijević, E., *J. Colloid Interface Sci.*, **24**, 47 (1967)

73. Matijević, E., Mangravite, F. J., Jr., and Cassell, E. A., *J. Colloid Interface Sci.*, **35**, 560 (1971)